SEMI-AQUATIC MAMMALS

SEMI-AQUATIC

Johns Hopkins University Press

BALTIMORE

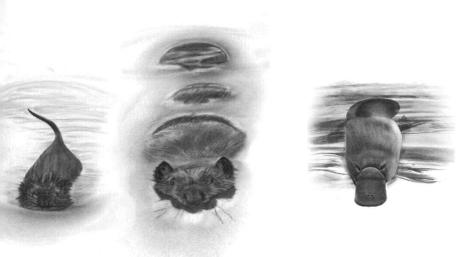

MAMMALS

Ecology and Biology

Glynnis A. Hood

Illustrated by Meaghan Brierley

© 2020 Johns Hopkins University Press
All rights reserved. Published 2020
Printed in the United States of America on acid-free paper
9 8 7 6 5 4 3 2 1

Johns Hopkins University Press
2715 North Charles Street
Baltimore, Maryland 21218-4363
www.press.jhu.edu

Library of Congress Cataloging-in-Publication Data
Names: Hood, Glynnis A., 1964– author.
Title: Semi-aquatic mammals : ecology and biology /
Glynnis A. Hood ;
illustrated by Meaghan Brierley.
Description: Baltimore, MD : Johns Hopkins University Press, 2020. |
Includes bibliographical references and index.
Identifiers: LCCN 2019056156 | ISBN 9781421438801 (hardcover) |
ISBN 9781421438818 (ebook)
Subjects: LCSH: Aquatic mammals. | Mammals.
Classification: LCC QL713 .H66 2020 | DDC 599.5—dc23
LC record available at https://lccn.loc.gov/2019056156

A catalog record for this book is available from the British Library.

Text designed and typeset by Amy Ruth Buchanan /3rd sister design.

*Special discounts are available for bulk purchases of this book. For more
information, please contact Special Sales at specialsales@press.jhu.edu.*

Johns Hopkins University Press uses environmentally friendly book
materials, including recycled text paper that is composed of at least 30
percent post-consumer waste, whenever possible.

For those whose years of
research and conservation
efforts give voice to these
remarkable mammals and
the habitats on which
they depend

9. The Prey: Predator-Prey Interactions   266

PART IV. REPRODUCTION

10. Mating and Offspring   291

PART V. CONSERVATION CHALLENGES & MANAGEMENT APPROACHES

13. Management Approaches — 379

FIGURES

TABLES

Freshwater semi-aquatic mammals live on every continent except Antarctica, and they range from high mountain streams in the Andes of South America to coastal wetlands of Australia. They are found in every major mammalian division (monotremes, marsupials, and eutherians) and vary in size from tiny water shrews to the massive common hippopotamus of sub-Saharan Africa. These are the mammals that have an obligatory daily dependence on freshwater habitats for almost all aspects of their ecology. It is a precarious ecological position that requires both aquatic and terrestrial ecosystems for daily survival. Given semi-aquatic mammals' intermediate state between land and water, there is a continuum on which physiological, morphological, and behavioral adaptations have evolved. Beavers, otters, and the platypus have modified tails and webbed feet to maximize swimming efficiency while still allowing effective movement on land. Other species, including the American and European mink, appear less physically adapted, yet demonstrate a complete dependence on freshwater systems. An intimate connection to increasingly imperiled freshwater ecosystems has also made many species of freshwater semi-aquatic mammals some of the most endangered mammals in the world.

This book addresses the ecology, conservation, and management of 140 species of freshwater semi-aquatic mammals, representing eight mammalian orders and twenty-two families. I do not discuss semi-aquatic mammals that live in marine environments or species of bats that hunt near or in freshwater habitats. I use the term "semi-aquatic mammal" to represent other commonly used terms in the literature (e.g., aquatic mammal, amphibious mammal, freshwater mammal, and riparian mammal). As a general rule, the book follows taxonomic divisions and ordering rules adhered to by Ronald M. Nowak (1999) in *Walker's Mammals of the World*, but it also incorporates updated taxonomies and ordering presented by Conner Burgin and his colleagues (2018) in their recent paper "How Many Species of Mammals Are There?" in the *Journal of Mammalogy*. Additional studies cited in the book also informed my approach to taxonomic nomenclature and ordering.

Tables follow those same taxonomic divisions and hierarchical ordering (i.e., always beginning with monotremes, followed by marsupials, and then the corresponding ordering of eutherians). New species (e.g., Sulawesi water rat, *Waiomys mamasae*) were discovered during the writing of this book, and others (e.g., *Baiyankamys*, *Dasymys*, and *Hydromys*) are regularly undergoing reclassification because of advancements in molecular biology. The naming and classification of these species reflect these new advances, and I have noted any ongoing deliberations. Some of these taxonomic designations might change with time. Science is about continual advancement and revision. Fortunately, no species went extinct during the writing of this book, but there are some in very precarious positions.

Along with common names, scientific names are provided for each species at their first mention in each chapter (rather than just when first mentioned in the book). This approach does not include tables within chapters, which contain both common and scientific names and are designed so they can stand alone if necessary. Common names reflect the names primarily used in the *Red List of Threatened Species* published by the International Union for Conservation of Nature (IUCN), or the common names used by various experts who have published them in the scientific literature. There is constant variation in the application of common names (e.g., there are many "water rats") in the literature, but I have used the same common names throughout the book. For Australian species, there is a move to rename species with their original indigenous names. *Hydromys chrysogaster* has many common names (e.g., rakali, rabe, water-rat, common water rat, golden-bellied water rat), but it is identified as the rakali in this book to follow current naming norms for species in Australia. There are also continental differences for *Myocastor coypus* (coypu or nutria). To stay consistent, I use the term "coypu" because of its global familiarity and immediate link to its scientific name.

What is notable in the existing research is the ongoing reassessment as to which species meet the criteria of being semi-aquatic and which ones do not. For example, the qualifying anatomical criteria in some studies disqualify American mink as semi-aquatic, while other studies and reviews consistently include it as specifically semi-aquatic as per the researchers' diagnostics relative to its behavior and ecology. The focus of each study often drives these criteria (e.g., anatomical versus behavioral). Other species (e.g., moose, *Alces alces*) have been occasionally noted as semi-aquatic, but I have not considered them as such given their heavy use of forested ecosystems without any necessary access to aquatic habitats for up to eight months of the year in northern environments. However, I have included the water chevrotain (*Hyemoschus aquaticus*) because it is never found farther than 250 m from water and is completely dependent on freshwater habitats on a daily basis throughout the year. There are other species that I originally included (e.g., riverine rabbit, *Bunolagus monticularis*), but after extensive review of the literature and species specialist reports, they ended up excluded from the final species list, sometimes well into the research process.

As described with beavers, otters, and platypuses, there is a range of characteristics that represent the diversity of physiological, morphological, and behavioral adaptations of these mammals that allows them to succeed in amphibious habitats. For the purpose of this book, a semi-aquatic mammal is a species that has, through some degree of evolutionary adaptation for at least part of its daily activities, a dependency on aquatic *and* terrestrial habitats over and above the basic need to consume water as part of its dietary requirements. To confirm the inclusion of species that I describe here, I used a combination of sources, including the academic, and at times nonacademic, literature and IUCN species specialist reports. The biology of some species is poorly known but I have included them because of strong associations with freshwater habitats in the limited studies available. No doubt, there are species that might warrant inclusion, yet are not represented in this book, and other species that might seem like unlikely candidates but were included. During research and writing, I made every effort to identify those species that require a combination of freshwater and terrestrial habitats for their daily activities and survival.

In the broadest sense, I have written this book for anyone who has an interest in semi-aquatic mammals and the habitats that support them. It may also be of interest to a wide range of scientists, including ecologists, biogeographers, mammalogists, paleobiologists, physiological ecologists, zoologists, animal behaviorists, conservation biologists, and natural resource managers. In short, it is written for those with both specific and broad interests in mammalian species that are specially adapted to freshwater systems. It is a book designed to be accessible to educated generalists, ecologists, and mammalian specialists. To highlight species-specific and ecological details, the text is accompanied by Dr. Meaghan Brierley's wonderful illustrations made specifically for this book. Her illustrations were created while spending an extensive amount of time working directly with specimens accessed from various collections. Additional figures from other sources are also included where necessary.

Scope and Organization

This book provides a review of the ecological diversity of semi-aquatic mammals, as well as their conservation and management. Books that present this group of mammals in their entirety are lacking, and ecologists and animal scientists often return to Alfred Brazier Howell's seminal book *Aquatic Mammals: Their Adaptation to Life in the Water*, written in 1930. I am grateful for his book in particular, because of its attention to detail and its comprehensive approach. The extensive amount of literature regarding individual studies on these mammals was also invaluable. I have aimed to provide a resource that brings these wide-ranging materials into one place. This is a book that provides a context for our current ecological knowledge of a global suite of semi-aquatic mammals.

The Introduction (Chapter 1) presents an overview of each of the semi-aquatic mammals included in the book. As such, the taxonomy, basic geographic distributions, and general characteristics of these species and their freshwater habitats help set the stage for the rest of the book. In this section, and throughout the book, specific scientific terminology is highlighted in bold at first mention and included in the Glossary at the back of the book. The Introduction is followed by the first section of the book, Part I, "Geographical Distributions," which begins with Chapter 2, "Paleobiology." Chapter 2 starts by discussing a broad evolutionary context of semi-aquatic mammals and mammal-like species relative to their early biogeography and phylogeny. Thereafter, I present a more detailed description of the early and present distributions and phylogeny of each of the extant taxonomic orders and families of living (and some long extinct) semi-aquatic mammals. Chapter 3, "Ranging across the Continents," describes the geographical distributions of semi-aquatic mammals within a global context. Along with the specific continental distributions of semi-aquatic mammals, habitat-specific details provide a fine-scale analysis of habitat selection. Chapter 4, the final chapter in Part I, describes ecological niches of semi-aquatic mammals, ranging from ecological engineering to trophic positions and special behavioral adaptations for extreme environments.

Part II addresses physical adaptations of semi-aquatic mammals. The first chapter of this section (Chapter 5, "Morphology") investigates the breadth of morphological adaptations for living in and moving in amphibious environments. There is an extensive overview of the key structural components of the bodies of semi-aquatic mammals and how they vary among species. Tables provide summaries of common morphological adaptations of the feet, tail, ears, eyes, and nose of several of these species. Next is Chapter 6, "Physiological Adaptations," which provides an examination of the physiological constraints and adaptations required for living in aquatic environments. Topics include aspects of thermoregulation, energetics, respiration, and digestion. The final chapter in this section is Chapter 7, "Locomotion and Buoyancy." As expected, it outlines the range of adaptations and strategies used by semi-aquatic mammals to maintain their ability to move on land while still moving efficiently through water, regardless of temperature or depth. Morphological and physiological considerations covered in the two previous chapters help set the context for Chapter 7.

In Part III, there is a focus on feeding ecology of semi-aquatic mammals, both as predators and as prey. In Chapter 8, "The Predators: Foraging Strategies and Niches," I provide an analysis of common foraging strategies and feeding niches (e.g., carnivores, herbivores, insectivores), as well as key prey items for various species. The categories used for feeding niches generally follow those presented in *Ecological and Envi-*

ronmental *Physiology of Mammals* by Withers et al. (2016). In Chapter 9, "The Prey: Predator-Prey Interactions," there is a switch in perspective that examines semi-aquatic mammals as prey instead of consumers. As such, this chapter discusses predator-prey ecology and predator avoidance strategies, as well as including an overview of parasites and diseases as a specialized form of predation.

In part, the broader division of mammals is based on their reproductive biology. All of the three divisions (monotremes, marsupials, and eutherians) have semi-aquatic representatives. Part IV, "Reproduction," describes the structural anatomy, physiology, and reproductive strategies of these mammals. There is only one chapter in this section (Chapter 10, "Mating and Offspring"). Along with a discussion on reproductive biology, there is also coverage of the sociality of reproduction and how these behaviors benefit reproductive success. I also describe the care and the timing of dispersal of offspring and how these behaviors influence recruitment of new individuals into the population. There is also a brief overview of captive breeding as it relates to conservation.

The final section of the book is Part V, "Conservation Challenges and Management Approaches." It begins with Chapter 11, "Status and Trends," which details the current conservation status and population trends of the 130 species for which conservation data exist. In this section I rely heavily on IUCN species reports, and I integrate current threats to these species at both local and global scales. From there, Chapter 12, "Introductions and Reintroductions," addresses the use of species translocations in both a historic and a conservation context. There is a series of case studies for four species of semi-aquatic mammals (North American beaver, coypus, muskrats, and American mink) that have been introduced into areas outside their natural distribution. In these cases, their ecological impacts are highlighted. Then a series of case studies describes the conservation reintroduction efforts of a number of semi-aquatic mammals and their varying successes. The book ends with a chapter examining management approaches (Chapter 13) and various challenges associated with wildlife management, as well as solutions to help ensure the persistence of semi-aquatic mammals. Appendix A provides a taxonomic list of the 140 species discussed in this book, while Appendix B suggests additional online resources that might be of interest to the reader.

My aim is to present a comprehensive resource for scientists and wildlife enthusiasts who are interested in semi-aquatic mammals and the freshwater habitats on which they depend. My hope is also to increase awareness of a group of mammals whose complete dependence on both aquatic and terrestrial habitats places them in the unique position of being ecological barometers in some of the most threatened ecosystems in the world. The book starts with their prehistoric beginnings and ends with their present status and potential for future survival in rapidly changing environments.

ACKNOWLEDGMENTS

Semi-aquatic mammals and their freshwater habitats have always been part of my life. Dr. Meaghan Brierley, the illustrator of this book, and I both grew up on the shores of Kootenay Lake in southeastern British Columbia, Canada. Our family cabins are still a short walk apart. The mountain rivers and streams, as well as the lakes and diverse wetlands of the Creston Valley, provided us with childhoods filled with wild encounters with mink, river otters, muskrats, beavers, and water shrews. Both Meaghan's and my parents, along with Meaghan's grandparents, encouraged a love of biology and the natural world. I still remember Meaghan's grandfather, Vaughan Mosher, watching a mink with me as it ate a fish on the rocky shoreline of Kootenay Lake. It was this type of early introduction to semi-aquatic mammals and their ecology that inspired me to pursue a lifetime of studying freshwater systems and the mammals they support. I hope the content of this book and Meaghan's remarkable illustrations will bring a broader understanding and respect for these incredible mammals and their freshwater habitats.

There are two people who were instrumental in bringing this book from an idea to reality. Dr. Ludwig (Lu) Carbyn was the first to suggest that I write another book, albeit on beavers, and put me in touch with Vincent J. Burke, Executive Editor of Johns Hopkins University Press. After Vince and I agreed that there were several beaver books already on the market, I remarked that we needed one book that included all of the semi-aquatic mammals. Vince asked me to submit a proposal, which became *Semi-aquatic Mammals: Ecology and Biology*, and with some excellent guidance from Vince the process began.

When I asked her, Meaghan Brierley immediately agreed to illustrate the book, and she has been an integral part of the project from its beginning. The writing process was more protracted than anticipated because of a series of unforeseen events, and Vince's patience and support were invaluable. As the project extended longer than anticipated, Meaghan was consistently patient with my delays and showed unwavering commitment to the book. It would not have happened without her.

Following Vince's departure from JHUP, Tiffany Gasbarrini took over as a patient and supportive editor. This book has been guided to completion by Tiffany and the rest of the team at JHUP, including Editorial Assistant Esther P. Rodriguez, Managing Editor Juliana McCarthy, Production Editor Hilary S. Jacqmin, and Publicity Manager Kathryn Marguy and the marketing team. Of course, a book is only as good as its copyeditor, and Carrie Love is one of the best. I am especially thankful for her guidance, knowledge, and remarkable attention to detail.

Dr. Robert Brooks provided a blind peer-review of the manuscript, and thereafter he kindly offered to openly discuss the manuscript further if needed. His edits and suggestions made for a much stronger manuscript. This book also benefited from a sabbatical granted by the University of Alberta, which was championed by my Departmental Chair, Dr. Peter Berg. In addition, I am very grateful to Dr. Robert Balyk, Dr. Ronald Henderson, and Gord Nadeau, who helped me return to writing after some unexpected delays. My colleague Dr. Anne McIntosh was always on standby for a cup of tea.

I am especially thankful for the early proposal edits and ongoing support of both Dr. Dee Patriquin and my late mother, Kathleen Hood. Dee was also there for photography adventures and provided constant assistance with template formatting, editing, and advice. Meaghan and I appreciate Lindsay Turner for his input regarding illustrations for the book. Lindsay also illustrated Figure 6.5, which depicts countercurrent circulation. Duncan Abercrombie, General Manager of Animal Damage Control and member of the Alberta Trappers Association, was very generous with his time so that Meaghan could have extensive access to various semi-aquatic mammals, along with detailed hands-on descriptions of their functional anatomy. His depth of knowledge of North American mammals is remarkable, and it ensured accuracy and detail in the illustrations. Bill Abercrombie, President of Animal Damage Control, also provided specimens to aid in the illustrations. Dr. Mark Edwards, Curator of Mammalogy at the Royal Alberta Museum in Edmonton, kindly allowed us to access the museum collections and made room for us in what was a busy schedule in a brand new museum building. He and his staff were very helpful in setting up an articulated beaver skeleton for us. University of Alberta Biology Technicians Dawn Lawson and Marian Forre, at the Augustana Campus, provided ongoing access to their mammal collections. Similarly, staff at the University of Calgary's Biology Museum provided welcome access to their collections. Dr. Heather Proctor from the Department of Biological Sciences at the University of Alberta was very generous in providing images of a beaver beetle (*Platypsyllus castoris*) on short notice. Without hesitation, Dr. Bruce D. Patterson, Curator of the Field Museum of Natural History, provided a photograph of an Ethiopian amphibious rat (*Nilopegamys plumbeus*) from a painting by Leon Pray that is housed at the museum. It is one of the only images we have of this illusive and critically endangered mammal.

Professor Christopher R. Scotese, Director of the PALEOMAP Project, kindly provided geographic information system (GIS) files for a complete set of maps for the Phanerozoic Eon. These files allowed for the production of a series of maps ranging from the Paleozoic to the Late Cenozoic for Figure 2.1. Dr. Scotese's generosity was invaluable and allowed me to present the geographic context for the paleobiology of semi-aquatic mammals. We appreciate his generosity and remarkable research. Similarly, the International Union for Conservation of Nature (IUCN) provided access to updated shapefiles that were used to create a series of species distribution maps in various chapters. These data files were helpful in various chapters, especially in Chapter 3, "Ranging across the Continents." The hard work and dedication of the IUCN and its staff and scientists is critical to the conservation of species and their habitats.

Finally, we thank our families, friends, and colleagues who provided patience and support during the time spent researching, illustrating, and writing this book.

Introduction

The evidences of evolution are most abundant upon every hand, and still we know so very little about them.

A. Brazier Howell, 1930, *Aquatic Mammals: Their Adaptations to Life in the Water*

What Is a Semi-aquatic Mammal?

In a world where water covers more than 70% of the earth's surface (Shiklomanov 1993), many species of mammals have evolved adaptations for fully aquatic or semi-aquatic specialization. Although most mammalian species are able to swim instinctively, with giraffes and apes being exceptions (Howell 1930; Crandall 1964), only a select group of species are semi-aquatic. Their specialized physiological, morphological, and behavioral adaptations for both aquatic and terrestrial habitats differentiate semi-aquatic mammals from other mammals, as does their obligatory association with water. In his seminal text, *Aquatic Animals: Their Adaptations to Life in the Water*, A. Brazier Howell (1930) states that to merit the term "aquatic" an animal must be able to swim, dive, seek food in the water, and close both nostrils and ears while submerged. In addition to these abilities, most semi-aquatic mammals have waterproof fur and shortened appendages, as well as some modification to their feet (e.g., webbing, fringes of stiff hairs) and tails (e.g., altered shape, compression) to aid locomotion.

More generally, a semi-aquatic mammal is a species that has, through some degree of evolutionary adaptation, a dependency on aquatic and terrestrial habitats for at least part of its daily activities. This dependency on aquatic habitats is over and above the basic dietary need to consume water. They are never called "semi-terrestrial" because their aquatic adaptations make these species stand out from other mammals. The degree of adaptation and dependence on aquatic habitats varies among semi-aquatic mammals, an ambiguity not generally seen in mammals completely adapted to either aquatic or terrestrial habitats.

There are also consistent trends relative to body size. Freshwater semi-aquatic mammals tend to be larger than their terrestrial counterparts (Wolff and Guthrie 1985), yet smaller than their fully aquatic (often marine) relatives, mainly because freshwater mammals are restricted to freshwater habitats (Fish 2000). Although one might think an American water shrew (*Sorex palustris*), weighing only 8 to 18 g, is small, it is one of North America's largest shrews. The capybara (*Hydrochoerus hydrochaeris*), a semi-aquatic species of South America, and the North American beaver (*Castor canadensis*) are the largest and second-largest rodents in the world, respectively. Jerry Wolff and R. D. Guthrie (1985) suggest that a larger body size in small semi-aquatic mammals helps reduce **predation** risk by aquatic predators, such as fish. It also increases an individual's competitive advantage in aquatic environments where food quality is high, but the food's abundance is lower and patchier than in many terrestrial habitats. The idea of a thermoregulatory advantage, they argue, is less likely, because there appears to be no relationship between body size and water temperature in these mammals (Irving 1973; Wolff and Guthrie 1985). Indeed, Wolff and Guthrie determined that small-bodied semi-aquatic mammals are often found in colder mountain streams, and larger mammals, such as South America's nutria (coypu; *Myocastor coypus*), which can be as large as 9 kg, maintain a southern distribution in North America because of intolerance to harsh, cold winters (Carter and Leonard 2002). The largest of the semi-aquatic mammals, the common hippopotamus (*Hippopotamus amphibius*), which weighs up to 1,500 kg, lives in the heat of sub-Saharan Africa (Fig. 1.1). It is also one of the only semi-aquatic mammals that lacks fur.

Many, but not all, of the semi-aquatic mammals inhabit surface freshwater systems, which cover less than 10% of the world's surface (Lehner and Döll 2004; Mitsch and Gosselink 2007). Of the twenty-nine currently accepted orders of mammals (Wilson and Reeder 2005;

Figure 1.1. Comparison of capybara and common hippopotamus.
M. BRIERLEY

Feldhamer et al. 2015), several contain semi-aquatic species that have
an obligatory dependence on freshwater habitats for many of their daily
activities (Veron, Patterson, and Reeves 2008). They are found in each
of the three major groupings of mammals: order Monotremata and cohorts Marsupialia and Placentalia (Fig. 1.2). As with their adaptations,
the degree of dependence on aquatic habitats also varies among species of semi-aquatic mammals. Animal physiologist, Frank Fish (2000:
683) refers to semi-aquatic mammals as "evolutionary intermediates
between ancestral terrestrial mammals and their fully aquatic descendants." Their choice of habitat, from wetlands to remote rivers, results
in many species being poorly studied (Bagniewska et al. 2013), which is
concerning. With the rapid loss and degradation of freshwater habitats
on a global scale (Mitsch and Gosselink 2007; Junk et al. 2013), many
semi-aquatic mammals represent some of the most endangered species
on earth (IUCN 2019).

The more commonly used term "semi-aquatic mammal" (or "semi-
aquatic mammal") is often replaced by "riparian mammals" (Dunstone
and Gorman 1998), "aquatic mammals" (Howell 1930; Estes 1989; Taylor
1989; Feldhamer, Thompson, and Chapman 2003), "amphibious mammals" (Irving 1973; Dunstone and Gorman 1998; Feldhamer, Thompson,
and Chapman 2003; Feldhamer et al. 2015), or "freshwater mammals"
(Feldhamer et al. 2015; Veron, Patterson, and Reeves 2008). To remove
ambiguity regarding fully aquatic mammals that spend their entire lives
in water (i.e., cetaceans and sirenians) versus mammals that spend time

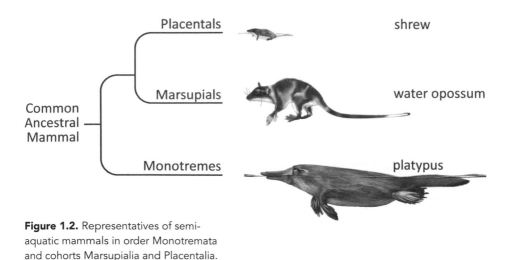

Figure 1.2. Representatives of semi-aquatic mammals in order Monotremata and cohorts Marsupialia and Placentalia. M. BRIERLEY

in water and on land, "semi-aquatic" is the term most often used for those species that still have an ecological tie to terrestrial habitats, no matter how limited. In marine habitats, seals and walruses, which are also considered semi-aquatic, still need solid surfaces on which to give birth and rest; however, their physical adaptations reflect a higher dependency on water for movement and feeding. In this book, I focus on semi-aquatic mammals that are preferentially within freshwater systems, despite there being some species able to exploit brackish or even saline waters from time to time. These mammals occupy a range of dependency, with some defined as riparian specialists elsewhere, rather than semi-aquatic specifically.

In preparing for this book, it became apparent to me that the definition of what qualifies a mammal as "semi-aquatic" varies throughout the literature. There will undoubtedly be some species that are included that represent the outer range of freshwater-dependent for some people, and others that are still being discovered that would deservedly belong in this book. To provide a cohesive approach, I have integrated definitions and examples from Howell (1930), Dunstone and Gorman (1998), Feldhamer et al. (2015), the International Union for Conservation of Nature (IUCN) Red List (IUCN 2019), and numerous academic papers and texts from experts within the field. From those definitions, combined with species-specific accounts from seminal texts (e.g., *Walkers Mammals of the World*; Nowak 1999), a deep body of scientific literature, and conversations with other biologists, a comprehensive species list emerged that represents a global suite of 140 mammals that are freshwater-dependent with semi-aquatic adaptations in varying degrees. For almost ninety years, Howell's 1930 book has been the most comprehensive resource for these fascinating species. As expected, our understanding, classification, and the status and habitats of these species has changed dramatically over time.

The Semi-aquatic Mammals

Freshwater semi-aquatic mammals are found on every continent, except for Antarctica (Fig. 1.3). Biogeographic isolation has also resulted in the absence of native freshwater mammals on some larger island nations, such as New Zealand, where terrestrial mammals are naturally absent or rare. Humans, however, have moved several species of freshwater mammals around the world, which has caused difficulties for both native species and their habitats. Despite this broad distribution, some semi-aquatic mammals are dramatically different in lineage but surprisingly similar in ecological adaptations.

These differences begin with a division of class Mammalia into two subclasses: Prototheria and Theria (Fig. 1.4). Prototheria represents the monotremes, which are the only mammals to lay eggs (**oviparous**). Of its five remaining species (Veron, Patterson, and Reeves 2008; Merritt 2010), the platypus (*Ornithorhynchus anatinus*) is its sole semi-aquatic representative. It is **monotypic**, and thus the only species of the family Ornithorhynchidae. Theria is further divided into two cohorts (previously called infraclasses): metatherians (marsupials) and the eutherians (often called the "placentals," although marsupials also have a placenta which differs in use). As with the monotremes, the marsupials have only two semi-aquatic species—the water opossum locally known as the yapok (*Chironectes minimus*) and the little water opossum also called the thick-tailed opossum (*Lutreolina crassicaudata*). The remaining species regularly categorized as semi-aquatic are within the eutherians.

As with their dominance within the mammals as a whole, the rodents represent the largest complement of the freshwater semi-aquatic eutherians, with more than eighty-one semi-aquatic species. They range in

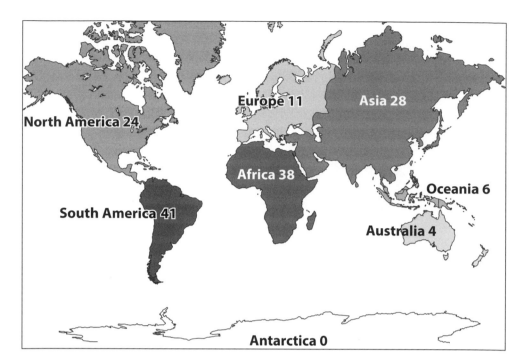

Figure 1.3. Global distribution of freshwater semi-aquatic mammals. Some species occur on more than one continent. Species-specific distributions are described in Chapter 3. M. BRIERLEY

size from two genera of tiny water voles to the large-bodied capybaras and beavers. Carnivores contribute twenty species within five families. Twelve species of otters dominate the semi-aquatic carnivores; however, mink, a mongoose, a genet, an otter civet, and two species of cats are also considered semi-aquatic by many mammalogists. The shrews, the star-nosed mole (*Condylura cristata*), and two species of desmans (order Eulipotyphla, "truly fat and blind") comprise eighteen species, while the hoofed mammals, a tenrec, otter shrews, and two species of rabbits provide the rest.

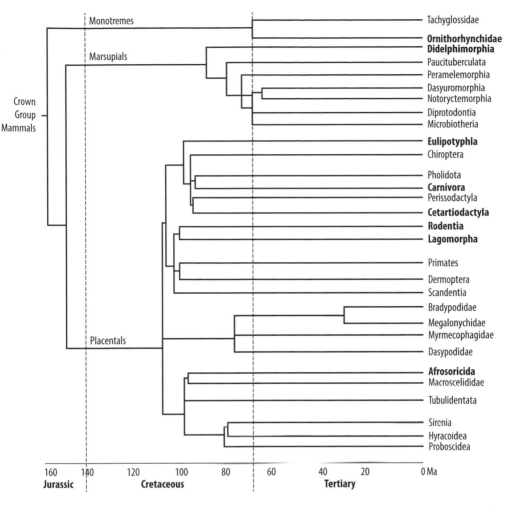

Figure 1.4. Generalized phylogeny of major living mammalian clades (based on Withers et al. 2016: Fig. 1.1). **Bold** indicates presence of semi-aquatic representatives. G. A. HOOD

The Monotremes

The prototherians, the oldest form of mammals, flourished in the Mesozoic, with a few surviving into the Cenozoic (Savage and Long 1989). Today only echidnas and the platypus remain. They are both egg-laying, toothless monotremes; the platypus is the only semi-aquatic species. The platypus is the only living member of the family Ornithorhynchidae (literally "bird snout") and is only found in Australia. Unlike most other mammals, monotremes have only one opening (**cloaca**) for both their excretory (feces and urine) and reproductive tract. In many ways, they retain features of their reptilian ancestors and produce small, soft-shelled, permeable eggs from which the young are born (Feldhamer et al. 2015). One of their distinctions as mammals, however, is the presence of mammary glands. The platypus, like all monotremes, lacks teats and instead secretes milk from pores on its belly (Griffiths et al. 1973).

Along with the rakali (common water rat, *Hydromys chrysogaster*) and the false water rat (*Xeromys myoides*), the platypus is one of the only aquatic mammalian predators in Australia (Fish 2000). When it comes to foraging, the platypus possesses electroreception in its specialized bill to detect and analyze electric fields in its environment to help search for prey items (Scheich et al. 1986). In the mammals, this is a rare adaptation shared with only three other mammals in the world. Although its purpose is not fully understood, the male platypus also has a sharp, venomous spur on its hind limb that might aid competition among males in breeding season (Feldhamer et al. 2015).

The platypus has been called the "paradoxical platypus" (Fish et al. 1997: 2651) because of its primitive anatomical structures for swimming, yet its deft ability to swim and dive with minimal energetic expenditure (Grant and Dawson 1978a). The platypus is the smallest known mammal (0.9 to 1.1 kg) to maintain its normal body temperature in cold water (Grant and Dawson 1978a). Remarkably, it can sustain a stable body temperature in water as cold as 5°C for up to three hours, despite having retained various reptilian features from their synapsid (reptilian) ancestors (Feldhamer et al. 2015). Its dense fur, however, reflects a functional convergence with other semi-aquatic mammals such as beavers and muskrats (*Ondatra zibethicus*) (Estes 1989). As a semi-aquatic species, the platypus's unique anatomy and physiology reveals links to its deep evolutionary past and its perfect suitability for current ecological conditions.

The Marsupials

Marsupialia consists of seven orders, twenty-two families, and 331 species (Wilson and Reeder 2005), yet its only truly semi-aquatic representatives are the water opossum (yapok) and the little water opossum (thick-tailed opossum) of South America. The water opossum ranges from southernmost Mexico as far south as northeastern Argentina, while the little water opossum is found from Guyana in the north to Argentina in the south. Both are from the order Didelphimorphia and family Didelphidae ("two wombs"), although these species have distinct differences. The water opossum has an abdominal pouch (marsupium) on both sexes that opens to the rear (Marshall 1978a), while the pouch is undeveloped, but present, in the little water opossum (Marshall 1978b). A combination of a well-developed sphincter muscle, fat, and fur allow the female water opossum to seal the pouch completely during a dive, thereby protecting the young within a watertight space. As with other marsupials, it is the distinct reproductive mode that separates the water opossum from the monotremes and the eutherians. The young in both the water opossum and the little water opossum are not fully developed when born and attach to the mammae within the marsupium after two weeks to continue their development until weaned two or three months later. The male's pouch is not completely waterproof and is used mainly to protect the male genitalia from colder temperatures while underwater. The birth of underdeveloped young and continued development during lactation is a major distinction of the marsupials from the eutherians.

Of the two species, the water opossum appears to be more exclusively semi-aquatic and has dense water-repellent fur, webbed hind feet, and a streamlined body. Some researchers consider the water opossum to be the only semi-aquatic marsupial (Galliez et al. 2009); however, the little water opossum is regularly described as an excellent swimmer and diver despite its ability to live in the savannahs and woodlands that are near the shores of the streams, lakes, and wetlands where it eats fishes and invertebrates (Nowak 1999). Regidor, Gorostiague, and Sühring (1999: 271) call the little water opossum "one of the least studied didelphid marsupial species," yet, when described, freshwater habitats are regularly a habitat of choice.

The Eutherians

Species in order Afrosoricida, which includes three semi-aquatic otter shrews in central and western Africa and one semi-aquatic species of tenrec on the island of Madagascar, were once represented along with more distantly related species in the previously recognized order Insectivora. Given recent molecular analyses, the insectivores have since been split into three orders, of which order Afrosoricida is one. Further classification of the tenrecs and otter shrews has a "tumultuous taxonomic history" (Everson et al. 2016: 890); however, current genetic analysis solidly places tenrecs and otter shrews into family Tenrecidae and family Potamogalidae, respectively (Wilson and Reeder 2005; Feldhamer et al. 2015; Everson et al. 2016). One semi-aquatic mammal exists in Tenrecidae—the webbed-footed tenrec (*Limnogale mergulus*); three are in Potamogalidae—the tiny Nimba otter shrew (*Micropotamogale lamottei*), the slightly larger Rwenzori otter shrew (*Micropotamogale ruwenzorii*), and the much larger giant otter shrew (*Potamogale velox*). All species are poorly studied and their phylogeny has been revised several times. The Rwenzori otter shrew (de Witte and Frechkop 1955) and Nimba otter shrew (Heim de Balsac 1954) were only formally documented in the twentieth century, although local peoples certainly knew of them previously. Many samples we have today are those individuals caught in local fish traps. Unlike other eutherians, the tenrecs and otter shrews have a single opening for their anus and urogenital tract (cloaca), much like the monotremes. Tenrecidae and Potamogalidae contain some of the world's most primitive mammals (Cornelis et al. 2014), although "primitive" belies their specialized biology (Olson 2013). The tenrecs and otter shrews represent a diverse group relative to body size, habitat use, and physical characteristics. The web-footed tenrec looks like a large shrew, while the otter shrews, as the name implies, resemble small otters to some degree. The four semi-aquatic species in Tenrecidae and Potamogalidae prey on aquatic animals and range from being visibly adapted to aquatic habitats to less specialized in appearance but still dependent on freshwater environments.

There are some distinct differences among the tenrecs and the otter shrews. These differences began twenty-five to forty-two million years ago (Ma) when a single early species of tenrec colonized Madagascar from Africa, likely on a raft of floating vegetation (Olson 2013). The remarkable diversity of tenrecs on the island of Madagascar, including the webbed-footed tenrec, evolved from that single species. The webbed-footed tenrec is often compared to a muskrat in appearance (Feldhamer et al. 2015), given its webbed hind feet and laterally flattened tail. However, its long muzzle is distinct. Unlike the vegetarian muskrat, the webbed-footed tenrec is a predator of fish, amphibians, and aquatic invertebrates. Its range is limited to the rivers, lakes, and marshes of the rapidly disappearing forests of eastern Madagascar. The remaining tenrecs in Africa eventually went extinct except for the otter shrews, which are now in a different family from Malagasy tenrecs (Feldhamer et al. 2015).

As with the webbed-footed tenrec, all three species of otter shrews are well adapted for aquatic life. Only the Rwenzori otter shrew and the giant otter shrew have webbed feet and conjoined second and third toes on the hindfeet (Guth, Heim de Balsac, and Lamotte 1959). However, despite less obvious physical adaptations, the Nimba otter shrew is an excellent swimmer and can remain submerged for up to fifteen minutes (Vogel 1983). One distinguishing feature of otter shrews is the lack of clavicle, which is generally cited when discussing the giant otter shrew, yet unconfirmed when the other two species are specifically described (Guth, Heim de Balsac, and Lamotte 1959). All species are excellent hunters of aquatic prey, such as crustaceans and aquatic invertebrates that they locate underwater through the use of highly innervated whiskers. Their waterproof fur and affinity for riverine dens with underwater entrances show specific adaptations to semi-aquatic environments; however, the paucity of research on the webbed-footed tenrec and otter shrews leaves them vulnerable to rapid habitat changes throughout their range.

Figure 1.5. Swamp rabbit. M. BRIERLEY

ORDER LAGOMORPHA

Of the two families within order Lagomorpha (pikas, rabbits, and hares), family Leporidae contains two species of rabbits within one genera (*Sylvilagus*) that are restricted to riverine and wetland habitats. Although not always considered to be semi-aquatic mammals, they are often included in assessments of semi-aquatic species (Veron, Patterson, and Reeves 2008). The critically endangered riverine rabbit (*Bunolagus monticularis*) is found in the central Karoo region of South Africa, but despite its name, it is not semi-aquatic. According to the IUCN (Collins, Bragg, and Birss 2019), the population of the riverine rabbit declined by at least 60% over the past seventy years, primarily due to extensive habitat loss and fragmentation from agricultural expansion. Unlike the riverine rabbit, the freshwater-dependent swamp rabbit, *S. aquaticus* (Fig. 1.5), and marsh rabbit, *S. palustris*, swim regularly, with the latter being the strongest swimmer of the two. Both are obligate wetland species in the United States and are generally considered to be common, except for the lower Keys marsh rabbit (*S. p. hefneri*), which is critically endangered due to pollution, vehicle strikes, and predation by domestic cats.

ORDER RODENTIA

Of all mammalian orders, Rodentia has the largest number of species. The British population geneticist J. B. S. Haldane posited that "the Creator would appear as endowed with a passion for stars, on the one hand,

and for beetles on the other" (Haldane 1949: 248) given the overwhelming number of beetles relative to all other species of insects, birds, and mammals. When it comes to mammals, there must have also been a passion for rodents; they are found on every continent except Antarctica and comprise approximately 42% of all mammalian species.

There are five suborders of rodents, thirty-six extant families, and at least 2,300 species (Feldhamer et al. 2015), of which as many as eight families contain over eighty-one species that are semi-aquatic. A distinguishing trait of all rodents is their single pair of upper and lower incisors and, as with lagomorphs, a distinct **diastema** (gap) between the incisors and premolars (where canine teeth and additional molars are found in many other orders of mammals). Their incisors, which grow perpetually, allow rodents to gnaw (*rodere* in Latin) on material as soft as herbaceous dicots and as hard as wood. In fact, while studying North American beavers in Alberta, Canada, I measured a beaver-cut stump of a trembling aspen tree (*Populus tremuloides*) that was 185 cm in diameter (Hood and Bayley 2008a). Because of its hard enamel, the anterior (front) side of a rodent's incisors is more resistant to wear than the posterior (back) side, which is composed of softer dentine. So, as rodents chew, this arrangement results in consistently sharp cutting teeth.

Beavers (family Castoridae) are among the best known semi-aquatic rodents. There are two species of beaver: the North American beaver and the Eurasian beaver (*C. fiber*). Beavers range in size to over 30 kg in weight and 120 cm in length, second only to the capybara. Both species of beavers are highly specialized for freshwater habitats and are known as "ecosystem engineers" because of their ability to modify aquatic habitats in dramatic ways. Their webbed hind feet and dorsoventrally flattened tails aid in propulsion, as further described in Chapter 7, "Locomotion and Buoyancy."

Two lesser known species of semi-aquatic rodents come from families Thryonomyidae and Capromyidae. Both species are dependent on freshwater habitats. The greater cane rat (*Thryonomys swinderianus*) is seldom found far from water and swims and dives without difficulty in the marshy reed beds of sub-Saharan Africa. The Cabrera's hutia (*Mesocapromys angelcabrerai*) lives in the swamps, marshes, and small islands off the west coast of Cuba. Hutias are generally considered to be terrestrial; however, the Cabrera's hutia shows a distinct dependence on wetlands for foraging and nesting. Very little is known about this endangered mammal, and more research is warranted given rapid habitat loss within its distribution.

The majority of semi-aquatic rodents are within the large superfamily Muroidea. Three families within Muroidea contain semi-aquatic rodents: Nesomyidae, Cricetidae, and Muridae. The only potential semi-aquatic species in family Nesomyidae is Delany's mouse (also known as Delany's swamp mouse), *Delanymys brooksi*, from the high-altitude swamps and shrublands of Burundi, the Democratic Republic of the Congo, Rwanda, and Uganda. With a body length of 5 to 6 cm (tail length 8.7–11.1 cm) and a maximum weight of 6.2 g, it is one of the world's smallest rodents. Given its rarity and the paucity of research on this species, there is debate as to its classification as semi-aquatic, but its dependence on increasingly fragmented freshwater systems is clear.

Cricetidae (voles and mice—nineteen genera containing forty-two semi-aquatic species) and Muridae (Old World rats and mice—fourteen genera containing thirty semi-aquatic species) are two of the most speciose families in order Rodentia. Once considered a subfamily of Muridae, molecular studies now place Cricetidae in its own family (Feldhamer et al. 2015). Muridae remains the largest family of mammals, despite the elevation of several subfamilies from it, while Cricetidae is the second largest. Together the two families contain more than 50% of all species of rodents. The broad diversity of these families still reflects an evolving taxonomic challenge; but there is a general biogeographic distinction between the Muridae and Cricetidae, with the former being native to Eurasia, Africa, and Australia and the latter to the Americas, Eurasia, and East Africa.

Semi-aquatic cricetids include many New World species of rats, voles, lemmings, and mice, as well as the round-tailed muskrat (*Neofiber alleni*) and muskrat. Eight of the semi-aquatic cricetid species are monotypic (Table 1.1). Members of Cricetidae range from the subarctic to the tropics and vary broadly in appearance. Of the **polytypic genera**, there are five with four or more species (*Holochilus, Ichthyomys, Nectomys, Rheomys*, and *Neusticomys*).

All four species of crab-eating rats within the genus *Ichthyomys* (*I. hydrobates, I. pittieri, I. stolzmanni*, and *I. tweedii*) are endemic to South America and frequent swampy areas and streams. Each species is of con-

Table 1.1. Semi-aquatic species in family Cricetidae

Species	Common name	Species	Common name
Amphinectomys savamis	Ucayali water rat	Neotomys ebriosus	Andean swamp rat
Anotomys leander	Ecuador fish-eating rat	Neusticomys ferreirai	Ferreira's fish-eating rat
Arvicola amphibius	European water vole	N. monticolus	montane fish-eating rat
A. sapidus	southern water vole	N. mussoi	Musso's fish-eating rat
Chibchanomys orcesi	Las Cajas water mouse	N. oyapocki	Oyapock's fish-eating rat
C. trichotis	Chibchan water mouse	N. peruviensis	Peruvian fish-eating rat
Deltamys kempi	Kemp's grass mouse	N. venezuelae	Venezuelan fish-eating rat
Holochilus brasiliensis	web-footed marsh rat	N. vossi	Voss fish-eating rat
H. chacarius	Chaco marsh rat	Ondatra zibethicus	muskrat
H. lagigliai	Lagiglia's marsh rat	Oryzomys couesi	Coues's rice rat
H. sciureus	marsh rat	O. palustris	marsh rice rat
Ichthyomys hydrobates	crab-eating rat	Rheomys mexicanus	Mexican water mouse
I. pittieri	Pittier's crab-eating rat	R. raptor	Goldman's water mouse
I. stolzmanni	Stolzmann's crab-eating rat	R. thomasi	Thomas's water mouse
I. tweedii	Tweedy's crab-eating rat	R. underwoodi	Underwood's water mouse
Lundomys molitor	Lund's amphibious rat	Scapteromys aquaticus	Argentine swamp rat
Microtus richardsoni	North American water vole	S. meridionalis	plateau swamp rat
Nectomys apicalis	western Amazonian nectomys	S. tumidus	Waterhouse's swamp Rat
N. palmipes	Trinidad water rat	Sigmodontomys alfari	Alfaro's rice water rat
N. rattus	common water rat	Synaptomys borealis	northern bog lemming
N. squamipes	South American water rat	S. cooperi	southern bog lemming
Neofiber alleni	round-tailed muskrat		

servation concern because of habitat loss. Body size ranges from 11 to 21 cm (tail 9 to 14 cm) and the largest of the species weighs up to 120 g (*I. tweedii*; Voss 1988). All four species are specialized freshwater predators. Their large hind feet are fringed with hair, their hind toes are partly webbed, and they have stiff hairs under their tails to aid in locomotion.

Similarly, the Neotropical water rats (genus *Nectomys*) of South America and Trinidad and Tobago (specifically, Trinidad water rat, *N. palmipes*) possess a swimming fringe under the tail and on their large, sparsely haired hind feet (Nowak 1999). There are four currently recognized species (*N. apicalis*, *N. palmipes*, *N. rattus*, and *N. squamipes*), all of which are semi-aquatic. The IUCN also lists *N. grandis* as a fifth species; however, it is sometimes considered to be a subspecies of *N. squamipes*, or synonymous with *N. magdalenae* (now *N. apicalis*). The partial interdigital webbing of the hind feet distinguishes the *Nectomys* from other members of the subfamily Sigmodontinae (New World rats and mice). *Nectomys* species are also relatively large compared to other species of Sigmodontinae. The South American water rat (*N. squamipes*) ranges in length from 16.0 to 25.5 cm, with a tail from 16.3 to 25.0 cm, and weighs between 160 and 420 g (Nowak 1999). Unlike *Ichthyomys* species, the Neotropical water rat populations are stable, although prior to its updated classification as a small-footed bristly mouse (*N. rattus*), the formerly known small-footed water rat (*N. parvipes*) was considered critically endangered by the IUCN (Nowak 1999). Similarly, the Magdalena nectomys (*N. magdalenae*) was considered data deficient by the IUCN prior to its reassessment as a western Amazonian nectomys (*N. apicalis*), thus highlighting the influence of taxonomic reassessment on conservation priorities.

All seven species of fish-eating rats (genus *Neusticomys*) live along streams in the dense tropical forests of South America (Velandia-Perilla and Saavedra-Rodríguez 2013). Although their feet are smaller than those of *Nectomys* species, their hind feet still have a slight fringe of stiff hair to increase surface area for swimming after aquatic invertebrates. Their ability to swim is evident in the name of their genus, which derived from the ancient Greek *neustikós* (able to swim) and μῦς (pronounced mûs "mouse").

As with any large taxon, the morphological diversity and degree of specialization within the cricetids is extensive. From the highly specialized Ecuadoran fish-eating rat (*Anotomys leander*) from Andes of Ecuador, with its tiny slit-like ear that closes via muscular contractions when submerged (Voss 1988), to the more northerly muskrat, which is the largest (body length 22.9–32.5 cm, tail length 18–29.5 cm) and heaviest (681–1,816 g) of the cricetids, these species highlight the diversity of freshwater semi-aquatic mammals. Such diversity and specialized adaptations are explored throughout this book.

Muridae, the largest of the rodent families, now has a worldwide distribution excluding Antarctica and some oceanic islands (Feldhamer et al. 2015). Murids share an enlarged keyhole-shaped infraorbital foramen (opening in the maxillary bone below the lower margin of the eye), a broad zygomatic plate to support strong chewing muscles, and a sciurognathus lower jaw (in which the angular process at the back of the jaw and incisors begin in the same area). Three murid genera have three or more semi-aquatic species (Table 1.2).

Genus *Dasymys* (Ancient Greek for "hairy mouse") is endemic to Africa and has recently been divided into eleven species, many of which were previously classified as the African marsh rat (*Dasymys incomtus*). Their taxonomy is still under review, but all eleven species are listed below. The shaggy swamp rats range in size from 9.0 to 11.3 cm (tail length 9.7 to 18.5 cm) and weighs 48 to 150 g (Kingdon 1974; Nowak 1999). Their common name is apt given the quality of their fur, ranging from long and straight to "shaggy" (Nowak 1999). Species within *Dasymys* are adept swimmers as reflected by their preferred habitats of marshes, bogs, and reed beds (Kingdon 1974).

Three genera of semi-aquatic water rats are endemic to Australo-Papuan freshwater ecosystems: *Crossomys*, *Hydromys*, and *Parahydromys*. They share a dorsally restricted cranium behind the orbits and simplified or obliterated occlusal molar patterns, of which the third upper and lower molars are severely reduced or absent altogether (Helgen 2005). Both the earless water rat (*C. moncktoni*) and the waterside rat (*P. asper*) are monotypic. Two species from genus *Baiyankamys* (*B. habbema* and *B. shawmayeri*) were once classified within genus *Hydromys*. Three *Hydromys* species (*H., hussoni*, *H. neobritannicus*, and *H. ziegleri*) are endemic to Papua New Guinea and nearby islands, while the distribution of the rakali (*H. chrysogaster*) extends to Australia. Also called "beaver rats" or "common water rats" (Nowak 1999: 1611), the rakali, which is quite large (body 11.9 to 35.0 cm, tail 10.3 to 35.0 cm, maximum weight

Table 1.2. Semi-aquatic species in family Muridae

Species	Common name	Species	Common name
Baiyankamys habbema	mountain water rat	Deomys ferrugineus	Congo forest mouse
B. shawmayeri	Shaw Mayer's water rat	Hydromys chrysogaster	rakali
Colomys goslingi	African wading rat	H. hussoni	western water rat
Crossomys moncktoni	earless water rat	H. neobritannicus	New Britain water rat
Dasymys capensis	Cape marsh rat	H. ziegleri	Ziegler's water rat
D. foxi	Fox's shaggy rat	Nesokia bunnii	Bunn's short-tailed bandicoot rat
D. griseifrons		Nilopegamys plumbeus	Ethiopian amphibious rat
D. incomtus	African marsh rat	Otomys lacustris	Tanzanian vlei rat
D. longipilosus		Parahydromys asper	waterside rat
D. medius		Pelomys fallax	creek groove-toothed swamp rat
D. montanus	montane shaggy rat	P. hopkinsi	Hopkins's groove-toothed swamp rat
D. nudipes	Angolan marsh rat	P. isseli	Issel's groove-toothed swamp rat
D. robertsii	Robert's shaggy rat	Rattus lutreolus	Australian swamp rat
D. rufulus	West African shaggy rat	Waiomys mamasae	Sulawesi water rat
D. shortridgei		Xeromys myoides	false water rat

300 g), prefers both **lotic** (moving) and **lentic** (still) aquatic habitats. As with many semi-aquatic mammals, the rakali was overharvested for its thick, waterproof fur until conservation measures were initiated in 1938 (Nowak 1999). The earless water rat is the most specialized for aquatic life, with its very large, webbed hind feet.

Two additional families of rodents include semi-aquatic species: Caviidae and Myocastoridae. Caviidae contains the capybara (*Hydrochoerus hydrochaeris*) and the lesser capybara (*H. isthmius*), and Myocastoridae contains the coypu (*Myocastor coypus*). Outside of its South American distribution, the coypu (also known as the nutria) is an invasive species on every continent except Antarctica and Australia. A relatively large rodent (body length 43 to 63.5 cm, tail length 25.5 to 42.5 cm, weight 5 to 17 kg), it often degrades wetland plant and animal communities through its concentrated feeding and burrowing behaviors (Carter and Leonard 2002). Caviidae are also from South America; however, the distribution of the lesser capybara extends as far north as the Panama Canal. Although the lesser capybara is listed as data deficient by the IUCN (2019), both the capybara and coypu are considered to be of least concern. At a body length of 100 to 130 cm and a weight between 27 and 79 kg, the capybara is the world's largest extant rodent. Unlike many other semi-aquatic mammals, its long, course fur is sparse; however, it is an excellent underwater swimmer and can dive with ease. All three species are targeted by hunters: the capybaras mainly for meat and hides, and the coypu for its thick underfur and, more recently, to reduce non-native populations.

ORDER EULIPOTYPHLA

As with tenrecs and otter shrews, semi-aquatic species of shrews, desmans, moles, and their relatives were once grouped in with all other "insectivores." However, recent molecular analyses placed them in their own order. Despite Eulipotyphla literally translating as "truly fat and blind," many species in this order are neither fat nor blind, although most species have poor eyesight, with the star-nosed mole being functionally blind. What they lack in eyesight they make up for with specialized nerve cells in the sensory organs in their whiskers (vibrissae) or, in the case of the star-nosed mole, twenty-two curious appendages creating a star surrounding its nostrils. The facial appendages of the star-nosed mole are composed of sensory organs called the Eimer's organs, which in turn are composed of specialized neural structures that allow the moles to detect prey at rapid speeds (Catania 1999). As a part of his fascinating research, Kenneth Catania used high-speed cameras and showed that the "star" moves so quickly it cannot be seen with the naked eye. These specialized appendages touch twelve or more locations per second to "see" food items, resulting in the star-nosed mole briskly consuming its prey. To satisfy its massive appetite, the star-nosed mole has further assistance in detecting prey: whiskers on the sides of its nose, eyes, ears, and fore-

Figure 1.6. Russian desman. M. BRIERLEY

feet (Merritt 2010). The star-nosed mole is the only semi-aquatic mole; other moles are mainly fossorial (living below ground).

Other members of Eulipotyphla also accommodate for poor eyesight to become very effective predators (Table 1.3). To compensate for their tiny eyes, American water shrews use their whiskers and "underwater sniffing" to locate prey. For underwater sniffing, the shrew expels then re-inhales air bubbles through its nose to detect the scent of its prey (Catania, Hare, and Campbell 2008). Its whiskers aid in prey detection and help it forage throughout the day and night to meet a metabolism that is much higher than other mammals of its size (Platt 1974).

There are two families within Eulipotyphla that contain semi-aquatic species: Soricidae (shrews) and Talpidae (moles and desmans). The common grouping of the shrews is deceiving, however, given the hundreds of species of shrews within three major subfamilies—the white-toothed shrews (Crocidurinae), the red-toothed shrews (Soricinae), and the African shrews (Myosoricinae)—of which more than a dozen are semi-aquatic in the Crocidurinae and Soricinae subfamilies. In Talpidae, one species of mole and both living species of desmans are semi-aquatic. As such, they are more restricted in geographic range than the shrews. The endangered Russian desman (*Desmana moschata*) is limited to the Volga, Don, and Ural rivers in Russia, Ukraine, and Kazakhstan (Fig. 1.6). Listed as vulnerable by the IUCN, the Pyrenean desman (*Galemys pyrenaicus*) is confined to the Pyrenees Mountains in Andorra, France, Spain, and northern Portugal. As with the shrews, desmans have long, pointed noses, although the desman's highly flexible, tactile nose is grooved and divided into two lobes at the tip to aid in detecting prey. Shrews, the star-nosed mole, and desmans diverged early on in evolutionary history (similar to the tenrecs), and they possess neural and physiological characteristics that make them highly specialized mammals.

Table 1.3. Semi-aquatic species in families Soricidae and Talpidae

Species	Common name	Species	Common name
Crocidura mariquensis	swamp musk shrew	*Neomys anomalus*	southern water shrew
C. stenocephala	Kahuzi swamp shrew	*N. fodiens*	Eurasian water shrew
Chimarrogale hantu	Malayan water shrew	*N. teres*	Transcaucasian water shrew
C. himalayica	Himalayan water shrew	*Sorex alaskanus*	Glacier Bay water shrew
C. phaeura	Bornean water shrew	*S. bendirii*	Pacific water shrew
C. platycephalus	Japanese water shrew	*S. palustris*	American water shrew
C. styani	Chinese water shrew	*Condylura cristata*	star-nosed mole
C. sumatrana	Sumatran water shrew	*Desmana moschata*	Russian desman
Nectogale elegans	elegant water shrew	*Galemys pyrenaicus*	Pyrenean desman

ORDER CARNIVORA

Within order Carnivora there is tremendous diversity in the five families of currently recognized freshwater semi-aquatic mammalian carnivores: Felidae (cats), Herpestidae (marsh mongooses), Viverridae (otter civets and the aquatic genet), Mustelidae (mink and otters), and Procyonidae (raccoons). Many authorities recognize the unique dependence of two species of wild cats (family Felidae) on freshwater habitats, though they are more riparian specialists (Pacini and Harper 2008; Veron, Patterson, and Reeves 2008).

All carnivores have a specialized fourth upper premolar and first lower molar known as carnassial teeth. These teeth are designed to cut or shear prey, but they vary in length and sharpness within the carnivores. Carnivores also have canine teeth that are longer than their other teeth. Their tooth development suggests a diet of meat and meat alone, but some members of Carnivora have herbivorous or omnivorous diets. Within the semi-aquatic carnivores, however, all species kill and eat other animals. The otter civet (*Cynogale bennettii*) and marsh mongoose

(*Atilax paludinosus*) occasionally eat fruit, although it is not a large portion of their diets.

Within Felidae, the fishing cat (*Prionailurus viverrinus*) of southern and Southeast Asia is recognized by some authors as a freshwater-associated or riparian mammal (Pacini and Harper 2008; Veron, Patterson, and Reeves 2008). It has an extensive range in Southeast Asia and is a wetland specialist that targets rodents, birds, and fish. Another wild cat, the flat-headed cat (*P. planiceps*), is also mentioned. It lives in the swamps, wetlands, and riverine habitats of Malaysia, where it hunts for fish, shrimp, birds, and small rodents. Both species are subject to habitat destruction and poaching. However, Howell (1930) excludes all cats from his assessment of aquatic (or semi-aquatic) mammals in Carnivora (p. 30). Therein lies the challenge with the designation of "semi-aquatic" for mammals and its varying imperatives. The inclusion of the fishing cat and flat-headed cat as riparian mammals by several authors is because of the almost exclusive reliance of these cats on freshwater habitats and because of the population decline caused by wetland draining and degradation throughout their range.

Two other families within suborder Feliformia (cat-like mammals) have semi-aquatic species: Viverridae (otter civets and the aquatic genet) and Herpestidae (marsh mongooses). Viverridae differ from Herpestidae by having longer tails, webbing between their five toes, retractile claws in most species, and longer and more pointed pockets (bursae) on the margins of their ears (Feldhamer et al. 2015). Differences in the auditory bullae and anal sacs, as well as genetic differences, resulted in the removal of the mongooses from Viverridae and their placement in their own family. The otter civet and aquatic genet (*Genetta piscivora*) are the only semi-aquatic viverrids from an otherwise very diverse and widespread Old World tropical and subtropical family (including Europe, Africa, the Middle East, and Asia). Both the otter civet and aquatic genet are species of concern and struggle with habitat loss. The Lowe's otter civet (*Cynogale lowei*) was once defined as its own species from a single specimen found in northern Vietnam in the winter of 1927, but Scott Robertson and his colleagues (2017) used multiple diagnostic techniques to determine that it was actually a juvenile Eurasian otter (*Lutra lutra*). Fortunately, resolving historic taxonomic errors allows for more focused conservation efforts on confirmed species.

The otter civet is a rare species that is especially suited to semi-aquatic environments. It has been listed as endangered since the mid-1990s, primarily through habitat destruction (IUCN 2019). Although rarely seen, its distribution includes the Malay Peninsula, Sumatra, the island of Borneo, and southern Thailand. With a maximum size of just under 90 cm from head to tail, and up to 5 kg in weight (Nowak 1999), it is an important predator and likely at the top of its trophic level, as are many civets (Veron et al. 2006). As its name suggests, it shares a common body shape with otters and, like many semi-aquatic mammals, has dense underfur, webbed feet, and the ability to close its nose and ears while underwater. As a predatory tactic, the otter civet sometimes remains underwater with only the tip of its nose above the surface (Pocock 1914). Most remarkable are its very lengthy and abundant whiskers at the end of its long muzzle and under its ears, which aid in detection of underwater prey (Pocock 1914; Nowak 1999).

One of the rarest of the African carnivores, the aquatic genet of the Democratic Republic of the Congo prefers shallow headwater streams and small brooks (Van Rompaey, Gaubert, and Hoffmann 2008; Van Rompaey and Colyn 2013). Very little is known about this species and its biology due to its scarcity and its solitary **crepuscular** and **nocturnal** behavior. Indeed, only thirty museum species exist globally and, in the wild, Bambuti pygmies and Bantu rarely capture aquatic genets in their ground snares (Hart and Timm 1978). Unlike other genets, the aquatic genet lacks spots and instead has a bushy black tail, a chestnut red to dull red body, and yellow to white spots on its forehead, muzzle, and cheeks (Nowak 1999; Van Rompaey and Colyn 2013). Adaptations that help them catch fish and other aquatic prey are bare palms and soles and sharp, narrow carnassial teeth. As with the somewhat larger otter civet, vibrissae on the aquatic genet's muzzle are very long. The aquatic genet attracts insectivorous fish by placing its vibrissae on the surface of the water (Van Rompaey and Colyn 2013).

Family Herpestidae is closely related to Viverridae and includes one semi-aquatic species: the marsh mongoose of sub-Saharan Africa (Baker 1992a), also known as the water mongoose. Of the species in family Herpestidae, the marsh mongoose is most closely linked to freshwater habitats. It is rarely far from water and mostly eats aquatic prey (especially crustaceans), often after searching underwater holes and crevices.

Its high consumption of aquatic crustaceans is unusual for herpestids (Baker and Ray 2013). As a prominent predator, the water mongoose plays an important role in the faunal community of papyrus swamps (Baker 1992a). However, of the semi-aquatic mammals, the water mongoose appears to be one of the least physically adapted to aquatic habitats (Estes 1989), despite being an excellent swimmer and diver (Taylor 1989). As such, it likely represents the outer limits of the definition of "semi-aquatic." The closely related long-snouted mongoose (*Herpestes naso*) is often associated with freshwater systems, but it is described as a terrestrial species by the IUCN (2019).

Family Mustelidae includes several semi-aquatic species in the two subfamilies: Lutrinae (otters) and Mustelinae (weasels, which includes mink). Globally there are eleven species of freshwater otters and two mink species. In addition, the marine otter (*Lontra felina*) does use some coastal freshwater habitats, but it is mainly found in coastal marine environments. The marine otter is included in this book to accommodate its rare use of fresh water. The closely related sea mink (*Neovison macrodon*) was hunted to extinction in the nineteenth century. Today the conservation status of most species of otters is of concern due to declining populations (Duplaix and Savage 2018; IUCN 2019).

Like many other mustelids, otters and mink have long thin bodies, with short limbs (Fig. 1.7). Their well-developed anal scent glands secrete strong smelling compounds used for chemical communication, defense, and the marking of territories (Feldhamer, Thompson, and Chapman 2003). In otters and mink the carnassial teeth are well developed to aid in hunting. Of the two subfamilies, otters are the most physically adapted for semi-aquatic environments (Estes 1989). So close is the otter's association with water that the Old English and Indo-European origin of the word "otter" shares the same root as the word for "water."

Semi-aquatic specialization in otters extends thirty thousand years into the fossil record (Chanin 1985). Otters are classified using three key characteristics: the shape of the baculum (penis bone), vocalizations, and male external genitalia (Davis 1978; Chanin 1985). These characteristics, combined with the analysis of the mitochondrial cytochrome *b* gene, divide otters into three clades: (1) the North American river, Neotropical, and marine otters, (2) the sea, Eurasian, spotted-necked, cape clawless, and small-clawed otters, and (3) the giant otter of South

Figure 1.7. North American river otter (*top*) and American mink (*bottom*). M. BRIERLEY

America (Koepfli and Wayne 1998). Of the seven genera of otters, six genera have freshwater species, although some venture into brackish waters in coastal habitats. They range in size from the 1 to 5 kg Asian small-clawed otter (*Aonyx cinereus*), which has a body length of 45 to 61 cm and a tail length of 25 to 35 cm, to the giant otter (*Pteronura brasiliensis*), which has a body length of 86 to 140 cm and a tail length of 33 to 100 cm (Nowak 1999) and is the largest species of the Mustelidae. Male giant otters weigh 26 to 34 kg and females weigh 22 to 26 kg—making them the same size or heavier than an adult Siberian husky.

There are three species within the genus *Aonyx*: the African clawless otter (*Aonyx capensis*; a.k.a. Cape clawless otter), the Congo clawless otter (*A. congicus*), and the Asian small-clawed otter. Although the Congo clawless otter was considered to be conspecific (same species) as the African clawless otter (Wozencraft 2005), the IUCN Otter Specialist Group considers it a distinct species (Jacques et al. 2015). Its rarity and remote habitats in Africa's central equatorial rain forest make it the least known of the African otters (Jacques, Duplaix, and Chapron 2004). For most individuals, the silver-tipped guard hairs on the Congo clawless otter and a four-sided black patch extending from below the eye to above the nose differentiate the two species in the field. Both species are "large and bulky" (Jacques, Duplaix, and Chapron 2004) and share similar weight and body lengths. Unlike the clawless otters, the Asian small-clawed otter has partial webbing between its front toes but, as its name suggests, has tiny claws that do not extend beyond the end of its toes. Its distribution, although shrinking in size, extends from the Himalayan foothills of northern India eastward to Indonesia.

The sole species within the genus *Lutrogale* is the smooth-coated otter (*Lutrogale perspicillata*). Its range extends in low-lying areas throughout south and Southeast Asia, with an additional isolated subpopulation in Iraq (de Silva et al. 2015). This thick-necked otter is the largest of the otters in Asia with a 7 to 11 kg weight, 65 to 79 cm head and body length, and 40 to 50.5 cm tail length (Nowak 1999). It has short, smooth fur and, although similar in appearance to the Old World otters (genus *Lutra*), the dorsoventrally flattened tail (flattened from upper and lower sides) is much more pronounced, and there are obvious keels on each side toward the end of the tail. Its physical similarities with the Old World otters once placed it into the same genus, but it was considered its own genus by Gray in 1865 (de Silva et al. 2015).

The Old World river otters include the Eurasian river otter (*Lutra lutra*) and the hairy-nosed otter (*L. sumatrana*). The Japanese otter (*L. nippon*) was officially declared extinct in 2012 after last being seen in 1979. The spotted-necked otter (*Hydrictis maculicollis*) was formerly included in the genus *Lutra*, but mitochondrial gene analyses by Klaus-Peter Koepfli and his colleagues (2008b) established its placement in genus *Hydrictis*. Old World otters weigh 5 to 14 kg, have a head and body length of 50 to 82 cm, and have a tail length of 33 to 50 cm (Nowak 1999). As with other mustelids, on average, the males are larger than the females. Typical of this genus, their short, wide neck supports a flattened and rounded head (Nowak 1999). The IUCN (2019) classifies the hairy-nosed otter as endangered and the spotted-necked and Eurasian otters as near threatened because of overhunting, ongoing habitat loss and degradation, and bioaccumulation of organic pollutants and heavy metals, among other anthropogenic threats (He et al. 2017; IUCN 2019).

New World river otters include four species: the Neotropical river otter (*Lontra longicaudis*), the southern river otter (*L. provocax*), the North

American river otter (*L. canadensis*), and the marine otter (*L. felina*). Only the Neotropical, southern, and North American river otters are freshwater species. As previously mentioned, the marine otter occasionally ventures into coastal fresh waters to hunt (Estes 1986; Nowak 1999). *Lontra* was considered to be the same genus as the Old World otters (*Lutra*), but *Lontra* and *Lutra* are now classified as **diphyletic** (derived from two different ancestral lines). All species of *Lontra* have fully webbed feet, streamlined bodies, and strong, agile tails that make them highly adapted to swimming, diving, and foraging underwater. Sharing a South American distribution with the Neotropical otter is the highly social giant otter, which lives in family groups of up to ten individuals. Its unique periscoping behavior (lifting its head and neck out of the water) reveals distinct chin and throat markings that allow for individual identification (Duplaix 1980). The giant otter is listed as endangered by the IUCN and critically endangered in Paraguay and Ecuador. It is extirpated in Uruguay and likely extinct in Argentina. Across its range, overhunting, habitat destruction, and naturally low reproductive rates are key factors in population declines (Groenendijk et al. 2015).

Sharing common habitat with the giant otter, the crab-eating raccoon (*Procyon cancrivorus*) is limited to coastline and riverbank habitats for much of the year, unlike its northern cousin the common raccoon (*P. lotor*), which is less dependent on freshwater habitats. As with the fishing cat and flat-headed cat, the crab-eating raccoon is more of a riparian specialist that exhibits a propensity for catching freshwater prey and living in close proximity to freshwater habitats (Gatti et al. 2006).

ORDER ARTIODACTYLA

Some modern phylogenists combine Artiodactyla and Cetacea (whales, dolphins, and porpoises) into the superorder Cetartiodactyla, thus highlighting the ecological link between hoofed mammals and cetaceans. The order Artiodactyla represents the even-toed ungulates and contains twelve species in five families (Suidae, Hippopotamidae, Tragulidae, Cervidae, and Bovidae) that are considered by the IUCN (2019) to be dependent on freshwater systems. As with carnivores, the degree of specialization within the ungulates varies broadly. The three species of *Babyrousa* are classified by the IUCN as terrestrial-freshwater specialists (IUCN 2019): hairy babirusa (*Babyrousa babyrussa*), Sulawesi babirusa

(*B. celebensis*), and Togian Islands babirusa (*B. togeanensis*). They are not included in a 2008 article by Geraldine Veron, Bruce D. Patterson, and Randall Reeves, "Global diversity of mammals (Mammalia) in freshwater," yet they are described as riparian mammals in Nic Pacini and David Harper's chapter ("Riparian, Semi-Aquatic, and Riparian Vertebrates") in David Dudgeon's 2008 book, *Tropical Stream Ecology* (Pacini and Harper 2008: Chapter 6). Given their clear dependence on wetlands and seasonally flooded swamps, the compromise would be to include *Babyrousa* as freshwater-dependent species that occasion upland habitats to a lesser degree. The same designation applies to the water chevrotain (*Hyemoschus aquaticus*; Family Tragulidae), which lives in seasonal swamps, mangrove swamps, papyrus stands, and other damp habitats. It is rarely found more than 250 m from water.

Of the ungulates, the common hippopotamus and, to a somewhat lesser degree, the pygmy hippopotamus (*Choeropsis liberiensis*) are well adapted to aquatic habitats. The semi-aquatic common hippopotamus exhibits many of the classic traits for life in water. As with many fully aquatic and semi-aquatic mammals, its ears, nose, and eyes are aligned on the top of its head, and both the ears and nose seal off when underwater. Although it does not often swim, per se, it uses its partially webbed feet to support itself on the bottom of waterbodies, where it pushes itself from the bottom in almost a gallop-like motion (Nowak 1999; Coughlin and Fish 2009). The faster the "gallop," the longer the hippopotamus stays suspended in the water column. The combination of its hefty weight (1,000–4,500 kg) and high bone density allows the hippopotamus to regain contact with the bottom to maintain forward momentum. The endangered pygmy hippopotamus, while still semi-aquatic, is less prone to swimming than its larger relative. Its larger nostrils and eyes set on the side of its more rounded head are less adapted to underwater activity, despite its high dependency on streams and wetlands.

Within family Cervidae, three species in three genera (*Blastocerus dichotomus*, *Elaphurus davidianus*, and *Hydropotes inermis*) are associated with freshwater habitats. Moose (*Alces alces*) are often associated with circumpolar wetlands since they are strong swimmers and are able to surface dive a few meters under the water. However, they are considered a terrestrial species by the IUCN (2019). Certainly in northern climates moose are forest specialists when waterbodies are covered in snow and ice for months at a time. At most they are seasonally associated with wetlands when they become very territorial at prime waterbodies. In contrast, marsh deer (*Blastocerus dichotomus*) are only found in marshes of the Pantanal region of central South America and the Gran Chaco natural region of south-central South America. The Père David's deer (*Elaphurus davidianus*) of the marshlands of subtropical China and the water deer (*Hydropotes inermis*) of China and the Korean peninsula fill a similar role as that of the marsh deer. All three species of deer are considered to be semi-aquatic, although they graze on riparian and grassland plants as well.

The two genera in family Bovidae (*Kobus* and *Tragelaphus*) contain three species that the IUCN (2019) considers terrestrial-freshwater specialists; however, their extent of obligatory use of freshwater habitats varies. Two of the southern lechwe (*Kobus leche*) of south-central Africa and the endangered Nile lechwe (*K. megaceros*) of Ethiopia and South Sudan are wetland specialists. The southern lechwe, is particularly adapted to wetland habitats and, similar to the sitatunga, has a water-repellent substance on the fur of its legs that aids rapid movement through water (Skinner and Chimimba 2005). The sitatunga (*Tragelaphus spekii*), a wetland antelope of central Africa, inhabits swamps and marshy habitats. In chatting with Camille Warbington about her fascinating doctoral research on sitatunga in Uganda, I was struck by the sitatunga's complete dependence on swamp habitats, as reflected by their long, wide feet (up to 16 cm in length) and water-repellent fur.

As mentioned before, the distinction between aquatic mammals, semi-aquatic mammals, riparian mammals, and similar terms presents challenges when definitively assessing what is and what is not distinctively semi-aquatic. Unlike a taxonomic scheme that follows specific morphological and genetic distinctions, the degree to which a mammal is semi-aquatic might be physically evident (e.g., the webbed feet and flatted tail of the platypus and beavers) or physiological determined (e.g., ability of a muskrat to hold its breath for up to fifteen minutes). In his book *Aquatic Mammals: Their adaptation to life in the water*, A. Brazier Howell stated the following:

> To merit the term aquatic, however, a vertebrate must be sufficiently at home in the water so that when near this element it will instinctively

seek it for concealment or escape. It must not only be able to swim with adequate speed but must be able to dive with facility and remain submerged. In almost all cases there is necessary the further corollary that it habitually seeks its food in the water. The ability to propel itself through the water does not mean that it has necessarily developed organs that are appreciably more highly specialized for such use than are found in its nearest, wholly terrestrial relative, but it does mean that such a vertebrate must have evolved some sort of valvular mechanism for closing both nostrils and ears while submerged. In other words an aquatic vertebrate must have the ability to close all orifices leading within the body. (Howell 1930: 4)

For some semi-aquatic species, Howell's criteria are easily met and are easily included in his book (e.g., beavers, otters), while others (e.g., marsh and swamp rabbits) are still counted, despite their lack of obvious morphological adaptations. Semi-aquatic species lie somewhere between the fully aquatic and the fully terrestrial—what Howell calls a "borderline existence" (Howell 1930: 5), where the degree of adaptation reflects specific niches within freshwater habitats. Of the species presented so far, approximately 13% currently fit the IUCN designation of data deficient, and an additional 4% of the species presented in this book have yet to be evaluated. As freshwater systems continue to be under threat, we are racing against the clock to assess the mammals that depend on them.

Freshwater Habitats

At the broadest level, surface freshwater systems are either lotic or lentic, with rivers and streams in the former category and wetlands, lakes, and reservoirs in the latter. Fresh water only comprises 2.5% of the world's water, while oceans make up the rest (Carpenter, Stanley, and Vander Zanden 2011). Of the global freshwater resources, 98.6% is in the form of groundwater or glaciers, which leaves only 0.26% available as surface water. Therefore, some of the world's rarest mammals are linked to one of the rarest resources on earth (Fig. 1.8). Of additional concern is that surface freshwater systems provide three-quarters of the water withdrawn for human use (Carpenter, Stanley, and Vander Zanden 2011).

Open-water pond, muskrat, *Ondatra zibethicus*

Small stream, Pyrenean desman, *Galemys pyrenaicus*

Beaver channels, North American beaver, *Castor canadensis*

Watering hole, common hippopotamus, *Hippopotamus amphibious*

Figure 1.8. Freshwater habitats and semi-aquatic mammals. M. BRIERLEY

Rivers and streams move water from a higher elevation to a lower elevation in some form of channelized way. There are various methods to classify them relative to stream order (relative size) and morphology (channel shape). The size of the catchment can influence stream order, with large rivers such as the Amazon classified as a twelfth-order stream, while a headwater stream is classified as a first-order stream. Common terms can also reflect the size of lotic waterbodies, progressing from a small brook to a creek, stream, and then river, although these terms are often interchangeable depending on local norms. The velocity of a stream or river depends on the stream gradient, channel depth and width, and the sheer volume of water in the system. Use of these systems by semi-aquatic mammals is directly associated with water velocity, depth, water clarity, stream substrate, prey abundance, and shoreline and within-stream vegetation, among other habitat features. While assessing fish habitats on Canada's west coast, I witnessed a small creek turn into a formidable stream in a matter of minutes during an impressive rainstorm. Semi-aquatic mammals, such as river otters and mink, must be adaptable to such changeable environments, especially when prey are similarly affected.

Lentic environments are broadly represented by wetlands, lakes, and reservoirs. As with lotic systems, there are various categories and classifications of lentic waterbodies. In William Mitsch and James Gosselink's seminal book, *Wetlands*, they provide thirty-five common terms used to describe wetlands around the world (Mitsch and Gosselink, 2007: Table 2.1). To avoid the varied uses of the same common terms, more formal systems exist to represent specific geographic areas and climatic influences. However, universally, the key definition of a wetland contains three critical components: (1) there is water present at some time of the year (excluding overly dry years), (2) soils reflect a predominately saturated state, and (3) there are **hydrophilic** (water loving) plants present. Marshes are wetlands dominated by herbaceous vegetation, while swamps and some bogs have woody vegetation, and wet meadows are as they seem. Wetlands are often divided into **ombrotrophic** (all water comes from precipitation) or **minerotrophic** (water comes from streams or groundwater); however, some wetlands receive water from both sources. Wetlands can also be **peatlands** (e.g., fens and bogs, some marshes and swamps) that form on thick accumulations of partly decayed plant material and **non-peatlands** (e.g., most marshes, open water ponds, most swamps) that form on mineral-based soils. There are entire textbooks based solely on wetland ecology, so a general description will suffice for this book, with more specific habitat descriptions provided while discussing individual mammals.

While wetlands are transitional features between land and lakes and rivers, lakes are generally larger than wetlands and, depending on size, can experience **stratification** (separation into distinct layers based on temperature and water density) if deep enough. As with wetlands, there are different categories of lakes, often based on how they were formed. For example, some lakes formed through glaciation are called tarns or cirque lakes, while those formed through volcanic activity are often called calderas. As with all waterbodies, the global distribution of lakes, from the tropics to the poles, will also influence their ecology and biodiversity.

Finally, reservoirs are anthropogenic waterbodies formed through the creation of a dam, lock, or diversion of a river or stream. They are sometimes referred to as reservoir lakes; however, ecologically they are very different. Because they were created for water storage, electrical production, or even flood abatement, the water levels in reservoirs can fluctuate dramatically. Such fluctuations create difficult ecological conditions for species that require some level of habitat stability. Despite these challenges, some species of semi-aquatic mammals have adapted successfully to these waterbodies. In the case of otters, reservoirs stocked for recreational fishing have provided new sources of prey.

Some species are habitat generalists and have evolved to monopolize on various lotic and lentic freshwater habitats. Other species are very specific in their habitat needs and, thus, are critically endangered because of their dependence on a specific type of threatened freshwater habitat. Regardless of species specialization, conservation efforts to protect freshwater systems will help protect the semi-aquatic species that call them home.

PART I
GEOGRAPHICAL DISTRIBUTION AND HABITATS

2

Paleobiology

It has been asked
by the opponents of
such views as I hold,
how, for instance,
a land carnivorous
animal could have
been converted into
one with aquatic
habits; for how
could the animal in
its transitional state
have subsisted?

Charles Darwin, 1859,
On the Origin of Species

Introduction

In 1803, Johann Friedrich Blumenbach named the platypus *Ornithorhynchus paradoxus*, roughly translated as "paradoxical bird snout" (Hall 1999). As a compromise to its original scientific name given by George Shaw in 1799, *Platypus anatinus* ("flat-foot duck"), and the conventions of scientific nomenclature, *Ornithorhynchus anatinus* ("bird snout duck foot") was assigned its current name. Despite the enduring nature of its new scientific name, this egg-laying mammal was the subject of great debate among European naturalists and anatomists of the time. It was questioned as a hoax: a new type of animal, and mammal-like, but reptilian as well. We now know that the platypus represents the monotremes, one of the three major groups of extant mammals. Despite these early debates, this unusual mammal provides a hint at the complexity of phylogenetic relationships among mammals and the many paths taken for species from all three groups—monotremes, marsupials, and eutherians—to find their niche in freshwater habitats. An exploration of the paleobiology of semi-aquatic mammals takes us to the heart of

convergent evolution, where unrelated species share similar characteristics to adapt to similar environments.

A brief overview of mammalian evolution begins with the synapsid lineage from the Late Paleozoic to the Early Mesozoic era, approximately 275 to 225 Ma. This group of amphibian-like tetrapods were well adapted to terrestrial habitats and, as **amniotes**, were distinguished by their self-contained amniotic egg, which eliminated reproductive constraints of aquatic habitats where fertilization of the egg by sperm was facilitated through water. From the Synapsida, the pelycosaurs (~358 to ~298 Ma) gave rise to the therapsids, which then diverged and became the dominant synapsid line approximately 250 Ma in the Middle Permian period while the supercontinent **Pangea** was still intact (Fig. 2.1). Given the distribution of the therapsids at temperate latitudes of Pangea, paleontologists believe the therapsids' ability to tolerate a seasonal climate reflected an ability to accommodate wider ranges of temperature and humidity, as is characteristic of modern mammals (Kemp 2017). There are two suborders of therapsids: Anomodontia and Theriodontia. The advanced suborder Cynodontia, derived from the theriodont group, survived beyond the Late Triassic epoch (~251 to ~201 Ma) to the Middle Jurassic epoch (~174 to ~163 Ma). They ultimately led to the evolution of today's mammals. The dentition, jaw structure, and hearing, among other traits, of the cynodonts served as transitional anatomical characteristics linking them to early mammals. By the Late Triassic, small, rat-sized cynodonts were now "extremely mammal-like" (Kemp 2017: 42). The cynodonts were extinct by the Middle Jurassic epoch, at which time several groups of mammals were already well established (Feldhamer et al. 2015). The mass extinction of reptiles 65 Ma (end of the Cretaceous) allowed a competitive release of mammals, which then diversified to fill in the niches (and size classes) previous dominated by reptiles and dinosaurs.

As a supercontinent, Pangea allowed mammals to move relatively freely from pole to pole. As with today's landscapes, however, mountains, deserts, and larger waterbodies would have presented some physical barriers to movement. As Pangea began to break-up, there were great **radiations** of terrestrial plants and animals as distance among land masses increased and long-range dispersal became more limited. As Pangea began to split apart in the Mid-Mesozoic, the large northern continent of Laurasia, composed of what is now North America and Europe, sepa-

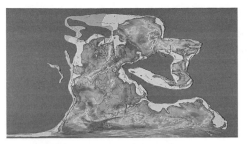
Permian-Triassic Boundary 250 Ma

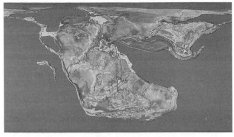
Triassic-Jurassic Boundary 200 Ma

Jurassic-Cretaceous Boundary 145 Ma

Cretaceous-Tertiary (K-T) Boundary 66 Ma

Modern Day

Figure 2.1. Continental distribution over time. Geographic information system (GIS) files courtesy of Prof. Christopher R. Scotese, Director of the PALEOMAP Project. G. A. HOOD

rated from the southern continent of Gondwana (also Gondwanaland). Gondwana contained present-day South America, Africa, India, Madagascar, Australia, Antarctica, New Zealand, Arabia, and New Caledonia. The resulting continents and subcontinents from each of these larger landmasses share similar species, often with common evolutionary origins.

Species in North America and Europe are similar to each other because of their shared evolutionary past in Laurasia, further connected intermittently with land masses between North America (Nearctic) and Europe (Palearctic) over time. Eutherians were dominant in the Holarctic (Nearctic and Palearctic combined), although marsupials originated in Laurasia and dispersed south to Gondwana when the connection between the two landmasses was intact. Mammals originating on Gondwana include the australosphenidans (including today's extant monotremes) and now extinct eutriconodonts (Chimento, Agnolin, and Martinelli 2016).

The formation of the Isthmus of Panama in the Late Pleistocene facilitated a southern migration of more successfully competitive placental mammals into South America after its separation from Gondwana, and a smaller northward migration of some of the marsupials from South America into North America. Today, approximately half of the contemporary species in South America are derived from North American ancestors, while only 10% of North American species originate from South America. Such connections among land masses occurred intermittently across the globe, thus aiding mammal migrations in the recent past. To the north, the **Bering land bridge** allowed migrations to and from North America to northeastern Europe. Detailed descriptions and timelines are covered extensively in geology and biogeography texts, although an abbreviated overview is provided in Table 2.1. Of interest in the exploration of the paleobiology of semi-aquatic mammals is how the species on these continents diversified into the unique forms and habitat preferences we see today.

The earliest mammals, distinguished through their primary jaw joint composed of the dentary (lower jaw bone) and squamosal bones (bones in the upper jaw), were relatively rare until the Cenozoic era (beginning 66 Ma; Table 2.2). This is the ball-and-socket jaw joint seen in mammals today. During the Mesozoic era (251–66 Ma), mammals and dinosaurs lived together in terrestrial habitats. Some paleontologists hypothesize that one group of the earliest known mammals (family Morganucodontidae) might have included the unknown ancestor of the monotremes (Feldhamer et al. 2015). There is debate about whether Morganucodontidae were true mammals, as defined by inclusion in the crown group of mammals, or whether they were mammaliaformes (mammal-shaped), a designation that includes the crown group of all extant mammals along with extinct (out-group) ancestors. Delving into this debate is beyond the scope of this book; however, it was the energetic, adaptable habits of the Morganucodontidae that set the stage for the mammalian radiations of the Cenozoic era. The fossil record reveals their adaptability

Table 2.1. Abbreviated geologic time scale

Era	Period	Epoch	Ma
Cenozoic	Quaternary	Holocene	0.012
Cenozoic	Quaternary	Pleistocene	2.68
Cenozoic	Neogene	Pliocene	5.333
Cenozoic	Neogene	Miocene	23.03
Cenozoic	Paleogene	Oligocene	33.9
Cenozoic	Paleogene	Eocene	56.0
Cenozoic	Paleogene	Paleocene	66.0
Mesozoic	Cretaceous	Upper	100.5
Mesozoic	Cretaceous	Lower	145.0
Mesozoic	Jurassic	Upper	163.5 ± 1.0
Mesozoic	Jurassic	Middle	174.1 ± 1.0
Mesozoic	Jurassic	Lower	201.3 ± 0.2
Mesozoic	Triassic	Upper	237.0
Mesozoic	Triassic	Middle	247.2
Mesozoic	Triassic	Lower	251.9 ± 0.024
Paleozoic	Permian	Lopingian	259.1 ± 0.5
Paleozoic	Permian	Guadalupian	272.95 ± 0.11
Paleozoic	Permian	Cisuralian	298.9 ± 0.15
Paleozoic	Carboniferous	Upper Pennsylvanian	307.0 ± 0.1
Paleozoic	Carboniferous	Middle Pennsylvanian	315.2 ± 0.2
Paleozoic	Carboniferous	Late Pennsylvanian	323.2 ± 0.4
Paleozoic	Carboniferous	Upper Mississippian	330.9 ± 0.2
Paleozoic	Carboniferous	Middle Mississippian	346.7 ± 0.4
Paleozoic	Carboniferous	Lower Mississippian	358.9 ± 0.4
Paleozoic	Devonian	Upper	382.7 ± 1.6
Paleozoic	Devonian	Middle	393.3 ± 1.2
Paleozoic	Devonian	Late	419.2 ± 3.2
Paleozoic	Silurian	Pridoli	423.0 ± 2.3
Paleozoic	Silurian	Ludlow	427.4 ± 0.5
Paleozoic	Silurian	Wenlock	433.4 ± 0.8
Paleozoic	Silurian	Llandovery	443.8 ± 1.5
Paleozoic	Ordovician	Upper	458.4. ± 0.9
Paleozoic	Ordovician	Middle	470.0 ± 1.4
Paleozoic	Ordovician	Late	485.4 ± 1.9
Paleozoic	Cambrian	Furongian	497.0
Paleozoic	Cambrian	Series 3	509.0
Paleozoic	Cambrian	Series 2	521.0
Paleozoic	Cambrian	Terreneuvian	541.0 ± 1.0

Table 2.2. Cenozoic era with major land mammal ages in North America and Europe

Epoch	North American Land Mammal Ages	Ma	European Land Mammal Ages	Ma
Holocene	Saintaugustinean	0.004	Aurelian	0.42
Holocene	Santarosean	0.014		
Pleistocene	Rancholabrean	0.3		
Pleistocene	Irvingtonian	1.4	Galerian	1.6
Pleistocene	Blancan	4.7	Villafranchian	3.5
Pliocene	Hemphillian	9.4	Ruscinian	5.3
Pliocene			Turolian	9.0
Miocene	Clarendonian	12.5	Vallesian	11.6
Miocene	Barstovian	16.3	Astaracian	16.0
Miocene	Hemingfordian	18.5	Orleanian	20.0
Miocene	Arikareean	29.5	Agenian	23.8
Oligocene			Arvernian	29.2
Oligocene	Whitneyan	31.8	Suevian	33.8
Oligocene	Orellan	33.9	Headonian	37.2
Eocene	Chadronian	37.0		
Eocene	Duchesnian	39.7	Robiacian	42.7
Eocene	Uintan	46.2	Geiseltalian	48.5
Eocene	Bridgerian	50.5	Grauvian	50.8
Eocene	Wasatchian	54.9	Neustrian	55.0
Eocene	Clarkforkian	56.2		
Paleocene			Cernaysian	55.9
Paleocene	Tiffanian	60.9	Thanetian*	59.2
Paleocene	Torrejonian	63.8	Selandian*	61.6
Paleocene	Puercan	66.0	Danian*	66.0

*Based on International Commission on Stratigraphy geologic timescale.

to various habitats, including amphibious ones. The three mammalian lineages of today arose during this time, with the monotremes (proto-therians) diverging earlier than the marsupials and eutherians. What is common within all three groups, and many of their mammal-like ancestors, was the secondary return to aquatic habitats to fill a vacant niche, an event that occurred independently for numerous species (Thewissen and Nummela 2008).

Paleobiologist Mark Uhen writes that *at least* seven separate lineages of mammals have returned to water either fully or as semi-aquatic species (Uhen 2007). Any retrospective assessment of mammalian evolution is limited by the mineralized nature of bone that allows for its preservation, and the ability to find that bone in the fossil record. Fossils require water transport and sediment cover (including tar pits and volcanic ash). Despite these challenges, paleontologists have discovered that secondary return to water by various mammals has deep roots in geologic history.

Extinct Mammal-like Semi-aquatic Taxa

Prior to the first mammals, and before pelycosaurs became extinct in the Late Permian, an early freshwater semi-aquatic pelycosaur (*Varanosaurus*) lived in rivers and lakes in what are now North America and Europe (Savage and Long 1989). It likely used its elongated snout, canine-like tusks, and long tail to its advantage when hunting fish (Fig. 2.2). A later mammal-like reptile from the Early Triassic, the *Lystrosaurus* (order Therapsida, infraorder Dicynodontia), has often been assumed to be semi-aquatic because of the high cortical thickness of the dorsal ribs, wide scapula, and paddle-like forearm (Ray, Chinsamy, and Bandyopadhyay

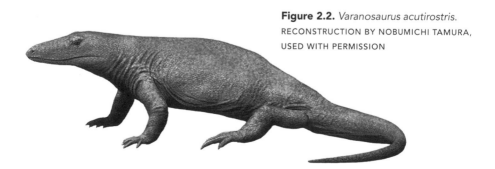

Figure 2.2. *Varanosaurus acutirostris.*
RECONSTRUCTION BY NOBUMICHI TAMURA, USED WITH PERMISSION

2005). However, as noted in Gillian King's article "The Aquatic *Lystrosaurus*: A Palaeontological Myth" (King 1990), there is ongoing debate about the semi-aquatic nature of this common vertebrate. A more recent analysis of bone microstructure by Dr. Sanghamitra Ray and her colleagues (2005) from the Indian Institute of Technology in Kharagpur provides some support for the case that at least one species (*Lystrosaurus murrayi*) was semi-aquatic, despite the fossorial nature of many species of *Lystrosaurus*. Some speculate that the downturned snout, short limbs, and well-developed tusks of a semi-aquatic *Lystrosaurus* helped it feed on the bottoms of rivers and lakes in what are now South Africa, India, Antarctica, China, and Russia (Savage and Long 1989). Given the origin of the fossils, and physical commonalities with distinctly fossorial *Lystrosaurus* species, the debate continues about the semi-aquatic nature of *Lystrosaurus murrayi* (Botha-Brink and Angielczyk 2010). Regardless, *Lystrosaurus* represents one of the few taxonomic groups that developed an ecological niche that allowed it to survive the massive extinctions at the end of the Permian period.

Debate continues around the habitat preferences of other early therapsids as well. The dinocephalian *Moschops capensis* is a large animal (~2.7 m long) from the Late Permian (~265 Ma) of the South African Karoo. The name "dinocephalian" means "terrible head," and it is with the head that some paleontologists find the link to aquatic habitats and to the vegetation *Moschops* consumed. A recent reconstruction places its head at an angle that suggests semi-aquatic behavior, much like today's hippopotamus (Benoit et al. 2017). The small clues that fossils present have traditionally provided brief glimpses into the overall ecology of early mammal-like species.

Another therapsid, *Haldanodon exspectatus* (order Therapsida, suborder Cynodontia) living from the upper Jurassic (~145 Ma) to the Cretaceous from Portugal, was a small, short, robust insectivorous animal. Given its short, stout limbs and humeri, it is often compared to modern desmans and other fossorial and semi-aquatic moles (Martin 2005, 2006; Luo 2007). Professor Thomas Martin (2005), from the University of Bonn, described it as more derived than *Morganucodon* because of the depression (**fossa**) below the scapular spine, a characteristic in monotremes but not in *Morganucodon* species.

More recently, a surprising find has given us a rare glimpse into how

Figure 2.3. *Castorocauda lutrasimilis.*
RECONSTRUCTION BY NOBUMICHI TAMURA,
USED WITH PERMISSION

the structure (**morphology**) of an early mammal-like organism matches its life habits (**ecomorphology**). This species from suborder Cynodontia shares an uncanny resemblance to the modern-day beaver and platypus. *Castorocauda lutrasimilis* (family Docodontidae) lived during the Middle Jurassic (~164 Ma) of Inner Mongolia (Ji et al. 2006). Its fossil remains were found in 2004 by Qiang Ji and an international team of researchers in northeastern China (Liaoning province). Unlike most fossils, its hair was preserved and consisted of dense guard hairs and underfur (Fig. 2.3). This was a particularly important discovery because it determined that fur was found in the immediate ancestors of modern mammals. *C. lutrasimilis* was approximately 425 mm long from snout to tail and had four incisors, one canine, five premolars, and six molars in its lower jaw—likely for feeding on aquatic invertebrates and fish, as is the case with otters (Ji et al. 2006). As with many modern semi-aquatic mammals, the caudal vertebrae 5 through 13 were dorsoventrally flattened in a butterfly pattern, much like caudal vertebrae 9 through 12 in modern beavers. Convergent with modern beavers, *C. lutrasimilis*'s tail had small carbonized scales next to caudal vertebrae 9 through 20. The forelimbs resembled those of the modern platypus in that their wide humerus was matched with wide forepaws. Its discovery opens up a new perspective on the diversity of mammalian ecomorphology.

Extinct Mammalian Orders

Traditionally, paleobiology primarily focused on fossilized skulls and teeth to aid in reconstructions of long-extinct species (Martin 2006). Teeth are often especially enduring over time and, along with associated skull bones, are sometimes all that remain of numerous mammalian fossils. Increasingly, the discovery of more complete skeletons and advanced analytical methods have aided our ability to investigate how Mesozoic mammals moved through the landscape via their preferred form of locomotion (Chen and Wilson 2015). For semi-aquatic and burrowing mammals, their shorter limbs and robust skeletons reflect a distinctly different presentation from the more slender, long-limbed, fully terrestrial forms, including climbers.

In their 2015 study, Meng Chen and Gregory Wilson from the University of Washington used six orders of mammals associated with semi-aquatic niches (Monotremata, Didelphimorphia, Carnivora, Lagomorpha, Rodentia, and Soricomorpha) to help quantify morphological indicators for habitat preferences of Mesozoic mammals. Using these metrics, they were able to identify two species from the now extinct order Eutriconodonta (once classified as "triconodonts") as semi-aquatic. The *Liaoconodon*, from the Early Cretaceous of eastern China's Liaoning Province, had a long body and paddle-like limbs. The second early mammal they identified was the *Yanoconodon*, from what is now China's Yan Mountains, about 300 km from Beijing. It was a small mammal, barely 13 cm long. After further examination, however, Chen's team determined it was not semi-aquatic, but rather terrestrial (Chen, Luo, and Wilson 2017). Such is the onerous task put to paleobiologists when trying to reconstruct biological history dating back over one hundred million years.

Another extinct order, Cimolesta, contained two species of nonplacental semi-aquatic mammals within family Pantolestidae: *Buxolestes minor* and *B. piscator*. Both species were otter-sized, fish-eating (piscivorous) mammals with broad, thick enameled teeth (Savage and Long 1989). *Buxolestes* fossils were found in the Middle Eocene deposits in Messel Germany (Fig. 2.4). A related species from the genus *Palaeosinopa* was found in the Early Eocene deposits of the Green River Formation of Wyoming (Rose and Von Koenigswald 2005). Its longer hind feet relative to the front feet and its long tail suggest that, like *Buxolestes*, it was a hind-limb paddler that relied on its tail while swimming. All three specimens of pantolestids appear to have been excellent burrowers and swimmers (Rose and Von Koenigswald 2005).

Within the now extinct family Desmostylidae (order Desmostylia), three species of genus *Desmostylus* filled either amphibious or fully aquatic niches (*D. hesperus*, *D. coalingensis*, and *D. japonicas*). Although their distribution was coastal (extending along the Pacific Rim from the Baja California peninsula up the Pacific coast, west to Sakhalin Island near Hokkaido, and then south to the Shimane Prefecture of Japan), there is strong evidence that these extinct herbivorous mammals from the Late Oligocene through to the Late Miocene (from 28.4 to 7.25 Ma) lived or spent a significant amount of time in freshwater or estuary ecosystems (Ando and Fujiwara 2016). Looking somewhat like a modern hippopotamus, they were 1.8 m long and weighed approximately 200 kg (Fig. 2.5). Their forward-facing tusks likely helped them forage for aquatic plants.

Figure 2.4. *Buxolestes.* RECONSTRUCTION BY GHEDOGHEDO, VIA CREATIVE COMMONS ATTRIBUTION-SHARE ALIKE LICENSE

Figure 2.5. *Desmostylus hesperus.* RECONSTRUCTION BY DMITRY BOGDANOV, VIA CREATIVE COMMONS ATTRIBUTION LICENSE

Paleobiology of Extant Orders

Order Monotremata (Monotremes)

It is generally accepted that the ancestor of the monotremes, including the platypus, diverged from other mammals as early as the Triassic, with echidnas and the platypus clades present by the Early Cretaceous (~150 Ma; T. Rowe et al. 2008). The presence of leathery eggs and a distinct lack of nipples sets them apart from therian species that evolved after the monotremes diverged, and acts as a diagnostic characteristic in paleomorphology. Compared to other mammalian groups, monotremes appear to have slower rates of diversification and morphological changes. An international study led by Professor Timothy Rowe and his research collaborators places the platypus-echidna divergence from 17 Ma to 80 Ma, with the research team favoring a more recent split within that timeline (T. Rowe et al. 2008).

FAMILY ORNITHORHYNCHIDAE (PLATYPUS)

Relative to most other continents, Australia has a limited fossil record, which makes it especially difficult to reconstruct the paleobiology of monotremes. However, a relatively recent analysis of an Early Cretaceous fossil from Australia, *Teinolophos trusleri*, places this species as the "most ancient member of the platypus clade" (T. Rowe et al. 2008: 1240). The findings of this study are compelling, including evidence of extensive electroreception in the fossilized beak, and firmly place this ancient species with other ornithorhynchids. Additionally, Rowe and his colleagues identified a younger Australian fossil, *Steropodon* (110 Ma), as a platypus (Fig. 2.6).

A recent discovery from the Two Trees site in the Riversleigh World Heritage Area, northwestern Queensland, Australia, revealed the largest known ornithorhynchid monotreme, *Obdurodon tharalkooschild* (Pian, Archer, and Hand 2013). This toothed mammal lived possibly during the Late Miocene or the Pliocene, which are time intervals without any representation of ornithorhynchids. Previously described species, *Obdurodon insignis* from central Australia (Woodburne and Tedford 1975) and *O. dicksoni* from the Riversleigh World Heritage Area in Queensland (Archer et al. 1992), are from the Late Oligocene and the Early to Mid-

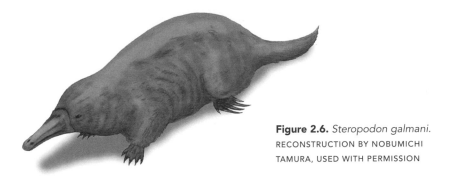

Figure 2.6. *Steropodon galmani.*
RECONSTRUCTION BY NOBUMICHI
TAMURA, USED WITH PERMISSION

Miocene, respectively. The only ornithorhynchid in South America is the Paleocene *Monotrematum sudamericanum* (Pascual et al. 1992). Interestingly, the modern platypus likely diverged earlier than the newly discovered giant platypus (*O. tharalkooschild*), as determined by their teeth. The modern platypus replaces its teeth with a keratinized pad as it develops into an adult, while *Obdurodon* retained their adult teeth and had an upturned bill for catching prey in open water (pelagic prey). The evolution of a hard pad, against which the modern platypus uses gravel and sediment to chew its food, might have accommodated the structures for electroreception through the enlarged infraorbital canal, thus leaving less space for roots of teeth (Asahara et al. 2016). As with Mesozoic monotremes, electroreception is used by the platypus in hunting aquatic prey in murky water.

Order Didelphimorphia (American Marsupials)

This order contains only one family, Didelphidae (the opossums), with most species living in Central and South America. The ancestor of opossums and the ancestor of the other extant marsupials diverged after the end of the Cretaceous period (Horovitz et al. 2009). The only North American species, the Virginia opossum (*Didelphis virginiana*) and the southern opossum (*D. marsupialis*), are a result of dispersal during the "great American biotic interchange" in the Late Cenozoic. Extant species are likely similar in appearance to their Cenozoic relatives. Unfortunately, the tropical regions where many didelphids lived have a relatively poor fossil record; however, the oldest recorded fossil of *Didelphis* (*D. solimoensis*) was discovered in Late Miocene deposits of Amazonia at the border of Peru and Brazil, along the upper Acre River (Cozzuol et al.

2006). Not only is it the oldest species of its genus, it is also the smallest. Given the limited amount of fossilized material, more questions remain regarding this early marsupial.

FAMILY DIDELPHIDAE (OPOSSUMS)

There are two extant species from family Didelphidae that are considered semi-aquatic, the water opossum (*Chironectes minimus*) and the little water opossum (*Lutreolina crassicaudata*), with *C. minimus* often being cited as the sole representative semi-aquatic marsupial (Marshall 1978a; Galliez et al. 2009). Both species became restricted geographically to the New World by the Mid-Miocene, despite a near global distribution of early marsupials. Although most New World marsupials have a fairly rich fossil record, the didelphids have a relatively poor one (Amador and Giannini 2016). Within the Didelphini clade, *Chironectes* and then *Lutreolina* were the first two genera to branch off from their common ancestor. *Chironectes* has a limited fossil record; however, a subfossil of a partial left jaw bone and some teeth were found near São Paulo, Brazil, and were determined to be Holocene in age. A Pleistocene fossil was found in cave deposits in Lagoa Santa, Minas Gerais, Brazil. The oldest find, from the Late Pliocene, was determined to be identical to the current species of water opossum (Marshall 1977, 1978a).

Recently, specimens from two biogeographically distinct populations of *Lutreolina crassicaudata* were reassessed through morphological and molecular analyses (Martiínez-Lanfranco et al. 2014). The specimens representing the western clade in the montane Yungas forests of Argentina and Bolivia were determined to be a distinct species, *L. massoia*. Despite current phylogenetic discoveries, the paleobiology of *Lutreolina* invites further analysis. The limited fossil record includes a fragment of the left part of a lower jaw dated to the Late Pleistocene in caves of Minas Gerais, Brazil; this species was assigned the name *Didelphys crassicaudata* by Lydekker in 1887 (Marshall 1978b). Additional Late Pliocene specimens were found in the Argentinian provinces of Buenos Aires and Catamarca, of which at least one might be *L. crassicaudata*.

Order Afrosoricida (Tenrecs and Elephants)

Order Afrosoricida is a recently recognized clade containing species previously classified in the formerly recognized order Insectivora. This new grouping, within the superorder Afrotheria, might seem an odd fit, given that Afrotheria also includes elephants and sea cows. However, genetic and molecular studies have confirmed their place in this diverse group (Stanhope et al. 1998), which is considered to be **monophyletic** (containing all descendants of a common ancestor). All species within Afrotheria evolved from a common ancestor following the Cretaceous-Tertiary (K-T) mass extinction approximately 66 Ma. Isolation is a powerful force in evolutionary biology, and Africa's isolation during the Early Cenozoic played an important role in the radiation of African mammals.

FAMILY TENRECIDAE (TENRECS)

Tenrec fossils on the continent of Africa date back to the Tertiary (66 to 2.6 Ma), but all extant tenrecs are found exclusively on Madagascar and are considered monophyletic (Everson et al. 2016). Fossils of *Parageogale*, *Protenrec*, and *Erythrozootes* are mainland representatives, yet there are no pre-Pliocene fossils from Madagascar (see Emerson et al. 2016). The colonization of Madagascar by tenrecs occurred 30 to 56 Ma (Oligocene and Eocene; Emerson et al. 2016). Given their similar niches, it is often assumed that the closest relatives of the web-footed tenrec (*Limnogale mergulus*) are the otter shrews of central Africa (*Potamogale velox* and *Micropotamogale* spp.). In actuality, aquatic tenrecs descended from ancestors of shrew tenrecs (*Microgale*), given their physical and molecular similarities (Asher and Hofreiter 2006).

The Nimba and Rwenzori otter shrews (*Micropotamogale lamottei* and *M. ruwenzorii*) and giant otter shrew (*Potamogale velox*) diverged early on; the three species are closely related and distinctly separate from other members of Afrosoricida (Kuntner, May-Collado, and Agnarsson 2011). A recent study places them in their own family Potamogalidae (Emerson et al. 2016); however, this change is not currently reflected in their classification by the IUCN. Given the strength of current molecular studies, it is likely that their classification in family Potamogalidae will become commonplace.

Order Proboscidea (Elephants and Extinct Relatives)

The distinguishing feature of mammals from Order Proboscidea is, not surprisingly, the presence of a trunk (proboscis), used for foraging and drinking. Family Elephantidae (elephants) is the only extant family, while other, now extinct, families ranged throughout Africa, Asia, Europe, North America, and South America. As a testament to their broad distribution, members of family Gomphotheriidae lived in South America from the early Middle Pleistocene until the Late Pleistocene (Prado et al. 2005). Early proboscideans originated in Africa during the Paleogene (66 to 23 Ma), with *Phosphatherium escuilliei* being the earliest proboscidean. Unlike its larger modern elephant relatives, *P. escuilliei* was less than a meter high at the shoulder.

FAMILY MOERITHERIIDAE (EXTINCT RELATIVE OF ELEPHANTS)

Like *P. escuilliei*, two species from genus *Moeritherium* (*M. lyonsi* and *M. trigodon*) were small, about 1 m at the shoulder. Unlike elephants, *Moeritherium* lacked a trunk and instead had a prehensile upper lip that was capable of grasping food (Fig. 2.7). Being semi-aquatic, they ate freshwater plants in swamps, as determined from stable isotope analysis of their teeth (Liu, Seiffert, and Simons 2008). Their fossil remains from 37 Ma (post-split of elephants and sirenians) were found in Egypt's Faiyum region, which fluctuated as a shallow estuary along the coast at that time.

Figure 2.7. *Moeritherium.*
RECONSTRUCTION BY DE HEINRICH
HARDER ~1920

Order Lagomorpha (Pikas, Hares, and Rabbits)

All species of Lagomorpha, including pikas, hares, and rabbits, share a common ancestor, despite distinct physical differences (Ge et al. 2013). Of interest in freshwater habitats are the two species of New World cottontail rabbits that are water-dependent: the swamp rabbit (*Sylvilagus aquaticus*) and the marsh rabbit (*S. palustris*), both of the United States. Lagomorphs arose in what is now China and Mongolia and then radiated to other continents. Unlike pikas, rabbits and hares are widely distributed in Eurasia, Africa, North America, and Central America. They have also been introduced to Australia and the southern regions of South America. Lagomorpha is a very old order that dates back to the Paleocene-Eocene boundary (55 Ma). A team of researchers from the American Museum of Natural History discovered the oldest fossil of a rabbit-like lagomorph, *Gomphos elkema*, during an annual paleontological trip to the Gobi Desert in Mongolia (Asher et al. 2005). This find not only established a starting point for what are today's lagomorphs, it also closed the debate supporting an earlier evolution of placental mammals at the K-T boundary of the Mesozoic, 65 Ma.

Fossils of one of the earliest known lagomorphs were found in west-central India and date to the Early Eocene (~53 Ma), thus identifying the time period when family Leporidae was diverging from other lagomorphs (Rose et al. 2007). There are several early genera of leporids (e.g., *Shamolagus*, *Lushilagus*, *Dituberolagus*, and *Strenulagus*) that gave rise to North American genera (e.g., *Mytonolagus*, *Megalagus*, *Paleolagus*) from the Middle Eocene well into the Oligocene (Ge et al. 2013).

FAMILY LEPORIDAE (HARES AND RABBITS)

Modern leporids diverged in the Early Miocene (~18.1 Ma), with *Nesolagus* of Sumatra and the Annamite Mountains of eastern Indochina and *Brachylagus* of North America being the earliest derived genera (Ge et al. 2013). Around 7.16 Ma, genus *Sylvilagus*, along with four other genera (*Oryctolagus*, *Caprolagus*, *Romerolagus*, and *Bunolagus*), formed a monophyletic group. The marsh rabbit and swamp rabbit, as sister taxa, are more closely related to each other than other cottontails and share a mean divergence time of 1.74 Ma, during the Pleistocene in North America. Their evolution into two distinct species is likely linked to geographic separation, or **vicariance** (Halanych and Robinson 1997). Despite the large number of extinct and extant leporids, these two species are the only rabbits identified as riparian-dependent.

Order Rodentia (Rodents)

Rodents enjoy a global distribution, except for Antarctica, New Zealand, and some oceanic islands. As the more speciose order of mammals occupying a broad range of habitats, classifying phylogenetic relationships among families and genera can be challenging. Molecular methods reveal complex and varied diversification rates, even within the same family, as influenced by biogeography and evolution (Fabre et al. 2012). Rodents first appear in the fossil record of the Late Paleocene (~57 Ma) to Early Eocene (~55 Ma) of Europe and North America. Of the six recognized clades of rodents (Fabre et al. 2012), three groupings include semi-aquatic mammals: castorimorphs (beaver-shaped), hystricomorphs (porcupine-shaped), and myomorphs (mouse-shaped). The oldest known rodents are in families Paramyidae, Alagomyidae, and Ischyromyidae, with *Paramys* often cited as the earliest known genus of rodent fossil (Feldhamer et al. 2015). Its skull was squirrel-shaped (sciuromorph), much like today's mountain beaver (*Aplodontia rufa*), a primitive fossorial rodent. Hystricomorph fossils appear in the Eocene of Asia, and then reflect a dispersal to Africa and then South America. South American hystricomorphs (caviomorphs) likely arrived by traveling on rafts of vegetation (Lacher et al. 2016). Globally, an explosive diversification of modern rodents began to appear in the Late Oligocene, and all of today's rodent families were established by the Late Miocene.

FAMILY CASTORIDAE (BEAVERS)

The first members of the beaver family originated in North America by the end of the Eocene. By the Early Oligocene (~33.5 Ma), the members of Castoridae expanded their range to Europe and Asia, where they underwent further diversification (Rybczynski et al. 2010). There are over thirty genera in family Castoridae in the fossil record; however, not all species were semi-aquatic, a trait that only evolved once in Castoridae (Rybczynski 2007). Semi-aquatic behaviors are evident in some species within two subfamilies: Castoroidinae and Castorinae. Canadian paleo-

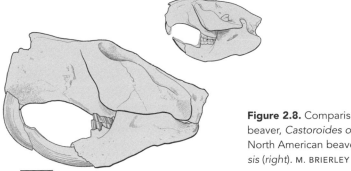

Figure 2.8. Comparison of skulls of giant beaver, *Castoroides ohioensis* (*left*), and North American beaver, *Castor canadensis* (*right*). M. BRIERLEY

5cm

biologist Dr. Natalia Rybczynski identified four extinct genera as semi-aquatic: *Castoroides, Procastoroides, Steneofiber,* and *Trogontherium.* As with modern semi-aquatic mammals, these taxa had shortened femurs, large hind feet that could support webbing, and specialized vertebrae in the tail. The European *Steneofiber eseri* and North American *Procastoroides* also possessed a grooming claw, which is used by modern beavers to groom their thick waterproof pelage (Shotwell, 1970; Hugueney and Escuillié 1995).

In North America, two large forms within subfamily Castoroidinae (*Procastoroides* and *Castoroides*) were much larger than modern beavers, with the Early Pliocene *Procastoroides* species being two-thirds larger than today's beaver species. The true giant beavers are from the Pleistocene, *Castoroides leiseyorum,* found in Florida, and its more northerly sister species, *Castoroides ohioensis,* ranging from the Yukon to the Great Lakes (Hulbert, Kerner, and Morgan, 2014). At over 2 m in length and up to 125 kg, they far exceeded the largest beaver found in the Late Pliocene of Europe (*Trogontherium cuvieri*), and they would have dwarfed today's beavers (Fig. 2.8). The European giant beaver was slightly larger than today's Eurasian beaver.

Extensive diversification in the Miocene resulted in at least nine semi-aquatic Castoridae genera in North America (*Anchitheriomys, Castor, Dipoides, Eucastor, Monosaulax, Nothodipoides, Priusaulax, Prodipoides,* and *Temporocastor*) and at least six in Europe (*Anchitheriomys, Castor, Chalicomys, Dipoides, Euroxenomys,* and *Steneofiber*), with the only remaining genera, *Castor,* originating in Europe (Rybczynski et al. 2010). Ancestors of the North American beaver (*C. canadensis*) likely crossed the Bering land bridge to North America in the Late Miocene (~7.6 to 8 Ma) (Horn et al. 2011). Recent molecular analysis by Liliya Doronina and her team (2017) supports a close relationship between Castoridae and Geomyoidea (kangaroo rat-related species) within a mouse-related clade.

FAMILY THRYONOMYIDAE (CANE RATS)

Family Thryonomyidae extends to the upper Eocene of Egypt, with many extinct species appearing in the Miocene fossil records of Africa and current-day Pakistan (López-Antoñanzas, Sen, and Mein 2004). Their geographic extent provides an excellent example of long-distance dispersal of rodents. Today there are only two closely related species, with the greater cane rat (*Thryonomys swinderianus*) closely associated with aquatic habitats of sub-Saharan Africa. The second species, the lesser cane rat (*T. gregorianus*) of sub-Saharan Africa, might belong to genus *Choeromys*, but more research is needed for verification (López-Antoñanzas, Sen, and Mein 2004). Presently, paleontologists consider *Gaudeamus aegyptius* the family's oldest ancestor (Late Eocene to Early Oligocene, 33.9 to 28.4 Ma of Egypt).

FAMILY CAPROMYIDAE (HUTIA)

The sole semi-aquatic representative of family Capromyidae is the Cabrera's hutia (*Mesocapromys angelcabrerai*), which is endemic to the mangrove swamps of Cuba. The molecular and morphological phylogenies of Capromyidae lack clarity, although a team led by Harvard's Pierre-Henri Fabre applied advanced molecular methods that inform these complex relationships (Fabre et al. 2014). *Zazamys* is the oldest capromyid and represents a diversification event dating to the Early Miocene (23 to 16 Ma). Given this date, hutias likely arrived in the Greater Antilles via **waif dispersal** from South America via rafts of vegetation. Fabre and his colleagues (2014) estimate that hutias and Atlantic forest spiny rats (Echimyidae) split in the Middle Miocene (14.8 to 18.2 Ma).

FAMILY CAVIIDAE (CAVIES, MARAS, AND CAPYBARAS)

Caviomorph rodents are broadly distributed throughout South America and comprise three highly divergent lineages: Caviinae (cavies), Dolichotinae (maras), and Hydrochoerinae (capybaras) (Pérez and Pol

2012). Evolution of South American caviomorphs dates to the Middle Eocene (45 Ma). They descended from hystricognath rodents that likely originated from Africa via waif dispersal. The only semi-aquatic species within family Caviidae are the capybara (*Hydrochoerus hydrochoerus*) and lesser capybara (*H. isthmius*), which are closely related to genus *Kerodon* (rock cavies). Modern capybaras are related to the extinct *Neochoerus*, which lived in both North and South America from 1.8 to 0.012 Ma. There were three known species: *N. aesopi*, *N. pinckneyi*, and *N. sirasakae*. Weighing over 90 kg, *N. pinckneyi* (Pinckney's new hog) was twice the size of the modern capybara. Along with many other New World megafauna, it went extinct approximately eleven thousand years ago (Ka).

FAMILY MYOCASTORIDAE (COYPU/NUTRIA)

Coypu (*Myocastor coypus*) is the only remaining member of family Myocastoridae. The fossil record indicates four other species in Myocastoridae (*Myocastor perditus*, *Myopotamus obesus*, *Myopotamus paranensis*, and *Myopotamus australis*), with coypu appearing in the Quaternary (0.781 Ma). *Myocastor perditus* fossils were recovered from the Quaternary of Bolivia (1.2 to 0.781 Ma). The coypu's closest relative is Cuvier's spiny rat (*Proechimys cuvieri*), a member of family Echimyidae (Lee et al. 2018). Recent molecular analysis places coypus in family Echimyidae, where it shares tribe Myocastorini with four other Neotropical genera: *Callistomys* (painted tree rat), *Hoplomys* (armored rat), *Proechimys* (spiny rats), and *Thrichomys* (punarés) (Fabre et al. 2017). This change is likely to become the norm in future rodent taxonomies. Pierre-Henri Fabre and his colleagues suggest that flooding might have resulted in the semi-aquatic condition of coypus relative to its arboreal relatives.

FAMILY CRICETIDAE (HAMSTERS, VOLES, AND LEMMINGS)

Family Cricetidae is one of the most specious and widely distributed families of rodents, thus presenting interesting phylogenetic challenges. It is commonly agreed that Cricetidae is ancestral to closely related family Muridae (Steppan, Adkins, and Anderson 2004). The earliest recorded cricetid is *Palasiomys* from the Middle Eocene (~37.2 to 33.9 Ma) in Central China (Rodrigues, Marivaux, and Vianey-Liaud 2010). Muroidea, a large superfamily of rodents that includes cricetids, began their extensive radiation in the Late Oligocene (~24.5 to 25.9 Ma) of the Old World (Schenk, Rowe, and Steppan 2013). Paleontologist Scott Steppan and his colleagues determined that, from the end of the Oligocene to the Early Miocene (~18.7 to 19.6 Ma), subfamilies Arvicolinae (voles, lemmings, and muskrats) and Cricetinae (hamsters) split from Sigmodontinae and Neotominae (the New World rats and mice). Shortly thereafter (~17.9 to 18.8 Ma) the lemmings and muskrats diverged from the hamster clade.

Subfamily Arvicolinae contains two semi-aquatic European water voles (*Arvicola amphibius* and *A. sapidus*) and several North American species, including a North American water vole (*Microtus richardsoni*), two bog lemmings (*Synaptomys borealis* and *S. cooperi*), the round-tailed muskrat (*Neofiber alleni*), and the muskrat (*Ondatra zibethicus*). The European genera date to the Early Pliocene, while the North American genera emerge in the Quaternary fossil record. However, of the four extinct species of genus *Ondatra* (*O. annectens*, *O. idahoensis*, *O. meadensis*, and *O. minor*), *O. meadensis* and *O. minor* both date to the Pliocene fossil record (~4.9 to 2.58 Ma) of the United States.

Subfamily Sigmodontinae includes several semi-aquatic species, all from the New World and most from the neotropics. The development of a geographic link and earlier waif dispersal between North and South America via Central America dramatically changed the distribution of sigmodontine rodents in the New World and changed their dispersal during Middle to Late Miocene (Leite et al. 2014). Once in the neotropics, species radiation was rapid relative to other muroids.

Despite some difficulties relative to fossil formation in the tropics and subtropics, two genera, *Holochilus* and *Lundomys*, have an extensive fossil history in South America; this history ranges from the Pleistocene to Holocene epochs (Pardiñas and Teta 2011). *Holochilus* and *Lundomys* are closely related to two other extinct South American sigmondontines: *Carletonomys cailoi*, found in the southeastern Buenos Aires Province of Argentina, and *Noronhomys vespuccii*, from the Fernando de Noronha archipelago of northeastern Brazil. *Carletonomys cailoi* was semi-aquatic (Pardiñas 2008). Its fossilized remains were mixed with other freshwater vertebrates, such as fish, amphibians, and coypu. Conversely, *Noronhomys vespuccii* likely descended from a semi-aquatic ancestor but evolved terrestrial habits once it rafted its way onto this relatively dry archipelago (Carleton and Olson 1999). It became extinct following European contact with South America because of the rats and mice that accompanied the ships.

FAMILY MURIDAE (TRUE MICE AND RATS, GERBILS, AND RELATIVES)

Family Muridae is the most speciose of the rodents and the most speciose of all mammals. As with cricetids, phylogeny is complex and challenging in this geographically and ecologically diverse taxon. Despite being more numerous than cricetids, there are fewer semi-aquatic murids. The distributions of the murids covered in this book range from the marshes of southeast Iraq (*Nesokia bunnii*) to sub-Saharan Africa (*Colomys goslingi, Dasymys, Nilopegamys plumbeus, Otomys lacustris,* and *Pelomys*), Indonesia (*Hydromys, Waiomys mamasae*), New Guinea (*Baiyankamys, Crossomys moncktoni, Hydromys, Parahydromys asper, Xeromys myoides*), and Australia (*Hydromys chrysogaster, Xeromys myoides, Rattus lutreolus*).

Murid rodents within the Australo-Papua region migrated from Southeast Asia to the island of New Guinea during the Late Miocene to Early Pliocene (~4 to 6 Ma) where they underwent rapid diversification (K. C. Rowe et al. 2008). Geographic separation resulted in a range of species and ecological specializations. All but the Ethiopian amphibious rat (*Nilopegamys plumbeus*) are within subfamily Murinae (Old World rats and mice), which diverged in the Early to Middle Miocene, between 15 to 20 Ma (Steppan and Schenck 2017). Of this group, subfamily Hydromyinae dispersed to Australia around 10 to 15 Ma, while members of Murinae followed in a second dispersal event approximately 2 Ma (Honeycutt, Frabotta, and Rowe 2007). As with other close oceanic islands, the key mechanism of dispersal was via waif dispersal.

FAMILY NESOMYIDAE (POUCHED RATS, CLIMBING MICE, AND FAT MICE)

Delany's swamp mouse (*Delanymys brooksi*) of east-central Africa is the only semi-aquatic representative of this exclusively African family of rodents. It is closely related (sister taxa) to the Old World Muridae and Holarctic/Neotropical Cricetidae (Steppan and Schenk 2017). Subfamily Delanymyinae diverged in the Early to Middle Miocene. However, based on extensive molecular and morphological analysis, Steppan and

Schenk (2017) united *Delanymys, Petromyscus* (rock mice), and *Mystromys* (white-tailed mouse) into subfamily Petromyscinae and abandoned the previous subfamily distinction. They also determined the age of divergence as Middle Miocene (14.7 Ma).

Order Eulipotyphla (True Shrews, Desmans, and Shrew-like Moles)

Previously grouped with Insectivora, the members of order Eulipotyphla diversified rapidly after the Eocene-Oligocene extinction event provided a competitive release from other insectivores (Ostende 1995; Douady and Douzery 2003). It was also the time of the **Grande Coupure**, or great break (Stehlen 1910), in which early paleontologists noticed that faunal extinctions appeared to result in a large turnover of European mammal species and diversification of others. Soricidae (true shrews) diverged at this time, along with the now extinct Heterosoricidae (Doby 2015). Although the fossil record of shrews and their relatives is more extensive in Europe and Asia than in North America, overall it is not as extensive as with other taxonomic groups (Moncunill-Solé, Jordana, and Köhler 2016).

Members of family Dimylidae, likely related to modern talpids, represent desman-like species from Oligocene and Miocene deposits from Europe in areas that would have had water at the time (Savage and Long 1989). Their double-rooted, broad, strong premolars hint at a diet of mollusks. In addition, skull morphology of *Dimyloides stehlini* from the Ehrenstein 4 site in Germany suggests that they are related to talpids (Schmidt-Kittler 1973). As more fossils from order Eulipotyphla are found, there will be increased insights into the taxonomic and ecological context of the members of Eulipotyphla.

FAMILY SORICIDAE (INCLUDING SUBFAMILY CROCIDURINAE)

The soricids have the greatest number of species of any of the insectivorous mammals and have a near global distribution, except for Australia, polar regions, and parts of South America. There are 250 species identified in the fossil record, 200 of which are extinct (Lopatin 2002). They are well documented in Europe as being present in the Lower to Middle Oligocene (~30 Ma), with one of the oldest European soricid fossils, *Srinitium marteli*, found in France, Bosnia and Herzegovina, and Germany (Hugueney 1976). Lopatin (2002) argues that *Soricolestes soricavus* from the Middle Eocene of Mongolia is the ancestor of all other shrews. One of the red-toothed shrews (Soricinae) of North America and Eurasia, *Paenelimnoecus crouzeli*, was found in Middle Miocene deposits in France (Reumer 1989). The fossil history of soricids continues to unfold and is progressing toward a more comprehensive assessment of soricid ecomorphology.

Extant genera of shrews with species that are predominately semi-aquatic (tribe Nectogalini) include the *Chimarrogale*, *Nectogale*, *Neomys*, and *Sorex* to some degree. For mammalogists, the question of whether use of aquatic habitats is ancestral in nature is of keen interest. An extensive genetic assessment of Nectogaline shrews determined that adaptations to aquatic habitats evolved independently multiple times in the *Neomys* and *Chimarrogale* + *Nectogale* lineages (He et al. 2010). The same is assumed true of North American shrews (*Sorex*), given that only three species of the 142 currently recognized species and subspecies are semi-aquatic.

FAMILY TALPIDAE (DESMANS AND SHREW-LIKE MOLES)

Desmans and the star-nosed mole are unusual in appearance and behavior, and they have odd tactile nasal morphology and hunting strategies. Despite a much broader distribution across Eurasia during the Miocene, the earliest European fossils were from the Oligocene. Within Desmaninae, there were four genera: *Dibolia*, *Mygalinia*, *Galemys*, and *Desmana*. Nine species were in genus *Dibolia*, four in *Galemys*, and six in *Desmana* (Rümke 1985). *Mygalinia hungarica* represents the sole known member of its genus, and it was found in the Ukraine in deposits dating from the Late Miocene to the Early Pliocene, 10 to 11 Ma and 13 to 14 Ma (Rzebik-Kowalska and Rekovets 2016).

Desmans had a brief appearance in North America from sites in Oregon, Nebraska, and New Mexico during the Middle Miocene (~14 to 16 Ma) to the Late Miocene (~4.9 to 10 Ma) (Martin 2017). One of the extinct North American species was the water mole, *Gaillardia*. James Martin cites *Gaillardia* as "one of the rarest fossil mammals of the Neogene of North America" (Martin 2017: 94). *Gaillardia thomsoni* (Matthew 1932) was found in the Snake Creek quarries of Nebraska, and then later in northern New Mexico. Not only are they the rarest fossil mammals of that time period, they are the oldest desman fossils in North America. More recent *Gaillardia* fossils found in Oregon were associated with fish, frogs and toads, shrews, and a prehistoric beaver (*Dipoides*), a testament to the animal's preference for aquatic habitats (Martin 2017).

North America's star-nosed mole is the only living species of genus *Condylura*. Fossils dating from the Middle to Late Pleistocene 11 Ka and Middle Pleistocene 700 Ka in the eastern United States were identified as the extant star-nosed mole (Guilday, Martin, and McCrady 1964; Gunnell et al. 2008). Older fossils from the star-nosed mole, dating back to the Middle Miocene (14 to 15 Ma) in Oregon, are considered the oldest of the genus (Gunnell et al. 2008; Sansalone, Kotsakis, and Piras 2016). Two other fossils, *C. kowalskii* and *C. izabellae*, from a Pliocene site in Poland (Skoczeń 1976) challenge the common belief that *Condylura* originated in North America. Similarly, an earlier fossil of *Condylura*, found in Kazakhstan in Middle Miocene (11.9 to 11.1 Ma), suggests a re-evaluation of *Condylura* distribution outside of North America and a re-evaluation of the age of the genus (see Sansalone, Kotsakis, and Piras 2016), a genus that likely evolved from a semi-fossorial shrew-mole.

Order Carnivora

The earliest carnivoran mammals, the viverravids and miacids from the superfamily Miacoidea, appear in the fossil record dating from 65 to 62 Ma (Flynn and Wesley-Hunt 2005). The paraphyletic Miacoidea precedes order Carnivora and instead is often classified in the clade Carnivoramorpha, which now includes modern carnivores. These extinct families (Viverravidae and Miacidae) both have origins in North America, although miacids likely arose in Europe as well. Modern carnivores diverged by the Eocene, with many modern clades obvious by the Oligocene (Flynn and Wesley-Hunt 2005). In particular, members of Felidae (cats), Viverridae (otter civet and genet), and Mustelidae (mink and otters) diverged during this time.

FAMILY FELIDAE (CATS)

Early felids appeared in the fossil record in Eurasia during the Oligocene (~21 Ma), with eight lineages of modern felids diverging by the Late Miocene (Werdelin et al. 2010). Fossils from genus *Proailurus* were

found in the Quercy region of France. One of the earliest fossils was identified as the now extinct *Proailurus lemanensis*, which lived in Europe
and Asia during the Mid-Oligocene. *Proailurus* is considered a direct
ancestor to modern felids. It weighed no more than a large house cat
and was approximately 1 m long from nose to tail, and was probably arboreal to some extent. The two species of interest to freshwater habitats,
Southeast Asia's flat-headed cat (*Prionailurus planiceps*) and the fishing
cat (*P. viverrinus*), are part of the leopard cat lineage, which diverged in
the Late Miocene (~6.2 Ma). Although there is generally a good fossil
history of the felids, some of the later radiations, including the leopard
cat lineage, are not as data rich (Johnson et al. 2006). Middle Pleistocene sites of Southeast Asia have revealed some fossils likely linked to
Prionailurus (Hemmer 1976). Felid fossils, in general, are not noted as
being associated with riparian or aquatic ecosystems.

FAMILY HERPESTIDAE (MONGOOSE)

Mongooses and other members of Herpestidae were often classified with
viverrids but are now recognized as their own family. As with viverrids,
their ancestors are within superfamily Miacoidea, which lived during
the Paleocene and Eocene (~66 to 33 Ma). Herpestidae have an Early
Miocene origin (Patou et al. 2009). As with the more modern felids, the
fossil record from Herpestidae is not very extensive (Peigné et al. 2005),
but it is linked to the Early to Middle Miocene of Europe (~18 Ma).
Leptoplesictis, an Early to Middle Miocene fossil of Europe and Africa,
is described as the "oldest fossil attributed to the Herpestidae" (Patou et
al. 2009: 70); however, this genus also shows genet-like qualities in the
morphology of its carnassial teeth. Only one modern herpestid is associated with freshwater habitats: the marsh mongoose (*Atilax paludinosus*).
The monotypic genus *Atilax* dates to Late Miocene sites in Africa (Hill
et al. 1985). The closest genetic relative to the marsh mongoose is the
long-nosed mongoose (*Herpestes naso*) (Patou et al. 2009), which is also
closely affiliated with freshwater habitats (Veron, Patterson, and Reeves
2008).

FAMILY VIVERRIDAE (OTTER CIVET AND GENET)

Ancestral viverrid lineages arose during the Late Eocene (~34 Ma) in Eurasia (Veron 2010), Asia, and Africa. As with many other carnivore fossils, viverrid fossils appear in the Eurasian Oligocene, ~23 Ma (Veron 2010; Feldhamer et al. 2015). As with the mongooses and their relatives, viverrids have a poor fossil record. The two earliest known fossils are *Herpestides* (Hunt 1996) and *Semigenetta* (Helbing 1927). By the Middle Miocene (~12 Ma), an ancestor of the African civets dispersed to Africa, followed by the ancestor of the genets approximately 11.4 Ma (Veron 2010). The otter civet (*Cynogale bennettii*), despite its unusual morphology, belongs to the Asian subfamily Hemigalinae (Asian palm civets) and is more closely related to the endangered Owston's palm civet (*Chrotogale owstoni*) and the near-threatened banded palm civet (*Hemigalus derbyanus*), both of which are forest species. Although considered equally related to the otter civet as the other two species, Hose's palm civet (*Diplogale hosei*) requires additional morphological work (Veron 2010). Very little is known of Hose's palm civet given the difficulty in finding this animal in the forests of Borneo.

The Democratic Republic of Congo's aquatic genet (*Genetta piscivora*) is a member of the subfamily Genettinae and is most closely related to Johnston's genet (*G. johnstoni*) of the Upper Guinean forests of Guinea, Sierra Leone, Liberia, and Côte d'Ivoire (Gaubert et al. 2004). The aquatic genet was originally placed in its own genus, *Osbornictis*, but is now confirmed within the genus *Genetta* (Veron 2010). Given the paucity of fossils for the otter civet and genets, emerging molecular methods have advanced the classification and evolutionary understanding of these species.

FAMILY MUSTELIDAE (MINK AND OTTERS)

Mustelids are thought to have evolved from the now extinct genus *Miacis* from family Miacidae. They are believed to be one of the oldest families within order Carnivora, with an emergence in Asia at the Eocene-Oligocene transition ~32.4 to 30.9 Ma (Sato et al. 2012). The earliest known musteloid fossils, *Mustelictis olivieri*, date to the Late Oligocene of France (~32.9 to 30.9 Ma). This insectivore was adapted for climbing trees (**scansorial**). The extinct *Plesictis plesictis* (24.7 to 23.3 Ma) is the earliest known stem mustelid (Sato et al. 2003). Family Mustelidae arose in the Early Miocene, approximately 16.1 Ma, with much of their diversification occurring in Asia, although there is still some debate about whether Asia or North America was its continent of origin (Sato et al. 2012). From their Eurasian roots, mustelids dispersed broadly to North and South America and Africa (Koepfli et al. 2008a). The North American dispersal, in the Early Miocene, followed the Bering land bridge.

Decoding the phylogeny of Mustelidae has presented various challenges. The relatively quick radiation of these species and repeated intercontinental dispersal events make it difficult to interpret the information, in much the same way it is difficult to interpret a rapidly changing weather system. As a result, numerous iterations of their evolutionary path have been presented. With improved molecular applications, many of the finer details are being resolved (Koepfli et al. 2008a). There are two recognized subfamilies within Mustelidae with extant semi-aquatic species: subfamily Lutrinae (*Aonyx*, *Enhydra*, *Hydrictis*, *Lontra*, *Lutra*, *Lutrogale*, and *Pteronura*) forms one monotypic lineage and is a sister clade to subfamily Mustelinae (*Mustela* and *Neovison*).

MUSTELINAE (MINK) The European mink (*Mustela lutreola*) and American mink (*Neovison vison*), despite their common names, have a more distant genetic relationship than first assumed (Koepfli et al. 2008a). The phylogenetic distance of these two species represents independent evolution of aquatic adaptations. Although *Neovison* still retains *Mustela* as closest relative, cytogenetic (genetics related to cell behavior), morphological, and biochemical data resulted in its placement into its own genus (Abramov 2000; Harding and Smith 2009). The only other member of *Neovison*, is the now extinct (~1880–1920) sea mink (*Neovison macrodon*). The North American origin of the American mink is supported by the fossil record. It likely diverged ~6 Ma along with its closest relative, the long-tailed weasel (*M. frenata*) (Koepfli et al. 2008a).

Interestingly, the European mink is closely related to the black-footed ferret (*M. nigripes*) of the North American plains (Harding and Smith 2009), although it is also closely related to the European polecat (*M. putorius*) via a possible earlier hybridization event (Davison et al. 2000). The European mink and black-footed ferret share an estimated divergence time of 1.1 to 1.3 Ma, thus reflecting the rapid diversification identified for these taxa. Some early fossils of *Mustela* were found in Late Miocene to Early Pliocene (~5.3 Ma) deposits in Europe. This is a time that evolutionary biologist, Dr. Klaus-Peter Koepfli and his team identified as the "Pliocene burst of diversification of mustelids" (Koepfli et al. 2008a), which followed a diversification of prey species and a shift to a cooling climate.

LUTRINAE (OTTERS) As with the Mustelinae, the otters (subfamily Lutrinae) have their origin near the Early to Middle Miocene boundary (Sato et al. 2012) and are closely related to the genus *Mustela* (Koepfli and Wayne 1988; Ginsburg 1999). Several species of otters diverged from a common ancestor approximately 11 to 14 Ma, a time often seen as the beginning of the evolution of carnivores (Table 2.3). North American otters (*Lontra*) are the only mustelids of non-Asian ancestry. The subfamily of all living otters (Lutrinae) is monophyletic (Sato et al. 2012) and, following the phylogenetic analysis of Klaus-Peter Koepfli and Robert K. Wayne (1998), can be grouped into three categories: (1) North American river, Neotropical, and marine otters, (2) sea, Eurasian spotted-necked, African clawless, and Asian small-clawed otters, and (3) the giant otter. Species from the genus *Lutra* started to appear in Europe during the Pliocene, while in North America, species of the genus *Lontra* appeared later in the Pleistocene, ~1.7 Ma (Willemsen 2006). It is generally believed that the now extinct species *Lutra licenti* made its way from China over the Bering land bridge to North America during that time, and that it was the origin of otters in North American (van Zyll de Jong 1972; Willemsen 2006). As with *Mustela*, thereafter the Eurasian otters (*Lutra*) and the North American otters (*Lontra*) experienced a rapid radiation and dispersal on a nearly global scale.

Members of genus *Aonyx* (African clawless otter *A. capensis*, Asian small-clawed otter *A. cinereus*, and Congo clawless otter *A. congicus*), diverged from a common *Lutra* ancestor 5 to 8 Ma (Koepfli and Wayne 1998), while the sole species of *Hydrictis* (spotted-necked otter *H. maculicollis*) has recently been distinguished from its former genus *Lutra*. The two remaining extant genera of otters, smooth-coated otter (*Lutrogale perspicillata*) and the giant otter (*Pteronura brasiliensis*), have a more complex phylogenetic relationship with other otters. Both otters have similar skull morphology and vocalizations, which has led to the hypothesis that they are more related than not (Duplaix 1980; Koepfli and Wayne 1998; Willemsen 1992). The giant otter is thought to have derived from the now extinct Hagerman's otter (*Satherium piscinarium*), identified from fossil remains in the United States from Idaho to Florida. Consequently, the giant otter is only distantly related to *Lontra* and *Lutra* species.

In 2016, Marco Cherin and his colleagues described a new species of otter, *Lutraeximia umbra*, from Early Pleistocene of Sicily, Italy. They placed this species into a monophyletic clade with the smooth-coated otter, one extinct otter (*Lutrogale cretensis*), and three Pleistocene otters from Italy (*Sardolutra ichnusae*, *Lutraeximia trinacriae*, and *Lutraeximia umbra*). Given the ever-emerging fossil record, the phylogeny of the spotted-necked otter and giant otter is a story that continues to unfold.

Unlike the mink, otters have a more extensive fossil record (e.g. *Mionictis*) that dates to Early Miocene deposits (~20 Ma) of both North America and Europe, shortly after otters diverged from other mustelids (Willemsen 2006). One extinct giant otter from the Late Pleistocene was *Megalenhydris barbaricina*, a very large otter with a slightly flattened tail (Willemsen and Malatesta 1987). It was one of four extinct species of Pleistocene otters from the island of Sardinia, Italy. Two of these Sardinian species, *Algarolutra majori* and *Cyrnolutra castiglionis*, were also found in Corsica. The fourth otter, *Sardolutra ichnusae*, was a small marine otter from Sardinia that likely descended from an otter in the genus *Lutra* (Willemsen 1992). Additional fossils continue to be found in the region.

In North American, a new discovery adds to the intrigue of how the story of otters unfolded. Usually, fossils of the medium sized (~16 kg) otter *Enhydritherium terraenovae* are associated with coastal habitats of California and Florida, thus reinforcing the assumption that it was a marine otter. However, a recent discovery in Mexico by Jack Tseng and his colleagues (2017) confirmed not only *E. terraenovae*'s presence in freshwater habitats but also its ability to completely live in freshwater systems as well as in near-coastal and marine environments (Tseng et al. 2017). The fossils were found in the state of Zacatecas, Mexico, which is nearly 600 km from the Gulf of Mexico and 200 km from the current Pacific coast—much farther than the previously recorded distance of 70 km from the Gulf of Mexico for sites in Florida (Lambert 1997). The 2017 study identified a possible overland route for this otter from California to Florida that dates back 5.59 to 6.95 Ma.

Table 2.3. Fossil records of extinct species of otters

Species	Age range	Location	Authority
Algarolutra majori*	2.588 to 0.012 Ma	Italy (Pleistocene)	Malatesta 1978
Aonyx antiqua	0.781 to 0.126 Ma	Netherlands (Quaternary)	Blainville 1841
Enhydritherium terraenovae*	13.65 to 4.9 Ma	USA (Pliocene, Miocene)	Berta and Morgan 1985
Lontra weiri	5.332 to 2.588 Ma	USA (Pliocene)	Prassack 2016
Lutra affinis	5.332 to 2.588 Ma	France (Pliocene), Greece (Miocene)	Gervais 1859
Lutra bravardi	2.588 to 1.806 Ma	France (Quaternary, Pliocene), Hungary (Pliocene), Slovakia (Pliocene)	Pomel 1843
Lutra bressana	2.588 to 0.012 Ma	France (Quaternary)	Deperet 1893
Lutra castiglionis	0.781 to 0.126 Ma	France (Pleistocene)	Pereira and Salotti 2000
Lutra fatimazohrae	3.6 to 2.588 Ma	Morocco (Pliocene)	Geraads 1997
Lutra hessica			Lydekker 1890
Lutra licenti		China (Pleistocene)	Teilhard and Piveteua 1930
Lutra lybica			Stromer 1914
Lutra palaeoleptonyx			Falconer 1868
Lutra simplicidens	0.781 to 0.126 Ma	Austria (Quaternary), Germany (Quaternary), UK (Quaternary)	Thenius 1965
Lutraeximia trinacriae	2.588 to 0.012 Ma	Italy (Pleistocene)	Burgio and Fiore 1988
Lutraeximia umbra	2.588 to 0.781 Ma	Italy (Pleistocene)	Cherin et al. 2016
Lutrogale cretensis*	0.126 to 0.012 Ma	Greece (Quaternary)	Symeonides and Sondaar 1975
Megalenhydris barbaricina*	0.126 to 0.012 Ma	Italy (Pleistocene)	Willemsen and Malatesta 1987
Nesolutra euxena	2.588 to 0.012 Ma	Italy (Quaternary)	Bate 1935
Sardolutra ichnusae*	2.588 to 0.0 Ma	Italy (Quaternary)	Malatesta 1977
Satherium piscinarium*	4.9 to 0.3 Ma	USA (Quaternary, Blancan, Pliocene)	Leidy 1873

Table 2.3. *continued*

Species	Age range	Location	Authority
Siamogale melilutra	11.1 to 4.9 Ma	Shuitangba, China (Pliocene to Miocene)	Wang et al. 2018
Siamogale thailandica			Ginsburg et al. 1983

Note: Data obtained from http://fossilworks.org. When information was available, all species were listed as scansorial insectivores. Given the emerging extent of the fossil record, this list is not exhaustive. Authority names are not in the references.

*Identified as "badger" on fossilworks.org.

FAMILY PROCYONIDAE (RACCOONS)

The raccoon family has a dynamic and changeable phylogenetic classification history. Evolutionary biologists have differed in their interpretations of morphological comparisons with similar taxa for many decades, but molecular genetics has helped clarify these relationships and divergence times (Koepfli et al. 2007). The oldest fossil for family Procyonidae is from genus *Pseudobassaris* from western European Late Eocene or Late Oligocene. By the Early Miocene, procyonid fossils appear in the North American fossil record, and subsequent diversification occurred, which resulted in today's genera (Koepfli et al. 2007). Genus *Procyon* appears by the Late Miocene or Early Pliocene in southern North America, with a possible extension into Central America (Baskin 2004). The raccoons are most closely related to the ring-tailed cats (*Bassariscus astutus* and *B. sumichrasti*) of the southern United States and Central America, respectively. They diverged approximately 11.4 to 12.8 Ma (Koepfli et al. 2007). The crab-eating raccoon of Central and South America's marshes and jungles is most closely related to the common raccoon (*P. lotor*). Fossils of *Procyon* do not appear until the Early to Late Pleistocene of South America; however, recent molecular analysis places their divergence from the common raccoon during the Early Miocene (~5 to 5.7 Ma), thus indicating the presence of a temporary land bridge prior to the great American biotic interchange (~3 Ma) (Koepfli et al. 2007).

Order Artiodactyla (Even-Toed Ungulates)

Hoofed herbivores likely originated in the Paleocene, with fossil forms occurring in Europe, Asia, and North America within the Eocene (~55 Ma). Perissodactyls (e.g., tapirs, rhinos, and horses) were more numerous than even-toed ungulates (Artiodactyla: pigs, deer, bovines, antelopes, etc.); however, artiodactyls have dominated since the Early Tertiary to today (Savage and Long 1989). There were two great radiations of artiodactyls. The first, in the Early Eocene (~54 Ma), involved pig-like species, while the second one (Late Eocene to Early Oligocene) led to the ruminants. From the northern continents, artiodactyls dispersed into Africa and South America.

The oldest known even-toed ungulate was the *Diacodexis*, which had an expansive range across North America, Europe, and Asia. This rabbit-sized ungulate lived from approximately 55.4 to 46.2 Ma. The expansion of grassland habitats allowed for many more artiodactyls that evolved after this tiny herbivore to succeed in a variety of habitats, including freshwater systems.

FAMILY SUIDAE (SWINE)

The oldest members of the Suidae first appear in the Late Eocene or Early Oligocene of Eurasia (~34.50 to 39.69 Ma) when they likely diverged from New World pigs (Tayassuidae; e.g., peccaries) (Frantz et al. 2016). Suidae colonized Eurasia and Africa by the Middle Miocene (~15 Ma), and by the beginning of the Pliocene (~5.33 Ma) all but the Suinae subfamily (appearing during Late Miocene, ~10 Ma) were extinct. One group of species, however, the riparian-dependent *Babyrousa*, adds a bit of mystery to the evolutionary picture. Paleobiologist Laurent Frantz and his colleagues call this genus "the enigmatic *Babyrousa*" (2016: 3.5). There is still active debate as to whether this genus is truly part of Suinae or in its own subfamily, Babyrousinae, that diverged from other suids in the Middle Miocene (~13 Ma).

The three species of "pig deer" (hairy babirusa *Babyrousa babyrussa*, Sulawesi babirusa *B. celebensis*, and Togian Islands babirusa *B. togeanensis*) live on geographically isolated islands of the Indonesian archipelago. They lack any extinct or living species that are closely related, which makes it difficult to map their phylogenetic history, and they are some-

Figure 2.9. *Anthracotherium magnum.*
RECONSTRUCTION BY DMITRY BOGDANOV,
VIA CREATIVE COMMONS ATTRIBUTION
LICENSE

times considered relic species of extinct Miocene suids. Their migration to these islands is also a mystery; they could have been brought with humans as people expanded into the area.

FAMILY HIPPOPOTAMIDAE (HIPPOPOTAMI)

Thanks to advances in molecular analyses, the hippos, classified in family Suidae for decades, are now firmly established in their own family (Boisserie 2005). In many studies, the physically similar artiodactyl family Anthracotheriidae was considered their closest relative; however, it is a complicated relationship given deficiencies in the fossil record (Boisserie 2005). Like hippos, members of Anthracotheriidae were semi-aquatic (Fig. 2.9), as evidenced by the habitats in which their fossils were found, and the morphology of the fossils themselves (Savage and Long 1989). Although found in Africa during the Oligocene, anthracotheres had a broad distribution that extended from Europe and Asia (Oligocene and Miocene) to North America (Oligocene). Genus *Merycopotamus*, the youngest genus of the family, died out in Asia by the Late Pliocene. Today, the closest relatives to hippos are the cetaceans (whales, dolphins, and porpoises). They shared a common ancestor in the crown group within suborder Whippomorpha (a blending of whale and hippo).

Family Hippopotamidae was a highly diverse and geographically extensive taxa, but only has two representative species today: the common hippopotamus (*Hippopotamus amphibius*) and the pygmy hippopotamus (*Choeropsis liberiensis*). The family's evolutionary history has created considerable debate over the years; however, in 2005 French paleontologist Jean-Renaud Boisserie published an extensive examination and review of all known species of hippopotami and their closest relatives. Through this review, he established that the pygmy hippopotamus represents an ancient lineage, thought to be most closely related to the extinct Chadian hippopotamids (e.g., *Saotherium mingoz* of the Pliocene). There are five recognized genera of hippopotami: true hippopotami (*Hippopotamus*, one extant and thirteen extinct species mainly from the Pleistocene to the Holocene), hexaprotodons or Asian hippopotami (*Hexaprotodon*, sixteen extinct species from the Late Miocene and Pliocene), *Archaeopotamus* (two extinct species from the Miocene), the pygmy hippopotamus (*Choeropsis*, one extant species), and the Chad hippopotamus (*Saotherium*, one extinct species from the Pliocene).

On Madagascar alone there were three species, one of which (*Hippopotamus lemerlei*) was more closely related to the common hippopotamus. Another small hippopotamus was initially classified as *Choeropsis madagascariensis* and then reclassified as *Hexaprotodon madagascariensis*; recently it has been suggested that it should revert back to *C. madagascariensis* (Boisserie 2005). In many ways it resembles the pygmy hippopotamus. The most recent Malagasy hippopotamus, *H. lemerlei*, went extinct approximately 1 Ka, although some accounts indicate a more recent extinction. Archaeological evidence indicates that some of these hippopotami were butchered by humans, which might have led to their extinction. Despite the extensive fossil record for hippopotami, their phylogeny remains complex as new species, yet to be identified, appear throughout their range.

FAMILY TRAGULIDAE (MOUSE DEER)

The tragulids have changed very little since their first appearance in the Oligocene (~34 Ma) when they were one of the first groups to diverge from the other ruminants (Wang and Yang 2013). The water chevrotain (*Hyemoschus aquaticus*) represents the sole representative of the African branch of Tragulidae that underwent extensive radiation during the Early Miocene (21.8 ffl 6.6 Ma) when they entered Africa across a temporary land bridge (Hassanin et al. 2012). All other members of this family are in genus *Tragulus* of Southeast Asia, which is where the family is thought to have evolved (Hassanin et al. 2012). Given their early emergence in the fossil record, tragulids retain more primitive features of suborder Ruminantia. As part of infraorder Tragulina, tragulids lack horns or antlers, have a plate formed by white fibrous tissue that ossifies to create an attachment point for the sacral vertebrae, have retained second and fifth digits, lack upper incisors, and have upper canines (Eisenberg 1981).

FAMILY CERVIDAE (DEER)

Like the Tragulidae, cervids are in suborder Ruminantia, but unlike the water chevrotain, they are within infraorder Pecora with all other ruminants (Hassanin et al. 2012). Members of the superfamily Cervoidea appear in the Miocene fossil record of Eurasia. Cervidae underwent a rapid diversification in the Middle Miocene and Plio-Pleistocene. Two subfamilies, Cervinae and Capreolinae, emerged in Asia around 7 to 9 Ma and 7.7 to 11.5 Ma, respectively. Current classifications place the marsh deer (*Blastocerus dichotomus*) of South America and the water deer (*Hydropotes inermis*) of China and Korea in subfamily Capreolinae (Groves 2016), although classification of the cervids remains dynamic. Unlike many other cervids, the water deer lacks antlers and has tusk-like canines, thus leading to confusion in its phylogeny until molecular methods clarified its lineage. A member of Cervinae, Père David's deer (*Elaphurus davidianus*) of China likely resulted from a hybridization event around the Late Pliocene or earlier (5.3 to 2.6 Ma) (Hassanin et al. 2012). Male Père David's deer have branched antlers with backward-facing tines.

FAMILY BOVIDAE (CLOVEN-HOOFED RUMINANTS)

Like the deer family, the bovids underwent rapid divergence at the Oligocene/Miocene boundary (~22.4 to 27.6 Ma), with extensive radiation beginning in the Miocene of the Old World (Hassanin et al. 2012). Africa supports the greatest amount of diversity, which is where all three species of interest in this book occur: southern lechwe (*Kobus leche*), Nile lechwe (*K. megaceros*), and sitatunga (*Tragelaphus spekii*). Tribe Tragelaphini diverged in the Middle Miocene of Africa between 15.9 to 16.4 Ma, as did the tribe representing the lechwe, tribe Reduncini (14.5 to 16.0 Ma). These radiations reflect a time with higher average temperatures

Figure 2.10. *Rodhocetus.*
RECONSTRUCTION BY PAVEL
RIHA, VIA CREATIVE COMMONS
ATTRIBUTION-SHARE ALIKE
LICENSE

(Middle Miocene Climatic Optimum), which favored the expansion of evergreen forests (Hassanin et al. 2012). Genus *Kobus* appears in the fossil record at the end of the Pliocene, with both species of lechwe appearing in the Pleistocene. The sitatunga has a slightly longer evolutionary history, with its origin in the Plio-Pleistocene of the Congo and Kenya.

Infraorder Cetacea (Whales and Relatives)

FAMILY PROTOCETIDAE

This diverse group of cetaceans had a near global distribution, including Africa, Europe, Asia, and North America. It is a fascinating group that represents the first stage of aquatic evolution for marine mammals in particular. During that evolution, however, some species lived in freshwater habitats. In particular, *Rodhocetus* is described as "otterlike in terms of its intermediacy on the terrestrial-aquatic axis" (Gingerich 2003: 446). Proportionally, however, it resembled a desman more than an otter, and it likely swam at the surface using alternate-pelvic paddling with additional buoyancy achieved through its non-wettable fur (Gingerich 2003). It lived during the Eocene (48.6 to 40.4 Ma) in what is now Pakistan (Fig. 2.10).

Figure 2.11. *Ambulocetus natans.*
RECONSTRUCTION BY NOBUMICHI TAMURA,
USED WITH PERMISSION

FAMILY AMBULOCETIDAE

One specific member of family Ambulocetidae, *Ambulocetus natans*, was also found in the area of modern-day Pakistan during the Early Eocene, 47.8 to 41.3 Ma (Fig. 2.11). As with *Rodhocetus*, it was a member of Cetacea, as indicated by the translation of its name "walking whale" (Thewissen, Hussain, and Arif 1994). Fossil analysis indicates that *Ambulocetus natans* used both freshwater and saltwater habitats associated with coastal swamps and shallow seas, thus representing one step closer to the evolutionary path of today's whales (Ando and Fujiwara 2016).

An overview of the paleobiology and phylogeny of today's freshwater mammals and their early relatives reinforces that aquatic habits and adaptations evolved independently several times in mammals. Some taxa, such as otters, reveal a consistent link between ancestral and modern forms in their adaptations to semi-aquatic habitats. Other taxa, including the rodents and hoofed mammals, reveal a diversity of locomotor styles within what are otherwise often monophyletic groups. The semi-aquatic mammals also demonstrate how competition and habitat availability drives convergent evolution.

3

Ranging across the Continents

Biogeography does more than ask Which species? and Where. It also asks Why? and, what is sometimes more crucial, Why not?

David Quammen, 1997, *Song of the Dodo: Island Biogeography in an Age of Extinctions*

Introduction to Distributions

How species distribute themselves locally, regionally, and globally have inspired some of the central theories in ecology today. As early naturalist scientists, such as Charles Darwin and Alfred Russel Wallace, explored the corners of the world, species new to Eurocentric scientists inspired investigations into broad-scale movements, specialized adaptations, and species interactions that helped form our current perceptions of an earlier world. These same investigations continue to help us understand the world of today, through the help of molecular genetics and an increasingly physically and technologically accessible world. Just as the previous chapter on paleobiology provides a glimpse into the past, this chapter offers an overview of current distributions of semi-aquatic mammals.

Species distributions are often described by range maps, lists of countries where a species is found, broad-based zoogeographic realms, habitat types, and seasonality, among others. Defining a distribution requires a holistic perspective to identify the myriad of factors that determine why a species chooses the geography it does. There is always an element of basic physical geography, especially for semi-aquatic mammals, whose reliance on the interface between land and water is critical. For example, Stolzmann's crab-eating rat (*Ichthyomys stolzmanni*), found in the Napo Province in eastern Ecuador and near the Andean city of Tarama in Peru, specifically requires clear, fast-flowing streams with rocky bottoms that support crabs and other invertebrates. To add to its specificity, the few individuals that have been documented live only in streams within primary forests. No other freshwater system, substrate, or forest appears to support Stolzmann's crab-eating rat. A simple change in water clarity from human-caused siltation might be enough to pose severe impacts on its population. The physical geography of an area at the smallest scales (rocky benthos) to the largest (oceanic separation, high mountain ranges), along with soils that allow for a favored plant to grow, help to define a species' range. Additionally, how species interact through competition, predation, and mutualism, for example, can change how organisms distribute themselves spatially. If crabs and invertebrates disappear, crab-eating rats will either need to find a new foraging area or, if anatomy and behavior allow, switch to a different food source.

There are native semi-aquatic mammals living on all continents except for Antarctica. They are also naturally absent from large oceanic islands, including New Zealand and remote island chains where dispersal was hindered by geographic challenges. Others have expanded their range with the help of humans, most often through intentional releases. Where there is fresh water, there are often semi-aquatic species specialized to take advantage of the habitats the fresh water provides. One glance at a world distribution map of semi-aquatic mammals highlights their successful colonization across the globe (Fig. 3.1). As expected, sub-Saharan Africa and South America are home to the greatest diversity of semi-aquatic mammals at just under 30% each, while Australia has only four species (~3%), with two of them also occupying expanded distributions farther north.

Asiatic species are the third-most numerous group at 21%, and North American species are the fourth-largest group with approximately 16%. Europe has just under 8% of the world's freshwater semi-aquatic mammals. These percentages, of course, exceed 100% because of the extended natural distributions of some species onto adjacent continents. In reality,

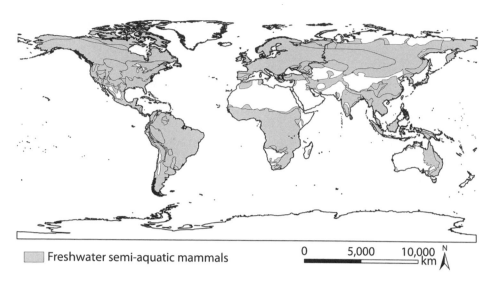

Figure 3.1. Global distribution of freshwater semi-aquatic mammals.
DISTRIBUTION DATA FROM THE IUCN

biogeographers are more inclined to categorize species to zoogeographic realms and regions to complement ecological similarities, rather than using continental boundaries. As with other organisms, semi-aquatic mammals reflect a broad range of distributions, from a singular location next to a particular stream on a remote tropical mountain, to an extensive range of habitats across diverse climatic zones. From their restricted ranges to continental distributions, semi-aquatic mammals provide a perfect example of evolutionary success regardless of habitat size.

Types of Distributions and Why

Species living within a specific geographic range are considered endemic to a particular area. Depending on scale, that range can include a whole continent or be limited to a particular lake or stream, as is the case of the recently classified Sulawesi water rat (*Waiomys mamasae*). In May 2012, three biologists were conducting night surveys in the western highlands on the island of Sulawesi, Indonesia, near the town of Mamasa (Rowe, Achmadi, and Esselstyn 2014). Unexpectedly, they caught an unknown species of semi-aquatic mammal by hand in a small, shallow, fast-flowing stream; it would not only be a new species to science but also a new genus. Local Mamasan people previously knew of its presence and called it the Mamasa Toraja's word "wai," for water, which was later incorporated into its scientific name. The distribution of endemic species is rarely random or uniform; they are usually clumped in regions. As with many endemic species, the isolation of island habitats creates the perfect environment for unique taxa. Several freshwater semi-aquatic mammals are considered as endemics within restricted ranges.

The island of New Guinea represents an ideal location for the evolution of endemic semi-aquatic mammals. It is an island of rugged mountain ranges and fast-flowing streams that make land expeditions into the interior difficult at best. The waterside rat (*Parahydromys asper*), although endemic, is widespread across the Cordillera Central and outlying ranges of the island from the northern Vogelkop Peninsula to the southeastern coast (Flannery 1995). However, three species of water rats from the genus *Hydromys* evolved in isolation from each other and show much narrower ranges. The western water rat (*H. hussoni*) lives in rivers, lakes, and streams of the montane forests in the isolated Wissel Lakes region of Indonesia's Papua Province, while Ziegler's water rat (*H. ziegleri*) inhabits streams and rivers in lowland tropical forests of the southern slopes of the Princess Alexandra range near Bainyik, Papua New Guinea. Living in further isolation, the New Britain water rat (*H. neobritannicus*) is endemic to the island of New Britain in the Bismarck Archipelago of Papua New Guinea, where it also resides in lowland streams, rivers, and wetland habitats. Conversely, the rakali (*H. chrysogaster*) is widely distributed from the island of New Guinea and its neighboring islands to the north to a variety of aquatic habitats as far south as Tasmania and southern Australia (Fig. 3.2).

Asiatic water shrews (*Chimarrogale*) exhibit endemism similar to the water rats of the island of New Guinea. The endangered Bornean water shrew (*C. phaeura*), from the Malaysian state of Sabah on the island of Borneo, lives in tropical forests of Mount Trus Madi, Mount Kinabalu, and the Crocker Range, where it forages for invertebrates. Over 3,000 km away on the Japanese islands of Honshu and Kyushu, the Japanese water shrew (*C. platycephalus*) lives in fast-moving, high elevation streams where it forages for invertebrates among boulders and large rocks. Another island species, the Sumatran water shrew (*C. sumatrana*), has a

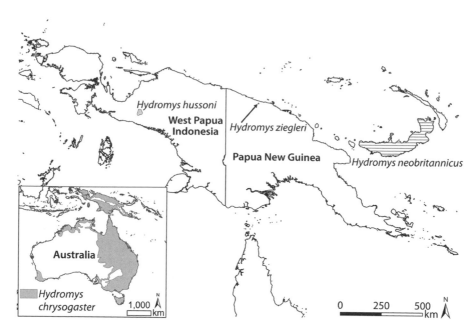

Figure 3.2. Distribution of *Hydromys* species.
DISTRIBUTION DATA FROM THE IUCN

very limited distribution. It has only been recorded in the Padang High-lands of southern Sumatra and remains a poorly documented species (Chiozza 2016). On the mainland, the Malayan water shrew (*C. hantu*) has a limited distribution but broader habitat preferences, including streams, wetlands, and rivers within intact primary lowland and mon-tane forests (Liat, Belabut, and Hashim 2013).

Another notable species that is endemic to its island habitats is the web-footed tenrec (*Limnogale mergulus*) of the eastern highlands and escarpment of Madagascar. It has only been found in approximately ten locations that support permanent, clean, and fast-flowing streams and rivers (Benstead, Barnes, and Pringle 2001). Not only is its range naturally restricted, its need for healthy freshwater systems makes this unique species vulnerable to extinction. Similarly, the Cabrera's hutia (*Mesocapromys angelcabrerai*), endemic to a small area along the coast and nearby islands of west-central Cuba, is now endangered because of wetland loss, by-catch in fishing nets, and predation by introduced black rats (*Rattus rattus*) (Soy and Silva 2008).

As with island species, species distributions on larger land masses are influenced by varied terrain and natural barriers, such as deserts, large bodies of water, and mountain ranges. These physical barriers help create isolated environments that either support species that evolved in geographic isolation (**autochthonous endemic**) or species that dispersed from their place of origin and then became the only survivors once the original populations went extinct (**allochthonous endemic**). Sometimes endemic species survive as taxonomic **relicts** of once diverse groups, or they are the remaining descendants of previously widespread taxa. Just as there are different species of the same genera living in similar aquatic niches on adjacent islands, comparable trends occur at the continental level as well.

The Amazonian region of South America boasts some of the highest biodiversity in the world, in part because glaciation did not slow evolution (which allowed competition among species) and habitat specialization promoted speciation over a longer time period. As islands do, Amazonia supports many endemic species that have restricted ranges, many of which belong to the ichthyomyine rodents (Voss 1988). The sole species within its genus, the Ucayali water rat (*Amphinectomys savamis*) is found close to the Ucayali River and near a lake in the Requena Province, Peru (Pacheco, Zeballos, and Vivar 2008). There are only two official records of this small rodent, and it remains a poorly understood species. Similarly, the endangered Ecuadoran fish-eating rat (*Anotomys leander*) has been recorded at three high elevation locations (2,800 m and 4,000 m above sea level) in the Andes of northern Ecuador (Voss 2015) and at only one locality in the Cordillera Central of Colombia (Marín and Sánchez-Giraldo 2017). It specializes on cold, fast moving streams surrounded by high elevation forest. Similarly, the Las Cajas water mouse (*Chibchanomys orcesi*) from southern Ecuador lives above tree line on the Cajas Plateau near Cajas National Park (Jenkins and Barnett 1997). It hunts for aquatic invertebrates and small fish in clear, cold, fast-flowing streams that support high oxygen levels. All seven species of fish-eating rats (*Neusticomys*) are endemic to the Amazonian zoogeographic region, with many living in relatively restricted ranges. Ferreira's fish-eating rat (*Neusticomys ferreirai*) is known only from a single lowland site where it was associated with a slow-moving stream in the state of Mato Grosso in south-central Brazil. With five known sites in Venezuela and Guyana, the increasingly vulnerable Venezuelan fish-eating rat (*Neusticomys*

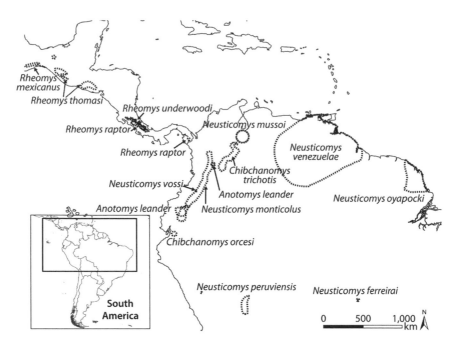

Figure 3.3. Ichthyomyine rodents from Mexico to South America. DISTRIBUTION DATA FROM THE IUCN

venezuelae) hunts invertebrates and small aquatic mammals in small streams within lowland forests. These species represent a suite of species unique to South America and share a common biogeographic distribution within tropical environments. Warm temperatures combined with precipitation produce the habitat diversity required to support a variety of species within relatively restricted geographic ranges.

Southern Mexico and Central America show similar trends relative to the more northern ichthyomyine rodents (Voss 1988). The water mice (*Rheomys*) of this region are also endemic to the area. The Mexican water mouse (*R. mexicanus*) hunts in pristine headwater streams in the province of Oaxaca, Mexico, the northernmost extent of *Rheomys*. Much of what we known about the Mexican water mouse dates to the late 1950s when biologist Thomas MacDougall secured four specimens from local fishers (Goodwin 1959). Given its dependence on intact forested riparian areas and intolerance to pollution, current deforestation and development activities in southwestern Mexico create difficult challenges

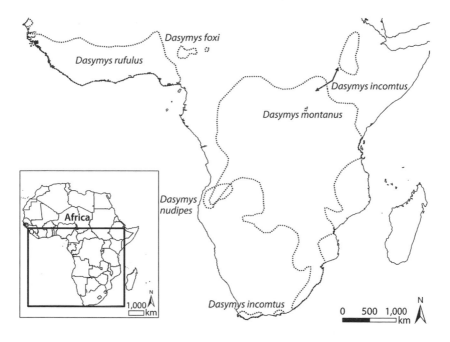

Figure 3.4. Distribution of *Dasymys*.
BASED ON IUCN DISTRIBUTION DATA AND
MULLIN, PILLAY, AND TAYLOR 2005

for this endangered rodent. The other three species (Goldman's water mouse *R. raptor*, Thomas's water mouse *R. thomasi*, and Underwood's water mouse *R. underwoodi*) also have limited ranges but are less isolated and cross more jurisdictional boundaries than the Mexican water mouse (Fig. 3.3.).

Sub-Saharan Africa supports a similar diversity of semi-aquatic mammals, including endemics with restricted ranges (Fig. 3.4). As molecular genetics provides clarity to taxonomic distinctions, species that were once thought widespread now comprise several species with much smaller individual ranges. Such is the case with *Dasymys*. Originally recognized as five distinct species (Musser and Carleton 1993), a more recent study identified eleven species (Mullin, Pillay, and Taylor 2005). Current data from the IUCN Red List still describes the original five species: Fox's shaggy rat (*D. foxi*) from Nigeria, African marsh rat (*D. incomtus*) from the Congo to South Africa, montane shaggy rat (*D. montanus*) from the Ruwenzori Mountains in Uganda, Angolan marsh rat (*D. nudi-*

pes) from the central highlands of Angola to Botswana, Namibia, and Zambia, and the West African shaggy rat (*D. rufulus*) from costal Senegal across through most of West Africa.

Taxonomic reclassification of one species into two or more species decreases the number of individuals per species and results in species-specific distribution maps covering smaller geographic areas. Such is the case for *Dasymys*. For example, Sarah Mullin and her colleagues suggest the Cape marsh rat (*D. capensis*), associated with intact rivers and wetland ecosystems in the Western Cape province of South Africa, possibly could be a relict population distinct from *D. incomtus* (African marsh rat) because of its geographic isolation and morphological differences. Once thought to be *D. incomtus* as well, *D. shortridgei* is limited in range to the Okavango Delta in Botswana, where its status is unknown. Similarly, *D. longipilosus*, a species seen as synonymous with the West African shaggy rat by some, lives only on Mount Cameroon. Another newly defined species about which very little is known, *D. griseifrons*, is found only in the mountains of Ethiopia. Still regarded as *D. incomtus* by some taxonomists, *D. medius*, has a more extensive range in the mountains of the Democratic Republic of the Congo (DRC), Burundi, Rwanda, Uganda, and Kenya. *D. robertsii* also occupies an area previously associated with *D. incomtus* in the northeastern corner of South Africa (Fig 3.5). This genus provides a good example of how species diversity and range contraction can influence our perception of distributions and subsequent conservation needs. Their taxonomy is still under review.

Endemic to the Albertine Rift valley in southwest Uganda, western Rwanda, Burundi, and extreme eastern DRC, the Delany's mouse (*Delanymys brooksi*) lives in high altitude marshes (2,000 m above sea level) within undisturbed stands of bamboo, cypress swamps, and montane forests. A good climber, this herbivore is adept at climbing on stems of marsh grass and similar plants with the help of its prehensile tail (Pacini and Harper 2008). Still endemic to the DRC, the aquatic genet (*Genetta piscivora*) lives within a patchy distribution associated mainly with shallow headwaters of streams from the right bank of the Congo River to the Rift Valley within the DRC. Harry Van Rompaey and Marc Colyn call it one of the rarest carnivores in Africa (Van Rompaey and Colyn 2013) and, despite being named in 1919, very little is known about this species.

Of even greater rarity is the Ethiopian amphibious rat (*Nilopegamys*

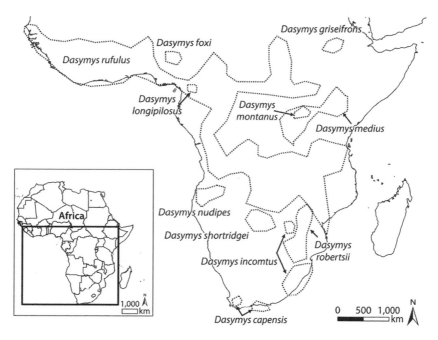

Figure 3.5. Revised distribution of
Dasymys species. BASED ON POINT MAP
IN MULLINS, PILLAY, AND TAYLOR 2005

plumbeus), known from a single specimen from a high elevation (2,600
m above sea level) mountain tributary in northwestern Ethiopia near
the source of the Little Abbai River (Nowak 1999). The river ultimately
flows into the Blue Nile. Although still listed as critically endangered on
the IUCN Red List, there is speculation that the Ethiopian amphibious
rat might now be extinct given the extensive agricultural expansion in
the area and lack of recent sightings.

Some endemic species with restricted ranges might also be relicts,
either because they are the sole survivors of a previously diverse group
(**taxonomic relicts**) or because they represent geographically restricted
descendants of a group that was originally widely distributed (**biogeographic relicts**). The Rwenzori otter shrew (*Micropotamogale ruwenzorii*) likely represents a relict population that evolved in the Albertine Rift
Mountains, which acted as a refugia during the Pleistocene (Nicoll and
Rathbun 1990). The Albertine Rift is part of the Rwenzori region of eastern DRC and western Uganda and is where most individuals have been

found near small rivers and streams in gallery forests, among other habitat types. More recently a Rwenzori otter shrew was recorded in Nyungwe Forest National Park, Rwanda (Stephenson 2016). The Nimba otter shrew (*M. lamottei*) is endemic to a small area in West Africa, including Mount Nimba and parts of the Putu range of Liberia, and limited locations in Côte d'Ivoire and Guinea. Its habitat associations are broader than the Rwenzori otter shrew. It too uses very small creeks and larger streams; however, it lives in small swamps in primary and secondary rain forests, and on cocoa and coffee plantations (Vogel 1983).

Several of the mammals in Australia are endemic due to geographic isolation over millions of years. Of the four semi-aquatic mammals, the Australian swamp rat (*Rattus lutreolus*) and the platypus (*Ornithorhynchus anatinus*) are geographically unique to Australia. The other two species, the rakali and false water rat (*Xeromys myoides*), have northern ranges extending to the island of New Guinea. The false water rat, although extending to the very southern tip of Papua New Guinea, has a disjunct distribution, with isolated pockets associated with mangroves and near-coastal freshwater swamps along Australia's northern coast, and more so along its eastern seaboard (Woinarski and Burbidge 2016). Disjunct distributions can occur when a species disperses through less desirable habitat until they settle in more suitable areas, or when a previously suitable range becomes geographically fragmented through either natural or anthropogenic means. Whatever the cause associated with the false water rat, its disjunct distribution leaves it vulnerable to extinction, especially when challenged with current degradation of mangrove and coastal wetland habitats.

The Australian swamp rat lives in wetland habitats and wet forests on the mainland and in a combination of moist forests, alpine areas, and moorlands in Tasmania and on the Bass Strait islands (Lunney 2008). As with the false water rat, it has a disjunct distribution farther northeast along the Queensland coast between the Atherton Tableland and Paluma (Burnett et al. 2016). The iconic platypus is also endemic to Australia, with a fairly broad distribution along the eastern coast in the states of Victoria, Tasmania, South Australia, Queensland, and New South Wales (Fig. 3.6). The only location in South Australia where the platypus can be found is the result of a translocation of a small population to the western end of Kangaroo Island (Carrick, Grant, and Temple-Smith 2008).

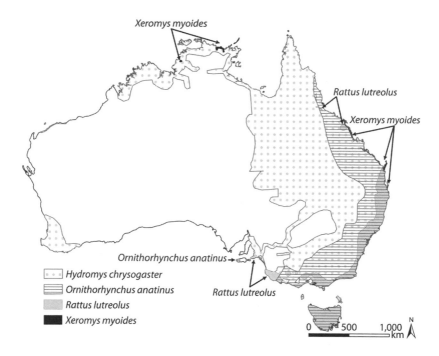

Figure 3.6. Distribution of semi-aquatic mammals in Australia. DISTRIBUTION DATA FROM THE IUCN

Endemic semi-aquatic mammals with restricted ranges in North America are markedly fewer than in the tropics. Despite several endemic species on the continent, many species have expansive distributions, with some ranging at a continental scale. In the far reaches of Alaska, however, the Glacier Bay water shrew (*Sorex alaskanus*) lives in bogs and streams on the flats around Gustavus and Bartlett Cove near Glacier Bay. Very little is known about this shrew, although it is thought to be similar to the closely related American water shrew (*Sorex palustris*) that is found throughout much of North America (Nowak 1999). As with many geographically restricted endemic species, Glacier Bay water shrews are hard to find and, therefore, difficult to study in detail.

At the other extreme, there are cosmopolitan species with a worldwide distribution encompassing most habitable landmasses. As with most species, no freshwater semi-aquatic mammals are naturally cosmopolitan, although some, such as the European water vole (*Arvicola amphibius*), occupy a broad range of aquatic habitats extending from the United Kingdom through continental Europe to the far reaches of Siberia. As with many broadly distributed species, the European water vole lives in a variety of freshwater habitats, from mountain streams to irrigation ditches. Its diet is equally as broad, ranging from aquatic and herbaceous plants to invertebrates, mollusks, small fish, and tadpoles. In winter it switches to tubers and roots (Nowak 1999). Two factors come into play with widely dispersed taxa: (1) they are capable of dispersing large distances, and (2) they have broad ecological tolerances. For example, the rakali, mentioned above, also has broad tolerances and was able to take advantage of various dispersal routes over time, yet its close relatives were not.

Similar to the European water vole, the Eurasian water shrew (*Neomys fodiens*) lives in a broad range of aquatic habitats from the Atlantic coast across Europe into the heart of Mongolia and China. As an insectivore, its diet is only slightly more restricted than the vole. It is an adept hunter across its range and is considered the most aquatic of all shrews (Hutterer et al. 2016b). Perhaps the species with the broadest range, however, is the Eurasian otter (*Lutra lutra*). It occupies a wide variety of aquatic habitats across three continents (Europe, Asia, and northern Africa), and it is one of the most widely distributed of the Palaearctic mammals (Roos et al. 2015). Regardless, the Japanese otter (sometimes described as a subspecies, *L. lutra whiteleyi*) was declared extinct on August 28, 2012, after a long period of overhunting and subsequent habitat degradation. Throughout its range, the Eurasian otter faces similar challenges and is classified as near threatened by the IUCN, despite its broad distribution (Fig. 3.7).

In sub-Saharan Africa, each of the three species of otters inhabit extensive ranges. Of the three, the African clawless otter (*Aonyx capensis*) is most widely distributed, with a distribution extending from coastal Senegal across to Eritrea and as far south as the southern tip of South Africa. In total, its range covers at least thirty-three countries. As with many otters, the African clawless otter can live at the interface of coastal brackish and freshwater habitats, as well as in arid areas of the Western Cape in South Africa (Nel and Somers 2007). Unlike most otters, it also shows some tolerance to pollution and modified landscapes (Somers and Nel 2003). The spotted-necked otter (*Hydrictis maculicollis*) is more typical of otters in its intolerance of pollution, including siltation. Despite a

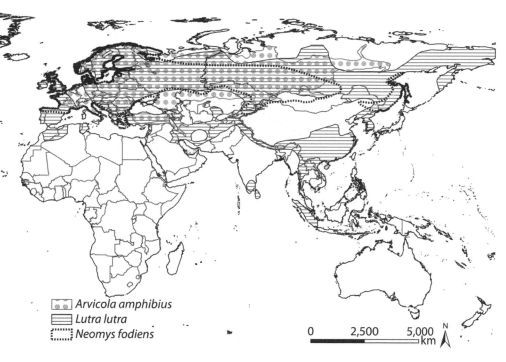

Arvicola amphibius
Lutra lutra
Neomys fodiens

0   2,500   5,000 km   N

Figure 3.7. Widely distributed Eurasian semi-aquatic mammals. DISTRIBUTION DATA FROM THE IUCN

relatively broad distribution, which on a map appears to be almost as extensive as that of the African clawless otter, it is most common in intact habitats associated with the large lakes of Central and East Africa, and in streams and rivers up to 2,500 m above sea level (Yalden et al. 1996). Its range appears to be decreasing, and it is now locally extinct (**extirpated**) in Burundi, Ghana, Lesotho, and Togo (Reed-Smith et al. 2014). The Congo clawless otter (*A. congicus*) is specific to diverse freshwater habitats in the rain forest of the Congo basin (Fig. 3.8).

The sitatunga (*Tragelaphus spekii*) shares a similar distribution as the Congo clawless otter, although it is dependent on seasonal swamps, riverine vegetation, reed beds, and stands of papyrus. Along with its distribution within twenty-four countries in a band extending from Ghana in West Africa through Central Africa and south to northern Botswana, there are disjunct populations in Angola, Chad, South Sudan, and along the western coast from Senegal to Guinea. It is extirpated in Niger and,

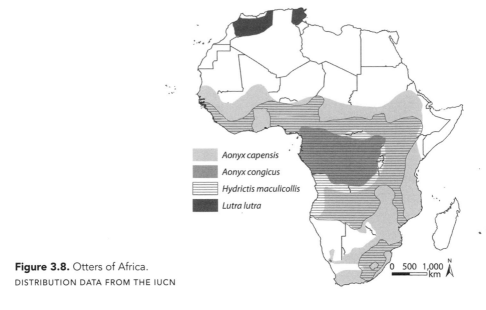

Figure 3.8. Otters of Africa.

DISTRIBUTION DATA FROM THE IUCN

Aonyx capensis
Aonyx congicus
Hydrictis maculicollis
Lutra lutra

as with the Congo clawless otter, perhaps Togo (May and Lindholm 2013), thus leaving a small, disjunct population in southeastern Ghana.

The common hippopotamus (*Hippopotamus amphibius*) is an iconic species found in rivers, lakes, and wetlands of sub-Saharan Africa. It lives in thirty-eight African countries and, surprisingly, one South American country. Its broad distribution belies its newfound vulnerability to poaching, habitat degradation, and the impacts of civil war (Fig. 3.9). The vulnerability of the common hippopotamus's skin to desiccation when not in water is balanced by its terrestrial forays to forage on grasses. Its distribution in Colombia is linked to drug lord Pablo Escobar. In the early 1980s, Escobar brought three pairs of common hippopotami from the United States to a private zoo on his ranch, Hacienda Nápoles (Vásquez 2012; Svenning and Faurby 2017). When Escobar was killed in 1993, the three females and one male that remained were left to fend for themselves. In time, they grew into a herd of twenty-eight animals. Some of these animals migrated via the Magdalena River to swampy areas of the Magdalena Medio Antioquia subregion (Vásquez 2012). There could be as many as forty wild hippopotami in Colombia. In contrast to the large range of the common hippopotamus, the endangered pygmy hippopotamus (*Choeropsis liberiensis*) has a limited range in the swamps and streambanks of lowland forests of Côte d'Ivoire, Guinea, Liberia,

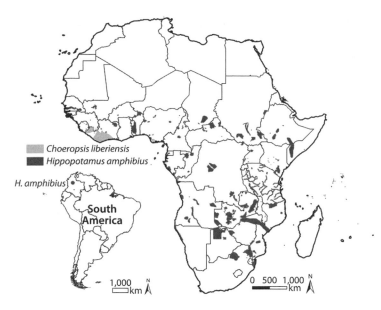

Figure 3.9. Hippopotamus distribution. DISTRIBUTION DATA FROM THE IUCN, EXCEPT FOR COLOMBIAN LOCATION

and Sierra Leone (Roth et al. 2004). Its range continues to decrease as deforestation continues. Unfortunately, this fascinating mammal is now extirpated in Nigeria.

Three South America species with extensive distributions are the capybara (*Hydrochoerus hydrochaeris*), the coypu (*Myocastor coypus*), and the endangered giant otter (*Pteronura brasiliensis*). The capybara uses varied habitats, from lentic habitats, such as marshes and estuaries, to rivers and streams from Colombia to the north to the Argentinian pampas in the south, and from the Andes to the coast of Brazil. The coypu has an equally expansive range and is rarely found more than 100 m from rivers (Bertolino, Perrone, and Gola 2005). Outside of South America, the coypu is an introduced pest species throughout Europe, the United States, the United Kingdom, and Japan. The damage it does to agricultural crops, riverbank stability, and native vegetation communities is costly and ecologically destabilizing (Carter and Leonard 2002). It also is a successful competitor with native semi-aquatic mammalian herbivores.

The original distribution of the giant otter extends from Colombia through Amazonia to the Misiones province of Argentina. Sadly, a suite of stressors has sent this wonderful species into rapid decline, specifically in parts of Brazil. It is likely extirpated from Uruguay and Argentina, and it occupies less than 2% of its former range in Paraguay (Groenendijk et al. 2015). Recent conservation efforts have offered some hope for specific populations; however, they continue to decline throughout their broad, yet fragmented range.

Similar to South America, North America also boasts several semi-aquatic mammals that occupy extensive ranges. The North American beaver (*Castor canadensis*) ranges from northern Mexico in the south to the Arctic Ocean in the north, and from the Pacific to Atlantic coasts. Within that extent, it is not found in Florida or some of the drier regions of the United States, and until a recent northern range expansion, it was not found on the Arctic tundra (Tape et al. 2018). It is one of the most wide-ranging mammals on the continent and is adapted to a variety of freshwater habitats and plant species. As a generalist herbivore, it has also successfully expanded its range in areas where it was intentionally introduced, either as a by-product of fur farming or for ecological goals when managers were not aware that it was a distinct species from the Eurasian beaver (*Castor fiber*). Consequently, introduced populations can be found in Argentina (Tierra del Fuego) and Chile in South America and in Belgium, Finland, Germany, Luxembourg, and Russia in Europe. Given that Eurasian landscapes coevolved with beavers, the impact of North American beavers was not as profound as in southern South America, where foraging, dam-building, and associated flooding has wreaked havoc on the southern forests and waterways (Anderson et al. 2009).

Two additional North American species, whose broad habitat preferences and native distributions have aided their success as introduced species, are the American mink (*Neovison vison*) and the muskrat (*Ondatra zibethicus*). The natural range of both species extends from the Beaufort Sea in the Arctic to the southern United States. After intentional introductions in Europe, they have achieved a broad distribution across Europe and have even spread as far as Japan. Similar to the North American beaver, they are also found in Chile and Argentina in South America. Unfortunately for the European mink (*Mustela lutreola*), competition with the American mink and its associated Aleutian disease has been a major factor in a dramatic range contraction—to the point that the European mink is now critically endangered. Muskrats, too, prove to be a superior competitor to the European water voles and the endangered Russian desman (*Desmana moschata*).

As seen with the Eurasian otter, the North American river otter (*Lontra canadensis*) lives in equally diverse habitat and ranges throughout much of North America (Cianfrani et al. 2018), with the exception of several states where it is rare or absent (Arizona, Hawaii, Indiana, Iowa, Kansas, Kentucky, Nebraska, New Mexico, North Dakota, Ohio, Oklahoma, South Dakota, Tennessee, and West Virginia). It was also extirpated from the province of Prince Edward Island in Canada by the early twentieth century because of habitat loss, although natural recolonizations appear to be occurring. As a young park naturalist on the west coast of British Columbia, Canada, I would watch otters playing in coastal brackish river deltas. Yet my formative years were spent watching them fish in the rivers and deep freshwater lakes of the interior mountains of British Columbia. As a top carnivore, they are adept at catching and consuming various prey species.

The American water shrew (*Sorex palustris*) reflects a similar equivalency to the Eurasian water shrew in habitat diversity and geographic extent, with a range from east-central Alaska to New Mexico and across Canada and much of the United States (Cook, Conroy, Herriges 1997).

Distributions across Broad Zoogeographic Realms

At the broadest level, semi-aquatic mammals occupy zoogeographic realms that represent large areas encompassing similar ecological and taxonomic characteristics. Examining endemic species living in restricted ranges, and more adaptable species with expansive distributions earlier in this chapter, helps identify the vulnerabilities and strengths of these mammals and their habitats. However, a quick glimpse into how species coexist in similar environments hints at past relationships, while providing insights into future distributions as well. A recent reassessment of Alfred Russel Wallace's zoogeographic regions by Ben Holt and his colleagues provides a fresh template for examining the global distribution of semi-aquatic mammals at the broadest of scales (Holt et al. 2013). As with Wallace before them, Holt's research team incorporated

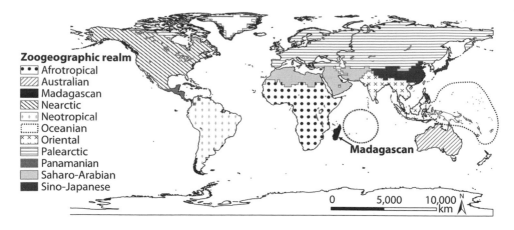

Figure 3.10. Zoogeographic realms. BASED
ON DATA AVAILABLE FROM HOLT ET AL. 2013
AND THE JOURNAL *SCIENCE*/AAAS.

species distribution data, but they also added data about phylogenetic relationships to increase our understanding of historical relationships among various regions. In total, they identified eleven zoogeographic realms, encompassing twenty zoogeographic regions (Fig. 3.10). The broader realms, described below, provide a global overview of the distribution of semi-aquatic mammals and allow for broad-scale comparisons.

Each of the tables represents only natural species ranges, not ranges advanced through species introductions. As with many taxa, semi-aquatic mammals can be widely distributed from a geographic viewpoint ("wide"), yet their populations can still be in decline and increasingly fragmented within those distribution. Other species might be restricted to a specific habitat type within a limited geography, regardless of population trends ("restricted"). Some restricted species are viable despite their limited distributions, while others are in rapid decline. Finally, some semi-aquatic mammals are known from only a handful of sightings or have been found at only one to three sites. These species are considered "isolated" in distribution. Other factors influencing population trends are covered to greater extent in Part V, "Conservation Challenges and Management Approaches."

Afrotropical

The Afrotropical realm is now distinct from the Saharo-Arabian realm, and it represents the unique taxonomic and phylogenetic nature of sub-Saharan Africa (Holt et al. 2013). It has thirty-six species of semi-aquatic mammals, all of which are unique to the Afrotropical realm (Table 3.1). The nature of this distribution speaks to the effectiveness of large deserts as barriers to dispersal for semi-aquatic species.

Australian

In 1876, Alfred Russel Wallace combined Australia, New Guinea, and many Pacific Islands into one faunal region as part of his broader development of global biogeographic regions. However, even then Wallace noticed that there was a distinct shift in species composition along an ecological boundary between what are now the Australian and Oriental realms. Now the Australian realm stands on its own, not just separated by Wallace's Line (a biogeographic boundary delineating a distinct taxonomic shift in distribution) but as a unique faunal area of its own (Holt et al. 2013). Of the four species of semi-aquatic mammals of the Australian realm, only two are endemic to the area: the platypus and Australian swamp rat (Table 3.2). The rakali and false water rat have distributions extending into the Oceanian realm.

Madagascan

The Madagascan realm is a phylogenetically distinct realm and is unlike the Afrotropical realm in species composition. There is only one species of semi-aquatic mammal on Madagascar: the web-footed tenrec (*Limnogale mergulus*). Its range is restricted to lotic habitats.

Nearctic

The Nearctic realm represents most of North America, except some parts of the high Arctic, and much of Mexico. There are eighteen species of semi-aquatic mammals in this realm (Table 3.3). The water opossum has a northern distribution that just overlaps the southern extent of the

Table 3.1. Distribution, range, and habitat associations of semi-aquatic species in the Afrotropical zoogeographic realm

Species	Countries	Extirpated	Range	Habitat
Nimba otter shrew, *Micropotamogale lamottei*	2		restricted	lotic
Rwenzori otter shrew, *Micropotamogale ruwenzorii*	3		restricted	lotic
giant otter shrew, *Potamogale velox*	~12		wide	lotic
greater cane rat, *Thryonomys swinderianus*	~8		wide	lotic
African wading rat, *Colomys goslingi*	11		disjunct	lotic
Cape marsh rat, *Dasymys capensis*	1		restricted	lentic & lotic
Fox's Shaggy Rat, *Dasymys foxi*	1		restricted	lentic & lotic
Dasymys griseifrons	1		restricted	lentic & lotic
African marsh rat, *Dasymys incomtus*	~9		wide	mainly lentic
Dasymys longipilosus	1		restricted	lentic & lotic
Dasymys medius	5		restricted	lentic & lotic
montane shaggy rat, *Dasymys montanus*	1		isolated	lentic
Angolan marsh rat, *Dasymys nudipes*	4		restricted	lentic
Robert's shaggy rat, *Dasymys robertsii*	1		restricted	lentic & lotic
West African shaggy rat, *Dasymys rufulus*	1		wide	coastal
Dasymys shortridgei	1		restricted	lentic & lotic
Congo forest mouse, *Deomys ferrugineus*	~7		wide	lentic & lotic
Ethiopian amphibious rat, *Nilopegamys plumbeus*	1	possibly extinct	isolated	lotic
Tanzanian vlei rat, *Otomys lacustris*	4		disjunct	lentic
creek groove-toothed swamp rat, *Pelomys fallax*	12		wide	lentic & lotic
Hopkins's groove-toothed swamp rat, *Pelomys hopkinsi*	~3		disjunct	lentic

Table 3.1. *continued*

Species	Countries	Extirpated	Range	Habitat
Issel's groove-toothed swamp rat, *Pelomys isseli*	2		isolated	lentic
Delany's mouse, *Delanymys brooksi*	4		restricted	lentic
swamp musk shrew, *Crocidura mariquensis*	8		wide	lentic
Kahuzi swamp shrew, *Crocidura stenocephala*	2		restricted	lentic
marsh mongoose, *Atilax paludinosus*	~44		wide	lentic & lotic
aquatic genet, *Genetta piscivora*	1		patchy	lotic
spotted-necked otter, *Hydrictis maculicollis*	18	Burundi, Ghana, Lesotho, Togo	wide	lentic & lotic
African clawless otter, *Aonyx capensis*	32		wide	lentic & lotic
Congo clawless otter, *Aonyx congicus*	10		wide	mainly lentic
pygmy hippopotamus, *Choeropsis liberiensis*	4	Nigeria	restricted	mainly lentic
common hippopotamus, *Hippopotamus amphibius*	38	Algeria, Egypt, Eritrea, Liberia, Mauritania	wide	lentic & lotic
water chevrotain, *Hyemoschus aquaticus*	~12	West Uganda	disjunct	lotic
southern lechwe, *Kobus leche*	5		disjunct	lentic
Nile lechwe, *Kobus megaceros*	2		restricted	lentic
sitatunga, *Tragelaphus spekii*	26		wide	lentic

Table 3.2. Distribution, range, and habitat associations of semi-aquatic species in the Australian zoogeographic realm

Species	Countries	Extirpated	Range	Habitat
platypus, *Ornithorhynchus anatinus*	1		wide	lentic & lotic
rakali, *Hydromys chrysogaster*	3		wide	lentic & lotic
Australian swamp rat, *Rattus lutreolus*	1		wide	lentic
false water rat, *Xeromys myoides*	2		disjunct	lentic

Nearctic, with much of its range extending well into South America. Coues's rice rat (*Oryzomys couesi*) is mainly a Panamanian species, but it extends north into southern Texas. The Neotropical otter, as its name suggests, has a distribution centered in the Neotropical realm, but it has a range extending into northern Mexico.

Neotropical

At forty-two species, the Neotropical realm is the most speciose of the zoogeographic realms (Table 3.4). At the continental scale, many of these species, such as the Ecuadoran fish-eating rat, are endemic and occupy a disjunct range in western South America. Similarly, the Lagiglia's marsh rat (*Holochilus lagigliai*) occupies an isolated habitat in western Argentina (Fernández et al. 2017). Other species, such as Coues's rice rat and the capybara, are broadly distributed. The Neotropical otter has an even broader distribution, which extends through the Panamanian realm into the Nearctic. As seen in other realms, many species of otters have extensive distributions that cover more than one realm. As relatively long-lived, large freshwater carnivores, large ranges help provide the food and resources needed to fill otters' ecological requirements.

Table 3.3. Distribution, range, and habitat associations of semi-aquatic species in the Nearctic zoogeographic realm

Species	Countries	Extirpated	Range	Habitat
water opossum, *Chironectes minimus*	21		wide	lentic & lotic
swamp rabbit, *Sylvilagus aquaticus*	1		wide	lentic & lotic
marsh rabbit, *Sylvilagus palustris*	1		restricted	lentic
North American beaver, *Castor canadensis*	3		wide	lentic & lotic
North American water vole, *Microtus richardsoni*	2		wide	lentic & lotic
round-tailed muskrat, *Neofiber alleni*	1		restricted	mainly lentic
muskrat, *Ondatra zibethicus*	2		wide	lentic & lotic
Coues's rice rat, *Oryzomys couesi*	10	Jamaica	wide	lentic
marsh rice rat, *Oryzomys palustris*	1	Pennsylvania	wide	lentic
northern bog lemming, *Synaptomys borealis*	2		wide	lentic
southern bog lemming, *Synaptomys cooperi*	2		wide	lentic
Glacier Bay water shrew, *Sorex alaskanus*	1		isolated	lentic & lotic
Pacific water shrew, *Sorex bendirii*	2		restricted	lentic & lotic
American water shrew, *Sorex palustris*	2		wide	lentic & lotic
star-nosed mole, *Condylura cristata*	2		wide	lentic
American mink, *Neovison vison*	2		wide	lentic & lotic
North American river otter, *Lontra canadensis*	2	Prince Edward Island	wide	lentic & lotic
Neotropical otter, *Lontra longicaudis*	20		wide	lentic & lotic

Table 3.4. Distribution, range, and habitat associations of semi-aquatic species in the Neotropical zoogeographic realm

Species	Countries	Extirpated	Range	Habitat
water opossum, *Chironectes minimus*	21		wide	lentic & lotic
little water opossum, *Lutreolina crassicaudata*	11		wide	lentic & lotic
capybara, *Hydrochoerus hydrochaeris*	11		wide	lentic & lotic
lesser capybara, *Hydrochoerus isthmius*	4		restricted	lentic & lotic
coypu, *Myocastor coypus*	6		wide	lentic & lotic
Ucayali water rat, *Amphinectomys savamis*	1		isolated	lentic & lotic
Ecuadoran fish-eating rat, *Anotomys leander*	1		disjunct	lotic
Las Cajas water mouse, *Chibchanomys orcesi*	1		isolated	lotic
Chibchan water mouse, *Chibchanomys trichotis*	3		restricted	lotic
Kemp's grass mouse, *Deltamys kempi*	3		wide	lentic
web-footed marsh rat, *Holochilus brasiliensis*	3		wide	lentic
Chaco marsh rat, *Holochilus chacarius*	2		wide	lentic
Lagiglia's marsh rat, *Holochilus lagigliai*	1		isolated	lentic
marsh rat, *Holochilus sciureus*	10		wide	lotic
crab-eating rat, *Ichthyomys hydrobates*	3		restricted	lentic & lotic
Pittier's crab-eating rat, *Ichthyomys pittieri*	2		restricted	lentic & lotic
Stolzmann's crab-eating rat, *Ichthyomys stolzmanni*	2		disjunct	lotic
Tweedy's crab-eating rat, *Ichthyomys tweedii*	2		disjunct	lotic
Lund's amphibious rat, *Lundomys molitor*	2		restricted	lentic & lotic
Western Amazonian nectomys, *Nectomys apicalis*	4		wide	lentic & lotic
Trinidad water rat, *Nectomys palmipes*	~2		restricted	lentic & lotic

Table 3.4. *continued*

Species	Countries	Extirpated	Range	Habitat
common water rat, *Nectomys rattus*	~6		wide	lentic & lotic
South American water rat, *Nectomys squamipes*	3		wide	lentic & lotic
Andean swamp rat, *Neotomys ebriosus*	4		wide	lentic
Ferreira's fish-eating rat, *Neusticomys ferreirai*	1		isolated	lotic
montane fish-eating rat, *Neusticomys monticolus*	2		restricted	lotic
Musso's fish-eating rat, *Neusticomys mussoi*	1		isolated	lotic
Oyapock's fish-eating rat, *Neusticomys oyapocki*	2		restricted	lotic
Peruvian fish-eating rat, *Neusticomys peruviensis*	1		isolated	lotic
Venezuelan fish-eating rat, *Neusticomys venezuelae*	2		restricted	lotic
Voss's fish-eating rat, *Neusticomys vossi*	2		isolated	lotic
Coues's rice rat, *Oryzomys couesi*	10	Jamaica	wide	lentic
Argentine swamp rat, *Scapteromys aquaticus*	2		wide	mainly lentic
Plateau swamp rat, *Scapteromys meridionalis*	1		isolated	lotic
Waterhouse's swamp rat, *Scapteromys tumidus*	4		restricted	lotic
Alfaro's rice water rat, *Sigmodontomys alfari*	7		wide	lotic
marine otter, *Lontra felina*	3		disjunct	marine & lotic
Neotropical otter, *Lontra longicaudis*	20		wide	lentic & lotic
southern river otter, *Lontra provocax*	2		restricted	marine, lentic & lotic
giant otter, *Pteronura brasiliensis*	~10		disjunct	lentic & lotic
crab-eating raccoon, *Procyon cancrivorus*	16		wide	mainly lotic
marsh deer, *Blastocerus dichotomus*	5		wide	lentic

Table 3.5. Distribution, range, and habitat associations of semi-aquatic species in the Oceanian zoogeographic realm

Species	Countries	Extirpated	Range	Habitat
mountain water rat, *Baiyankamys habbema*	1		restricted	lotic
Shaw Mayer's water rat, *Baiyankamys shawmayeri*	1		restricted	lotic
earless water rat, *Crossomys moncktoni*	2		restricted	lotic
rakali, *Hydromys chrysogaster*	3		wide	lentic & lotic
western water rat, *Hydromys hussoni*	1		isolated	lentic & lotic
New Britain water rat, *Hydromys neobritannicus*	1		restricted	lentic & lotic
Ziegler's water rat, *Hydromys ziegleri*	1		isolated	lotic
waterside rat, *Parahydromys asper*	2		wide	lotic
false water rat, *Xeromys myoides*	2		disjunct	lentic
hairy babirusa, *Babyrousa babyrussa*	1		disjunct	lentic & lotic

Oceanian

Traditionally, the Oceanian realm was merged with the Australian realm, but as a separate realm it represents the unique features of New Guinea and the Pacific Islands (Holt et al. 2013). There are ten species of semi-aquatic mammals in this realm (Table 3.5). Several are endemic species, including the hairy babirusa (*Babyrousa babyrussa*), the waterside rat, and all water rats (*Baiyankamys*, *Crossomys*, *Hydromys*) excluding the false water rat and the rakali, which are also found in the Australian realm.

Oriental

The Oriental realm covers Indonesia, Malaysia, Indochina, and India. It also includes the Indonesian island of Sulawesi. There are fourteen species of semi-aquatic mammals in this realm (Table 3.6). The presence of

Table 3.6. Distribution, range, and habitat associations of semi-aquatic species in the Oriental zoogeographic realm

Species	Countries	Extirpated	Range	Habitat
Sulawesi water rat, *Waiomys mamasae*	1		isolated	lotic
Malayan water shrew, *Chimarrogale hantu*	~1		restricted	lentic & lotic
Himalayan water shrew, *Chimarrogale himalayica*	8		wide	lotic
Bornean water shrew, *Chimarrogale phaeura*	1		disjunct	lentic & lotic
Sumatran water shrew, *Chimarrogale sumatrana*	1		isolated	lotic
elegant water shrew, *Nectogale elegans*	4		wide	lotic
flat-headed cat, *Prionailurus planiceps*	3	Likely Thailand	disjunct	mainly lentic
fishing cat, *Prionailurus viverrinus*	~8		wide	mainly lentic
otter civet, *Cynogale bennettii*	3		wide	lotic
hairy-nosed otter, *Lutra sumatrana*	5	India; Myanmar, possibly Brunei	restricted	lentic & lotic
smooth-coated otter, *Lutrogale perspicillata*	15	Isolated in Iraq	wide	lentic & lotic
Asian small-clawed otter, *Aonyx cinereus*	17		wide	lentic & lotic
Sulawesi babirusa, *Babyrousa celebensis*	1	Locally in Indonesia	disjunct	lentic & lotic
Togian Islands babirusa, *Babyrousa togeanensis*	1		restricted	lentic & lotic

large isolated islands creates an excellent context for speciation. Both species of babirusa, the Sulawesi and Togian babirusa, are endemic to this realm. Similarly, the Bornean and Sumatran water shrews, the otter civet, the flat-headed cat, and the Sulawesi water rat are unique to the Oriental realm.

Palearctic

Ben Holt and his colleagues (2013) extended the Palearctic boundaries to the northern reaches of the Western Hemisphere to account for phylogenetic similarities. There are fourteen species of semi-aquatic mammals in this expansive realm (Table 3.7), of which three species are more commonly found in the newly created Sino-Japanese realm. The elegant water shrew has a marginal range in the very south of the Palearctic realm in northcentral China, and the Himalayan water shrew has a distribution extending into northeastern China. The distribution of the Chinese water shrew extends only into a small part of the southern extent of the Palearctic range in central China.

Panamanian

The Panamanian realm is newly defined (Holt et al. 2013), and it serves to represent a phylogenetic link between North and South America. There are twenty-one species of semi-aquatic mammals in this realm; however, only three species are unique to the region: the Mexican water mouse, Thomas's water mouse, and Cabrera's hutia (Table 3.8). The other species have distributions extending northward from South America, with some continuing into the Nearctic realm.

Sahara-Arabian

The Sahara-Arabian realm acts as an intermediary between the Afrotropical and Sino-Japanese realms (Holt et al. 2013). However, as seen by the overlap of European and South Asian species, it also represents a connection with Europe (Table 3.9). For example, it is possible that the European water vole and Eurasian otter might range into the northeastern extent of the Sahara-Arabian realm. If so, the European water vole would be found in the northern reaches of Iraq, the Islamic Republic of Iran, and Afghanistan. The southern water vole is also primarily European in distribution. From the Sino-Japanese realm, the smooth-coated otter (*Lutrogale perspicillata*) is mainly found on the Indian subcontinent, but it has also been recorded in isolated locations in Iraq. The Himalayan water shrew is more associated with India, Southeast Asia, China, and Ja-

Table 3.7. Distribution, range, and habitat associations of semi-aquatic species in the Palearctic zoogeographic realm

Species	Countries	Extirpated	Range	Habitat
Eurasian beaver, *Castor fiber*	24	Moldova, Portugal, Turkey	wide	lentic & lotic
European water vole, *Arvicola amphibius*	48	Locally in Turkey, Georgia	wide	lentic & lotic
southern water vole, *Arvicola sapidus*	3	Central Spain	wide	lentic & lotic
Himalayan water shrew, *Chimarrogale himalayica*	8		wide	lotic
Chinese water shrew, *Chimarrogale styani*	2		wide	lotic
elegant water shrew, *Nectogale elegans*	4		wide	lotic
southern water shrew, *Neomys anomalus*	24		wide	lentic & lotic
Eurasian water shrew, *Neomys fodiens*	42		wide	lentic & lotic
Transcaucasian water shrew, *Neomys teres*	7		unknown	lotic
Russian desman, *Desmana moschata*	3	Belarus	restricted	mainly lentic
Pyrenean desman, *Galemys pyrenaicus*	4		restricted	mainly lotic
European mink, *Mustela lutreola*	6	19 countries*	restricted	mainly lotic
Eurasian otter, *Lutra lutra*	~80	Japan	wide	lentic & lotic
water deer, *Hydropotes inermis*	1	Locally in China, extinct Korea	restricted	riparian

*Austria, Belarus, Bulgaria, Croatia, Czech Republic, Finland, Georgia, Germany, Hungary, Kazakhstan, Latvia, Lithuania, Moldova, Montenegro, Netherlands, Poland, Serbia, Slovakia, and Switzerland.

Table 3.8. Distribution, range, and habitat associations of semi-aquatic species in the Panamanian zoogeographic realm

Species	Countries	Extirpated	Range	Habitat
water opossum, *Chironectes minimus*	21		wide	lentic & lotic
little water opossum, *Lutreolina crassicaudata*	11		wide	lentic & lotic
Cabrera's hutia, *Mesocapromys angelcabrerai*	1		restricted	lentic
capybara, *Hydrochoerus hydrochaeris*	11		wide	lentic & lotic
lesser capybara, *Hydrochoerus isthmius*	4		restricted	lentic & lotic
Ecuadoran fish-eating rat, *Anotomys leander*	1		disjunct	lotic
Chibchan water mouse, *Chibchanomys trichotis*	3		restricted	lotic
marsh rat, *Holochilus sciureus*	10		wide	lotic
crab-eating rat, *Ichthyomys hydrobates*	3		restricted	lentic & lotic
Trinidad water rat, *Nectomys palmipes*	~2		restricted	lentic & lotic
common water rat, *Nectomys rattus*	~6		wide	lentic & lotic
montane fish-eating rat, *Neusticomys monticolus*	2		restricted	lotic
Musso's fish-eating rat, *Neusticomys mussoi*	1		isolated	lotic
Venezuelan fish-eating rat, *Neusticomys venezuelae*	2		restricted	lotic
Coues's rice rat, *Oryzomys couesi*	10	Jamaica	wide	lentic
Mexican water mouse, *Rheomys mexicanus*	1		restricted	lotic
Thomas's water mouse, *Rheomys thomasi*	4		disjunct	lotic
Underwood's water mouse, *Rheomys underwoodi*	2		restricted	lotic
Alfaro's rice water rat, *Sigmodontomys alfari*	7		wide	lotic
Neotropical otter, *Lontra longicaudis*	20		wide	lentic & lotic
crab-eating raccoon, *Procyon cancrivorus*	16		wide	mainly lotic

Table 3.9. Distribution, range, and habitat associations of semi-aquatic species in the Sahara-Arabian zoogeographic realm

Species	Countries	Extirpated	Range	Habitat
European water vole, *Arvicola amphibius*	8		wide	lentic & lotic
Bunn's short-tailed bandicoot rat, *Nesokia bunnii*	1		restricted	lentic
Himalayan water shrew, *Chimarrogale himalayica*	8		wide	lotic
southern water shrew, *Neomys anomalus*	24		wide	lentic & lotic
Eurasian otter, *Lutra lutra*	~80	Japan	wide	lentic & lotic
smooth-coated otter, *Lutrogale perspicillata*	15	Isolated in Iraq	wide	lentic & lotic

pan. Of the seven species of semi-aquatic mammals in this realm, Bunn's short-tailed bandicoot rat (*Nesokia bunnii*) is endemic to the marshlands of southeastern Iraq at the confluence of Tigris and Euphrates Rivers. More research is required to fully assess its distribution.

Sino-Japanese

The Sino-Japanese realm was also created in the reassessment by Ben Holt and his colleagues (2013). There are twelve species within this realm, of which there is some inclusion of predominately Palearctic species (e.g., Eurasian otter and Eurasian water shrew) and species from the Oriental realm (e.g., smooth-coated otter, hairy-nosed otter, Asian small-clawed otter, Himalayan water shrew, and elegant water shrew). The Japanese water shrew and Père David's deer are endemic to the realm, although the latter is mainly restricted to captive populations (Table 3.10).

The distribution of semi-aquatic mammals is dependent on the distribution of freshwater and riparian habitats. Easily defined by two categories, lentic or lotic, freshwater systems represent a substantial variation relative to vegetation cover, water clarity, depth, temperature, and food resources. Despite the global diversity of semi-aquatic mammals, they are

Table 3.10. Distribution, range, and habitat associations of semi-aquatic species in the Sino-Japanese zoogeographic realm

Species	Countries	Extirpated	Range	Habitat
Himalayan water shrew, *Chimarrogale himalayica*	8		wide	lotic
Japanese water shrew, *Chimarrogale platycephalus*	1		restricted	lotic
Chinese water shrew, *Chimarrogale styani*	2		wide	lotic
elegant water shrew, *Nectogale elegans*	4		wide	lotic
Eurasian water shrew, *Neomys fodiens*	42		wide	lentic & lotic
fishing cat, *Prionailurus viverrinus*	~8		wide	mainly lentic
Eurasian otter, *Lutra lutra*	~80	Japan	wide	lentic & lotic
hairy-nosed otter, *Lutra sumatrana*	5	India; Myanmar, possibly Brunei	restricted	lentic & lotic
smooth-coated otter, *Lutrogale perspicillata*	15	Isolated in Iraq	wide	lentic & lotic
Asian small-clawed otter, *Aonyx cinereus*	17		wide	lentic & lotic
Père David's deer, *Elaphurus davidianus*	1	In wild	captive	lentic
water deer, *Hydropotes inermis*	1	Locally in China, extinct Korea	restricted	riparian

only as healthy as the freshwater systems on which they depend. Species especially vulnerable to extinction are those with restricted or isolated distributions, or those facing increased habitat fragmentation within otherwise broad ranges. Many species described in this chapter require immediate research on the very basic aspects of habitat associations and distributions to allow them to continue to range across habitats extending from coastal estuaries to high mountain streams.

4

Ecological Niches

The history of life
on earth has been a
history of interaction
between living
things and their
surroundings.
—Rachel Carson, 1962,
Silent Spring

Introduction

In the early part of the twentieth century, two main approaches
were established to describe a species' place or role in the environment, otherwise known as its **niche**. On the surface, it
seems a simple idea: what does a species need, and what does
a species do? But of course species-environment interactions
are much more complex, especially when bridging terrestrial
and aquatic environments. Also, in many cases we know so
little about a species that our scale of analysis complicates interpretations. How science has conceptualized a species' niche
has progressed over time and sometimes concurrently. In 1917,
American zoologist Joseph Grinnell published his study "The
Niche Relationships of the California Thrasher," in which he
described a niche as how species *respond* to resources. Soon
thereafter, in 1927, British zoologist Charles Elton provided his
own definition in his classic book, *Animal Ecology*. Rather than a
species' response to resources, Elton focused on the *impact* that
a species has on the environment. With time, both biotic and
abiotic resources were incorporated into G. Evelyn Hutchinson's niche concept outlined in 1957 (Hutchinson 1957; Devictor et al. 2010). Rather than the niche being a component of the
environment, Hutchinson identified it as an attribute of a species or population as it relates to the environment (Colwell and
Rangel 2009). As such, no matter where a particular species is,

its attributes follow it, which has implications in recolonization, reintroductions, and management of non-native species, as seen in Chapter 12, "Introductions and Reintroductions." In ecology, a niche is often described through a mathematical model, but the variables come from the species or population. It is important to understand how it interacts with other species and the physical environment to fully understand the ecosystem as a whole.

Species Interactions

In 1933, Alexander H. Leighton published a manuscript in the *Journal of Mammalogy* entitled "Notes on the Relations of Beavers to One Another and to the Muskrat." Over the course of the paper, he recounts his observations of both captive and wild beavers near his home in Digby County, Nova Scotia, Canada. He named one of the beaver kits Paddy and named a yearling Tony. Soon thereafter Leighton caught a muskrat, which he named Jerry, and added him to the beaver group to study what type of **symbiotic** relationship might develop. This was unconventional research by today's standards, but his paper is still cited. Leighton's account provides some insights into species interactions.

> On August 11 I caught a male muskrat (hereafter to be called Jerry) and put him in the pen with Tony and Paddy where he remained until September 27. He took up his living quarters right away in the sod house along with the beavers. On August 14 I discovered that he had dug a little hole for himself at the back of the beaver chamber and from that time onward made it his bed. (p. 33)

By September 9, five muskrats were living with Jerry and the beavers in the beaver lodge, without any help from Leighton. Leighton described the interaction as **commensalism**, a symbiotic relationship where one species benefits and the other is unaffected. In this case, beavers are the neutral party. For years, I have seen similar behaviors where muskrats enter occupied beaver lodges until freeze-up, where they then live with the beaver family presumably until spring. In discussing this species association with colleagues, I have found the main consensus was that it is likely **mutualism**, where both species benefit. Muskrats provide extra body heat in the lodge and beavers keep muskrats warm. Temperatures within lodges can be 35°C warmer than outside air temperatures (Stephenson 1969). However, in a camera-based study, hours of film footage helped establish that the relationship is actually exploitative on the muskrats' part (Mott, Bloomquist, and Nielsen 2013). They gain housing and steal food, while the beaver loses some of its larder. Despite the exploitative behavior, it is not quite to the extent of **amensalism**, where the benefits to one species negatively impact another species. Instead, this relationship appears to be similar to Leighton's conclusion years before. Cy Mott and his colleagues suggest that one benefit for muskrats might be to avoid muskrat lodges elsewhere, where cannibalism can occur at high population densities. The minor loss of food likely does not affect overwinter survival of the beaver family, given the lack of interspecific aggression seen toward muskrats, even inside the lodge.

A poorly studied aspect of commensalism relative to semi-aquatic mammals is the relationship between their latrine sites and **coprophagous** (feces-eating) insects. Various freshwater mammals have latrines where they regularly defecate and urinate (e.g., raccoons, mink, otters). A study in Madagascar that examined habitat use of the web-footed tenrec (*Limnogale mergulus*) in forested and deforested areas found more fecal pellets in zero-canopy streams than in forested riparian habitats, thereby suggesting that coprophagous insects using latrine sites as food sources were less abundant in deforested areas shared with the webbed-footed tenrec (Benstead, Barnes, and Pringle 2001). Increased deforestation not only would cause population declines for the tenrecs, but might also indirectly impact insects associated with their latrine sites. Similarly, a 1998 study along the Weber River in northern Utah determined that the leaf beetle (*Chrysomela confluens*) benefited from cottonwood stumps left behind by foraging activities of North American beavers (Martinsen, Driebe, and Whitham 1998). Using their upper incisors as an anchor and lower incisors as the main cutting teeth, beavers can quickly cut several trees in a localized area (Fig. 4.1). In the case of the Weber River study, once beavers felled a tree, the stumps produced new shoots, which provided forage for the beetles. As with the Madagascar study, indirect effects of species interactions can positively influence ecological communities.

Many examples of commensalism can be found in freshwater habitats; however, mutualism is less obvious in freshwater environments

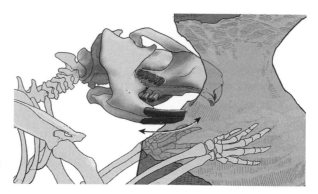

Figure 4.1. Jaw motion of North American beaver when chewing. M. BRIERLEY

Figure 4.2. Beaver beetle (*Platypsyllus castoris*), dorsal view. H. PROCTOR, USED WITH PERMISSION

(Dodds and Whiles 2010). A classic example of an interaction that benefits both parties is the ectoparasitic beaver beetle (*Platypsyllus castoris*) that only lives in the pelage of both species of beavers (Peck 2006; Fig. 4.2). Beetle densities can range from 0 to 192 adults per beaver (Janzen 1963). Adults feed on epidermal tissue, as well as on secretions from skin glands and wounds. In this case, the beetle gains nutrients from the shed skin and fluids, while the beaver benefits from a personal cleaner. There is no record that the extensive amount of grooming done by beavers has any effect on beetle presence. Only one other animal was found with a beaver beetle: a single North American river otter that likely picked up the beetle while in a beaver lodge, or after eating a beaver kit (Peck 2006). As with many cases of mutualism, beavers and beaver beetles are exclusively associated.

There are cases where the mutualistic relationship is not exclusive. Interspecific cleaning is often linked to feeding behaviors of one of the species. The association between capybara (*Hydrochoerus hydrochaeris*) and a bird, the wattled jacana (*Jacana jacana*), in the Gatun Lake region of Panama provides an excellent example of mutualism (Marcus 1985). In this case, the bird eats ticks that are parasitizing the capybara, thus obtaining a nutrient-rich meal, while the capybara reduces its infestation of **ectoparasites**. The Panamanian study was unique in that such behavior was unusual for the wattled jacana, and this particular association with capybara was not previously documented. With capybaras expanding into the area at the time of the study, this marsh-loving bird quickly adapted to this new feeding opportunity. A similar relationship exists between the common hippopotamus (*Hippopotamus amphibius*) and oxpecker birds (*Buphagus* spp.) in sub-Saharan Africa, although oxpeckers will also parasitize their hosts by creating wounds on an animal to get blood meals directly.

Some species associations are often equally as important as geographic barriers in how organisms distribute themselves spatially as well as temporally. Predation of one species by another sharing the same habitat can result in one of the species becoming nocturnal, using a different subhabitat, or going extinct. Similarly, **competition** for the same resources often results in niche separation, where the broad niche of a species (**fundamental niche**) retracts into a more restricted niche (**realized niche**) because of **competitive exclusion**. In this case, the more successful competitor completely excludes the less successful one from common resources and habitat. If not entirely excluded from an area, one species might change its use of resources to reduce competition with the more successful competitor. For example, in some parts of the Rocky Mountains of North America, beavers and North American elk (*Cervus elaphus*) compete for the same riparian woody plants, which has led to a localized loss of beavers (Baker et al. 2005). There is also the risk of extinction if adequate foraging alternatives are lacking. However, in a similar study in Elk Island National Park in western Canada, Suzanne Bayley and I determined that, rather than being excluded, beavers selected different size classes of preferred species of woody plants, and more nonpreferred forage species when faced with high-ungulate densities (Hood and Bayley 2008a). This switch in resource use, rather than habitat abandonment, is a form of **competitive exploitation**.

A good example of symbiosis and outcomes of interspecific competition in freshwater mammals lies with the otters. As seen in Chapter 3, species of the subfamily Lutrinae are found across the globe, except for Antarctica, Australia, New Zealand, and several other oceanic islands (Fig. 4.3). In many areas there is geographic overlap of more than one species. In sub-Saharan Africa, the range of the spotted-necked otter (*Hydrictis maculicollis*) overlaps with the Congo clawless otter (*Aonyx congicus*) in the Congo River basin and the African clawless otter (*A. capensis*)

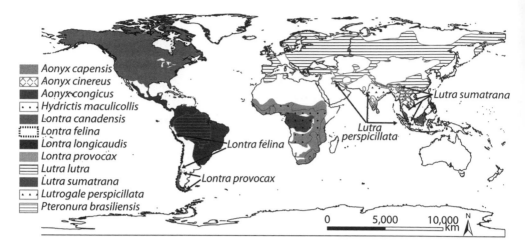

Figure 4.3. Global distribution of otters found in freshwater habitats. DISTRIBU-
TION DATA FROM THE IUCN

elsewhere. Interestingly, there are distinct size differences among the three species of otters (Nowak 1999). The spotted-necked otter is the smallest at approximately 4 kg in weight and ~100 cm in length includ-ing its tail. Fish are its primary prey. Conversely, the Congo clawless otter is ~140 cm long and weighs up to 25 kg; it primarily eats small terrestrial vertebrates, frogs, and eggs. The third otter, the African clawless otter, is somewhat smaller (~130 cm, 22 kg) and specializes on crabs, although it also has a varied diet.

David Rowe-Rowe, a South African wildlife biologist with more than thirty years of experience with otters, conducted a study on dietary dif-ferences among spotted-necked otters, African clawless otters, and the marsh mongoose (*Atilax paludinosus*) in their overlapping ranges in the Natal Province of South Africa (Rowe-Rowe 1977a). Fish and crabs domi-nated the diets of the spotted-necked otter and African clawless otter, respectively, while water mongoose mainly ate birds and mammals. De-spite all species' inclusion of fish, frogs, birds, mammals, and crabs in their diets, their main food choices dramatically reduce competition. Spatially, the spotted-necked otter avoids marine or estuarine habitats and prefers unsilted streams and water bodies bordered by dense ripar-ian vegetation. The broad distribution of the African clawless otter, on

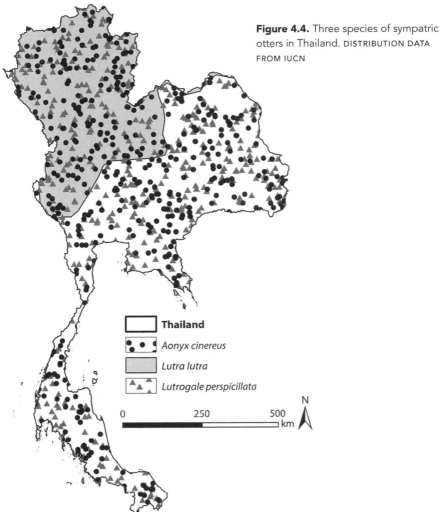

Figure 4.4. Three species of sympatric otters in Thailand. DISTRIBUTION DATA FROM IUCN

Thailand
Aonyx cinereus
Lutra lutra
Lutrogale perspicillata

0   250   500 km
N

the other hand, allows it to benefit from coastal environments to inland habitats as high as 3,000 m in Ethiopia (Yalden et al. 1996).

A similar study was conducted along the Huai Kha Khaeng River in western Thailand, where three other species of otters share overlapping ranges: the Eurasian otter (*Lutra lutra*), the smooth-coated otter (*Lutrogale perspicillata*), and the Asian small-clawed otter (*Aonyx cinereus*) (Kruuk et al. 1994; Fig. 4.4). In this case, there was dietary separation along with spatial separation. The Eurasian otter was mainly found in rapidly flowing waters in upper areas of the drainage, the smooth-coated otter preferred meandering sections of the river near a hydroelectric dam, and the Asian small-clawed otter selected habitats in the middle sections of the river. As with the African clawless otter, the Asian small-clawed otter specialized on crabs, especially *Potamon smithianus*. Both species of river otter (*Lutra* and *Lutrogale*) primarily ate fish, but the larger of the two otters, the smooth-coated otter, ate larger fish and hunted cooperatively, while the Eurasian otter was solitary (Kruuk and Moorhouse 1991).

Nicole Duplaix (1980) observed a similar situation in Suriname where two South American otters have overlapping ranges. In this case, the smaller, solitary Neotropical otter (*Lontra longicaudis*) eats smaller species of fish, while the highly social giant otter (*Pteronura brasiliensis*) consumes larger fish and often eats as a group. Along the southwest coast, the marine otter (*L. felina*) is almost exclusive to coastal marine habitats, while the southern river otter (*L. provocax*) uses some coastal habitats but is mainly a freshwater species. Competition between semi-aquatic mammals is not restricted to otters. In Europe, the southern water shrew (*Neomys anomalus*) choses slow-flowing or lentic habitats, in part due to competition with the larger Eurasian water shrew (*N. fodiens*), which is a much stronger swimmer (Hutterer et al. 2016a). A similar relationship occurs with the Himalayan water shrew (*Chimarrogale himalayica*) and the elegant water shrew (*Nectogale elegans*) where their habitat overlaps in parts of Asia. The webbed front and hind feet of the elegant water shrew give it a distinct advantage in aquatic habitats where it is able to catch small fish and aquatic invertebrates, while the less adapted Himalayan water shrew hunts insects and aquatic spiders (Nowak 1999). Competition is a major factor in spatial and temporal separation of species that share similar niches. When introduced species interact with native species, there is no time for evolutionary niches to settle out, which can result in dire consequences.

A final category of species interactions involves the role of **keystone species** (species whose actions have a disproportionate effect on the environment, over and above their own needs). A keystone species is likened to the keystone in an arch; once it is removed, the arch crumbles. In ecology, the loss of a keystone species often results in ecosystem declines and loss of diversity. The term is often overused and loosely applied;

however, in its truest sense, it relates to top-down effects on trophic levels. Robert Paine coined the term while studying food-web communities on the Pacific coast and how the removal a top predator, the starfish *Pisaster ochraceus*, ultimately led to a system with fewer species and dramatically altered food webs (Paine 1966, 1969). Various freshwater semi-aquatic mammals have been called keystone species, including beavers, muskrats, and hippopotami, because of their disproportionate impact on vegetation and wetland configurations, which then drives community structure and food-web function. For example, through the creation of "hippo lawns," the common hippopotamus provides an example of how a top herbivore improves forage selection for other species.

On the West African savanna, the range of the common hippopotamus and the kob antelope (*Kobus kob kob*) overlap, often in areas with nutrient-poor soils (Verweij et al. 2006). Through extensive grazing of aboveground biomass, new, more digestible plant growth is produced. In keeping vegetation heights down, hippopotami create prime growing conditions that produce more nutritious forage than in ungrazed swards. Mesoherbivores, such as the kob, rely on hippo lawns to augment an otherwise less nutritious food supply common to many parts of West Africa. Similarly, both species of beavers not only influence the expanse of aquatic vegetation through the creation of pond networks, they also dramatically influence riparian forests within 100 m of the pond edge, which is approximately the maximum forage distance for beavers (Hood and Bayley 2008a). Once large poplar trees are felled, understory plants flourish, and smaller poplars and less competitive species of trees (e.g., spruce, birch) can grow. The term "keystone species," which is linked to food-web dynamics, is often applied to how species change their physical environment; however, this phenomenon is better served with a different term in ecology (e.g., ecosystem engineer).

Species as Engineers

All species modify their environments to some degree; however, some species markedly change the landscape in enduring ways. Of the freshwater semi-aquatic mammals, there are several species whose day-to-day activities dramatically alter the structure and function of ecosystems. In 1994, a seminal paper by Clive Jones and his colleagues brought the concept of ecosystem engineering into a broader scientific discussion (Jones, Lawton, and Shachak 1994). The idea of organisms physically altering their environment was not new, but providing a theoretical context for their role in environments allowed for a more systematic understanding of their impacts at varying scales.

Burrows, Channels, and Trails

In freshwater systems, there are several ways that mammals physically change the environment. Many species of semi-aquatic mammals construct extensive burrow systems, likely a link to their phylogenetic connections to fossorial mammals. Others excavate the bottoms and shorelines of waterbodies and create deep hollows and distinct trail networks that then enhance aquatic connectivity and nutrient exchange. Two of the best known of these engineers are beavers and the common hippopotamus. The construction of dams by beavers and digging of channels perpendicular to the shoreline completely transform landscapes, while the creation of deep trail networks ("hippo highways") by common hippopotami expand swamp systems and link rivers to wetlands, as discussed below. Many additional species build nests or lodges that create refuges for other species and facilitate nutrient exchange. These innate behaviors in several orders of semi-aquatic mammals enhance dynamic interactions in biogeochemical cycles, water flows, and freshwater communities as a whole.

At the most basic level in freshwater systems, ecosystem engineering often involves moving and reconfiguring soil. Several species of semi-aquatic mammals, including water shrews (e.g., *Chimarrogale*, *Nectogale*, and *Sorex*) and muskrats (*Neofiber alleni* and *Ondatra zibethicus*), create extensive burrow systems that connect aquatic habitats with adjacent terrestrial habitats. Displacing soil alters drainage patterns, soil chemistry, and soil composition. Anaerobic environments become aerobic, thus altering chemical pathways and soil productivity. Mammals often bring organic materials into their burrows, and these materials then augment existing biochemical processes and microbial and invertebrate communities. In burrows with latrine areas, the feces increase nitrogen at local scales.

The Rwenzori otter shrew (*Micropotamogale ruwenzorii*), of the Ruwenzori mountain range of eastern Africa, digs tunnels into the bank of upland streams or small rivers, while ensuring that the entrance remains below the waterline. After staying in the burrow during the day, it emerges at night to hunt in the water and along the shoreline. Within the burrow it constructs a grass nest in the sleeping chamber (Vogel 1983). The closely related giant otter shrew (*Potamogale velox*) also constructs burrows with a nest chamber. The entrance is usually below the water, but it is sometimes at or above the waterline. Unlike the Rwenzori otter shrew, the giant otter shrew's burrows are regularly changed, thus leaving habitats for other species (Nicoll 1985). Several other semi-aquatic mammals in sub-Saharan Africa have similar behaviors, some of which vary even within the same species. The creek groove-toothed swamp rat (*Pelomys fallax*) is known to burrow in some areas; however, burrows of the African marsh rat (*Dasymys incomtus*) are much more extensive. This heavyset rodent constructs burrows up to 2 m long and 20 cm deep that are excavated parallel to the swampy edge of a river (Delany 1975). Aboveground, they build a domed nest that immediately links to a subterranean refuge burrow, which then leads directly to stream. They also have a surface entrance to the nest and distinct runways leading from their feeding areas (Skinner and Chimimba 2005). Unlike others of its genus, the Tanzanian vlei rat (*Otomys lacustris*) constructs extensive burrow systems, although less is known of its ecology and behavior. Farther east on the island of Madagascar, the web-footed tenrec (*Limnogale mergulus*) builds main burrows and temporary burrows that they use to rest while foraging at night (Benstead, Barnes, and Pringle 2001).

Three Australian mammals associated with freshwater, the rakali (*Hydromys chrysogaster*), platypus (*Ornithorhynchus anatinus*), and the false water rat (*Xeromys myoides*), excavate burrows adjacent to waterbodies. Burrowing and foraging activities of the rakali likely enhance the redistribution of nutrients in freshwater systems through the addition of **allochthonous** material, which is material transferred from one habitat to another (Ballinger and Lake 2006), as opposed to **autochthonous** material, which is created within the habitat itself. All semi-aquatic mammals facilitate movement of materials from terrestrial to aquatic systems, and vice versa, thus aiding nutrient exchange.

The platypus constructs a fairly simple burrow outside the breeding season that provides a simple shelter for both sexes. During breeding

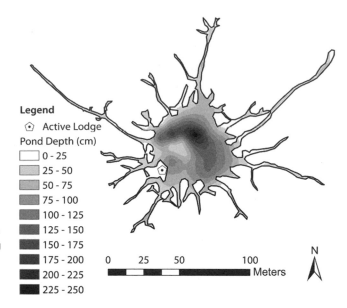

Figure 4.5. Bathymetry map of beaver pond in Miquelon Lake Provincial Park showing channels and excavation of pond bottom.
G. A. HOOD

Legend
⊙ Active Lodge
Pond Depth (cm)
☐ 0 - 25
25 - 50
50 - 75
75 - 100
100 - 125
125 - 150
150 - 175
175 - 200
200 - 225
225 - 250

0   25   50   100
                        Meters

N

season, the female digs a more complex burrow that opens just above the water and extends over 10 m into the bank, where the nesting chamber is located. This burrow helps protect the young, which is aided by a plug of soil that the female uses to close off the burrow when she leaves the young unattended (Nowak 1999; Thomas et al. 2018). An additional benefit of this burrow is that, as the female squeezes into the burrow on leaving the water, water is removed from her fur and absorbed by the soil, thus accelerating the fur-drying process (Nowak 1999).

American and Eurasian beavers also construct extensive burrows, especially along river systems where changeable water levels require flexibility in entrance locations. Above ground, they dig long channels away from the pond edge to facilitate foraging, predator avoidance, and territorial expansion (Hood and Larson 2015). In Miquelon Lake Provincial Park in Alberta, Canada, an estimated 1,700 m^3 per km^2 of soil can be displaced by beavers during channel excavations alone (Hood and Larson 2015; Fig. 4.5). Unaccounted for in that number is the extensive amount of digging beavers do along the pond bottoms as well. Normally flat pond bottoms are transformed into a network of troughs and ridges. Adjacent to dams and lodges, a tremendous amount of soil is required to construct and fortify both structures. Each spring as the rains wash away some of the existing soil on these structures, beavers need to excavate more of the pond substrate to maintain the structure of their dams and the structure and thermal integrity of their lodges. As such, the deepest location of the pond is usually at the main entrance to the lodge and immediately behind the dam. A key advantage for deep water at the lodge entrance is that much of the food cache remains accessible below the ice throughout the winter.

Wetland and muddy riparian soils often present ideal conditions for burrowing. Their high moisture content and buildup of soft sediments allow for easy digging adjacent to the water's edge. The star-nosed mole (*Condylura cristata*) is dependent on such soils and constructs a network of narrow tunnels (3.3 to 7.6 cm wide) that are up to 60 cm deep and up to 270 m long with either ground-level or underwater entrances. Some deeper parts of the tunnels can fill with water. Despite these elaborate tunnels, star-nosed moles are much less fossorial than other moles (Banfield 1974). During excavations, the star-nosed mole packs the soil into cylinders, which it then pushes to the surface to create large molehills. Renowned wildlife biologist, A. W. Frank Banfield, provides an excellent description of their digging behavior in his book *The Mammals of Canada* (1974). While digging, the tentacles of the mole's Eimer's organs on its nose, fold over the nares to prevent dirt from entering its nostrils; it uses the same strategy while diving.

Aboveground, well-defined trails also change riparian ecosystems. They can alter movements of other organisms and, in the case of the common hippopotamus, create broad networks of deep trails that fill with water and enhance aquatic connectivity and change the distribution of water (McCarthy, Ellery, and Bloem 1998). Pathways up to 1 m deep and 5 m wide extend from the pools that comprise the center of daytime activities for hippopotami, and also to upland grazing areas at night and to other waterbodies throughout the day. A study in the Ngorongoro Crater of Tanzania documented a network of ~100 m long trails radiating out from daytime wallowing pools (Deocampo 2002). Their edges form mud levees as the bodies of hippopotami push soil toward the edge of the channels. This soil then dries to create berms along the sides of the channel. Similar recessed trail systems form on lake bottoms as hippos move from their central domicile areas within a lake to local islands. According to a ten-year study in the Okavango Delta in Botswana, ecological engineering by the hippopotamus is "pivotal in determining the ecosystem's response to changing conditions" (McCarthy, Ellery, and Bloem 1998: 45). Generations of hippos use the same paths, which stir up sediment and restrict growth of aquatic plants. During droughts, hippo trails help create lagoons and pools that are used by fish (Mosepele et al. 2009).

The pygmy hippopotamus (*Choeropsis liberiensis*) also creates well-defined trails, or tunnel-like paths and canals, through thick vegetation (Nowak 1999). The pygmy hippopotamus is a solitary, primarily nocturnal species, so its paths are not as distinctive as those of the common hippopotamus. Several individuals will use the same path, but they tend to avoid each other outside the mating season (Robinson 1970). The sitatunga (*Tragelaphus spekii*), a somewhat less solitary ungulate, creates numerous branched pathways through the swamps and reedbeds in which it spends the majority of its time (Games 1983). The sitatungas' pathways through papyrus swamps are especially distinct. Along with their own trails, sitatungas also use trails created by hippos in shallower areas.

Lodges and Nests

Central to many of these burrows and pathways are lodges, dens, and nests. In their simplest form, nests are made from vegetation, such as the grass nest constructed by Delany's swamp mouse (*Delanymys brooksi*) of the Albertine Rift valley in central Africa. A larger version of a straw nest is made by the hairy babirusa (*Babyrousa babyrussa*) of Indonesia. Yet some species adopt nesting strategies that almost resemble avian counterparts. The Cabrera's hutia (*Mesocapromys angelcabrerai*), although dependent on wetlands, is also considered arboreal. It constructs a 1 m diameter communal nest in mangrove trees from mangrove branches and leaves. Much like some riparian songbirds, the Coues's rice rat (*Oryzomys couesi*) constructs a globular woven nest of ~46 cm in diameter in the reeds at a height approximately 1 m above the ground or water (Lowery 1974). Lund's amphibious rat (*Lundomys molitor*) of Brazil and Uruguay builds nests that are 30 cm across and 20 cm above the water in reedbeds. The marsh rats of South America (*Holochilus*) build even more elaborate nests that can be 3 m above the ground. As with Lund's amphibious rat, they are made of reeds but also sometimes cane and leaves to create a tight weave. The lower section of the nest is filled with plants, while the upper half of the nest contains the living area and entrances. They are relatively large nests and can weigh 100 g and have diameters reaching 40 cm (Nowak 1999).

Other species of semi-aquatic mammals combine mud and vegetation to build their nests. The false water rat builds a dome nest from mud and can be associated with a set of burrows. At 60 cm high, the dome nest is distinctive. There is an opening at the top to a narrow tunnel that leads to a nest chamber and a series of interconnected tunnels (Nowak 1999).

As mentioned previously, beavers are adept at using mud and sticks to build formidable structures. Their lodges are the central location for a beaver family and can endure for decades with regular maintenance. They often begin with a bank den, consisting of a burrow into a riverbank or shoreline of a pond. In self-contained ponds, beavers can also pile up mud and sticks to form a construction platform. Over a period of a few days, members of the family pile sticks and mud around the entrance until a fully formed lodge is made. Within the lodge, there is a platform that allows the beavers to stay out of the water while resting and additional entrances to allow more than one escape route from predators. Although there is a distinct air vent at the top of the lodge, during the coldest days of winter it is obvious that there are other parts of the structure that allow airflow. Examining the outside of a beaver lodge in winter reveals various areas where frost forms as warmer air escapes through the walls of the occupied lodge. To prevent cold air from entering the lodge, beavers increase the size of their lodge by reinforcing the walls with thick mud as winter approaches. Dietland Müller-Schwarze describes a lodge that was 6 m in diameter at the surface and over 2 m high (Müller-Schwarze 2011).

Muskrat lodges are less enduring and last for only one season. They too are made of mud and vegetation, but the muskrats' dominant use of cattails ensures that spring rains and rising water levels wash away all evidence of their lodges. Despite the seasonal nature of muskrat lodges, they provide warm shelters even in the coldest months. In winter temperatures ranging from 7°C to –39°C, internal lodge temperatures range from –4.5°C to 20.0°C (MacArthur and Aleksiuk 1979). When muskrats were present, temperatures inside the lodges averaged 9.4°C higher than abandoned lodges (MacArthur 1977). A thick cover of snow also allows for additional insulation during cold winter months.

Special Environments

In his 2013 book, *Life in the Cold: An Introduction to Winter Ecology*, Peter Marchand presents three strategies for an organism to survive the rigors of winter: migrate, hibernate, or resist. Of the 140 species of mammals I have included in this book, none of them are true hibernators, despite some living in higher latitudes dominated by continental climates. The high mountain habitats occupied by several of these species present challenges like frozen waterbodies and limited resources during winter. Lack of **hibernation** is not surprising, however, given that most northern mammals do not hibernate (Marchand 2013). However, **torpor** (a state of decreased physiological activity) can occur in some species, as suggested by Novakowski's research on beavers living in Canada's subarctic (1967). In addition to not hibernating, none of the semi-aquatic mammals undertake seasonal migration. For some species, such as the coypu (*Myocastor coypus*), temperature is limiting, and their invasive distribution is generally restricted from areas with extended winter temperatures below freezing (Carter and Leonard 2002). The survival strategies of species that do take on the challenges of winter provide a fascinating look into behavioral and physical adaptations.

Along with limited food supplies, freezing temperatures present a major challenge to semi-aquatic mammals in winter. Even in summer months, spending too much time in water results in excessive heat loss and energetic compromises. In the winter, water under the ice remains at a near-constant 3.98°C, at which water is at its maximum density. The thermal conductivity of water is higher than air, and despite thick waterproof fur of some species, trapped air is displaced in water as water compresses the pelage, thereby countering fur's insulative properties. Heat loss and hypothermia are counteracted to some extent by the waterproofing and the structure of individual hairs. In addition to morphological adaptations, there are physiological adaptations, including countercurrent circulation in the extremities of some species. Chapters 5 ("Morphology") and 6 ("Physiological Adaptations") provide more insight into these adaptations; however, there are many behavioral attributes that complement other processes to help semi-aquatic mammals succeed in cold climates. Unlike species that are solely adapted to water or land, semi-aquatic mammals must employ a multi-faceted approach to foraging in the frigid water under the ice, while still functioning on land upon emergence. Beavers, muskrats, river otters, mink, and more-northerly water shrews are the main species living in colder regions across the Holarctic, and they have developed behavioral strategies to obtain food and adequate shelter for overwinter survival.

Beavers

Although different species, the North American and Eurasian beavers are functionally equivalent in many ways, including their overwintering strategies. Well before winter arrives, beavers begin to construct winter food caches of preferred forage species in front of the main underwater entrance to their lodge. They begin by imbedding a stick into the pond bottom and continually adding to their larder until ice cover covers the entire waterbody for the rest of the winter. To ensure that most stems remain accessible under the ice, they often put larger pieces of felled trees on top of the cache to help sink the mass of stems farther (Fig. 4.6). In his MSc studies in Elk Island National Park, Canada, Doug Skinner es-

Figure 4.6. Beaver lodge with food cache. G. A. HOOD

tablished that the most nutritional species that would leach nutrients the fastest (e.g., willows) are often placed at the bottom of the cache, while less palatable or more chemically stable species are higher in the cache and help weigh down the raft (Skinner 1984). Beavers then access the woody stems, along with submerged aquatic vegetation when available, from under the ice throughout the winter. In central Alberta, beavers begin constructing food caches as early as mid-August. In 2016, I was taken by surprise to see food caches started by early July in Newfoundland, Canada. From my experience, food cache construction appears to coincide with the first frost. In areas farther south, food cache construction begins as late as October or November (Müller-Schwarze 2011). During the months before freeze-up, a beaver family will cut hundreds of woody stems, which they bring back to the lodge to build the winter food cache. Typically the areas in front of the lodge where the cache is located are the last parts of a pond to freeze in the winter and the first to open up in the spring, by an average of eleven days, thus providing increased access to open water for a broad array of birds and mammals (Bromley and Hood 2013). Canada geese preferentially use beaver lodges as nesting platforms and will actively defend them from other geese. The lodge and cache areas also attract benthic invertebrates, fish, and amphibians at a disproportionate density to the rest of the pond because of the addition of coarse woody debris and altered substrates (France 1997).

The presence of a food cache immediately outside the main entrance to the lodge provides a ready food source for beavers over the winter, until the pond ice melts in the spring. Although the food cache provides much of their overwinter food supply, beavers also swim under the ice to find submerged aquatic vegetation more than originally assumed. While conducting a winter study on beavers in central Alberta, Canada, I tracked a radio-tagged beaver that swam from its lodge across the width of the pond several times as I quietly stood on the snow-covered ice above it. It was late December, and the air temperature was −25°C. If there is enough water depth below the ice, beavers can be quite mobile. In 1967, Nicholas Novakowski discovered that beavers in the Mackenzie River area of northern Canada only leave their lodges once every two weeks. However, other studies have noted daily excursions of up to forty minutes, depending on location (Dyck and MacArthur 1992). It appears that there is a latitudinal effect on winter activity levels. In times of drought, some ponds freeze to the bottom (freeze-outs), thus trapping beavers inside their lodges where they could starve. During an extreme drought in 2002 in Elk Island National Park, beavers regularly chewed their way out of the side of their lodges as winter progressed (Hood and Bayley 2008b). It was not uncommon to find the skeletal remains of these beavers on the ice adjacent to their lodges because coyotes regularly took advantage of the easy prey. The risk was necessary, however, because of an inability to access stored food, which had since frozen into the ice.

Muskrats

Unlike beavers, muskrats do not build food caches, and forage under the ice throughout the year. Their dependence on submerged aquatic vegetation creates a reliance on relatively shallow water that facilitates growth of aquatic plants but is deep enough not to freeze to the bottom. Emergent vegetation, especially cattails, are combined with mud to create domed huts (sometimes called lodges) and feeding shelters that are half the size of huts (MacArthur and Aleksiuk 1979). Muskrats also construct push-ups, which are piles of submerged aquatic vegetation pushed up through a crack in the ice or through a plunge hole. Each structure serves a different purpose to ensure muskrats survive the winter. Semi-aquatic mammals have two key challenges that are exacerbated in winter: staying warm and foraging under the ice without the immediate ability to surface and access oxygen as needed. Muskrats present unexpected strategies to reduce heat loss and extend their access to oxygen while foraging under the ice.

Average minimum winter ambient air temperatures across the Canada prairies, where muskrats are common, range between −12.5°C and −20.5°C (McGinn 2010). Even farther north, in Canada's subarctic and arctic, where muskrat trapping by indigenous trappers was the heart of the economy for centuries, it is not unusual to experience a month or more below −40°C, yet dry fur and countercurrent circulation provide muskrats with effective remedies for such exposure. Under the ice, water temperatures range from near freezing to ~4°C, and once wetted, a muskrat's fur provides less insulation, thus physiological and behavioral strategies are critical. Once away from its hut, a muskrat must use stra-

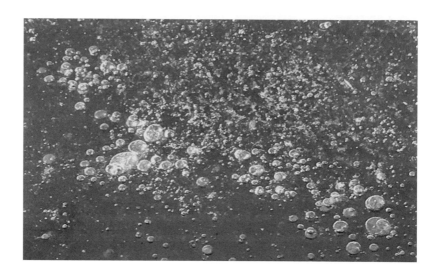

Figure 4.7. Bubbles within pond ice.
G. A. HOOD

tegic construction of feeding shelters and push-ups to provide waypoints that allow enough airspace and opportunity for it to breathe, as well as to groom and restore insulative properties in its fur (MacArthur 1989a). These structures allow the muskrat to forage away from its lodge for more than forty minutes (MacArthur 1979). Robert A. MacArthur, one of the foremost experts on muskrats, observed muskrats hyperventilating prior to diving into their plunge holes in the ice, thus producing "predive hyperthermia" (MacArthur 1979). This behavior, coupled with some other unusual strategies, help counter rapid heat loss once in water.

A strategy related to oxygen stores involves air bubbles frozen into the ice after being released by muskrats and by plants that photosynthesize under the ice (Fig. 4.7). While swimming under the ice, muskrats replenish their air supply by breathing from trapped air bubbles. Rather than working from anecdotal evidence, MacArthur set up an elaborate laboratory experiment involving a Plexiglas dive chamber that could vary a muskrat's access to air bubbles. Without access to these bubbles, excursion times were cut in half (MacArthur 1979). Marchand (2013) suggests that beavers might use the same strategy in winter to extend forays under the ice.

As with all semi-aquatic mammals, freeze-outs are a threat to muskrats, especially in winters with cold temperatures and little to no snow.

Droughts can leave waterbodies vulnerable to freezing down to the bottom. For muskrats, bank burrows provide some reprieve to the high mortalities experienced by muskrats that are completely dependent on huts during drought years (Messier and Virgil 1992). Locations of bank burrows in rivers and streams are similar to those of muskrat huts on ponds. They are often excavated near areas with herbaceous and emergent vegetation (Brooks 1985) and in banks bordered by open meadows and agricultural land (Brooks and Dodge 1981). Muskrats tend to avoid constructing burrows in banks lower than 0.2 m, with a slope < 10°, possibly to avoid flooding during fluctuations in water levels (Brooks 1985). Banks composed of > 90% sand and gravel would result in structural instability for the burrows and are avoided. On average, a muskrat family (e.g., two adults and several young) uses two burrows, much like muskrats that construct and use more than one hut (Brooks and Dodge 1986).

For muskrats that must leave their huts daily, foraging on vegetation above the ice is possible, although the nutritional content of dried sedges and cattails is poor. Muskrats can also eat parts of their huts over the winter as an additional source of food. During warm spells in the coldest winter months, or during dry years, I have regularly seen muskrats hiding in the cattails or walking across the snow. As with beavers, muskrats are at an increased risk of predation during freeze-outs; aerial and terrestrial predators, including mink, are threats to muskrats (Errington 1963).

River Otters

Compared to beavers and muskrats, there appears to be a distinct difference in habitat use and overwinter survival strategies for mink and river otters, both of which are carnivores. Unlike many semi-aquatic rodents and shrews, neither mink nor river otters construct shelters, but they are very efficient at using structures built by beavers and muskrats. In North America, river otters are particularly reliant on beaver-modified waterbodies, dens, and lodges. Extensive radio-tracking conducted on North American river otters in northern Idaho determined that beaver-impounded streams provide important habitats for otters during the winter months (Melquist and Hornocker 1983). In a similar study conducted by D. G. Reid and colleagues (1994) in the boreal region of

Alberta, Canada, river otters were so strongly associated with beaver ponds that the northern distribution of otters was suspected to be directly linked to beaver distributions. Beavers provide potential dens sites for otters in the form of bank burrows and beaver lodges. Of the beaver lodges studied in Alberta, 72% were not occupied by beavers. For those lodges that are occupied by beavers, resident beavers face threat of predation, especially of kits. Otters distinctly avoid sites with gradually sloping shorelines made of sand or gravel because they are poor-quality locations for burrow construction by other semi-aquatic mammals (e.g., muskrats and beavers). The large foraging ranges of otters necessitate several temporary shelters within their territories. In Idaho, thirty radio-tagged otters used eighty-eight different shelters in sixteen months (Melquist and Hornocker 1983).

An interesting trait of otters is their digging of passages through beaver dams to lower water levels in ponds in winter (Reid, Herrero, and Code 1988). This strategy allows otters to travel under the ice with additional airspace. In some cases, young otters also use the space under the ice for denning. A northern trapper once told me that otters will break beaver dams to draw the adults out of the lodge to fix the dam while the otter predates on the young inside the lodge. Otters' main prey item in winter is fish, which are easier to catch in winter given lower metabolic rates and movement in many prey species. Winter food scarcity makes river otters solo foragers with high levels of mobility. As such, habitat characteristics, such as resting sites and access under the ice, are critical for overwinter survival.

Reduced movement rates of otters were common in the Alberta study, especially for males, although overlapping winter home ranges were rare (Reid et al. 1994). Many otters stayed on the same waterbody for weeks and made most movements from den to den under the ice, unless moving to another waterbody. Otters also selected dens sites that had access to air and water under the ice, which made the steeper shorelines an important habitat feature. Despite their thick pelage, with an equivalent density to the northern fur seal (*Callorhinus ursinus*), it takes effort for an otter to restore the air layer in its fur. Otters use a number of strategies to restore the air layer, including rolling in snow, oiling their fur, shaking, piloerection, and pulling at loose skin (Tarasoff 1974).

Mink

River otters and American mink share many behavioral strategies in winter. However, rather than having a dominant focus on beaver dens and lodges, mink focus on abandoned muskrat burrows and huts. Surprisingly, given sometimes heavy predation rates on muskrats, mink are known to coexist with muskrats in larger burrow systems (Eberhardt 1973). During freeze-out events, mink can kill numerous muskrats, which they then cache for later consumption (Errington 1963). Mink are marginally adapted to aquatic life, yet increase aquatic foraging in winter, especially for fish (Gerell 1968). Their pelage, although thicker than their terrestrial counterparts, does not provide insulation equivalent to that of otters and muskrats. Unlike otters, mink hunt from the shore and then plunge into the water to catch fish, thus reducing time in water (Dunstone and O'Connor 1979). In a laboratory experiment published in 1976, Trevor Poole and Nigel Dunstone determined that mink minimize their time in water to an average of five to twenty seconds at a time, and they were most effective at catching smaller fish and fish that tended to shoal together. To reduce submersion time, mink located fish while on land, prior to entering the water. Mink also use two other thermoregulatory strategies to help them adapt to winter: lowering their resting body temperatures when denning and, like the otter, maintaining high levels of activity (Marchand 2013).

Water Shrews

Water shrews, the smallest of the semi-aquatic mammals, are at a distinct disadvantage in winter. America water shrews (*Sorex palustris*) have a faster cooling rate than any other endotherm (mammals, birds, and some fish). When swimming, shrews cool at a rate of −2.1°C per minute in water ranging from 10°C to 12°C (Calder 1969). Water shrews are larger than terrestrial shrews, which helps them to counter heat loss to some degree (Wolff and Guthrie 1985) but physiology plays a major role. Water shrews, along with muskrats (Fish 1979) and rakali (*Hydromys chrysogaster*) (Fanning and Dawson 1980), can tolerate basal temperatures as low as 28.5°C to 32°C with limited impacts and fast recovery rates. A combination of physiology and air in the water shrew's pelage helps counter the challenges of cold-water immersion, especially in winter (MacArthur 1989a). Given the high metabolism of shrews, obtaining energy from ready access to high energy food is equally critical.

In 1976, Polish biologist Krzysztof Wołk published an extensive study on the winter diet of the Eurasian water shrew (*Neomys fodiens*), the Palearctic equivalent to the America water shrew. In the heart of the Białowieża Primeval Forest, he constructed a series of elaborate structures of pipes (wells) associated with drainage ditches that he then augmented with "food table rafts" on which shrews would feed. Over two winters, Wołk collected food remains on the wooden rafts and compared them to similar samples obtained in autumn. His sample size was impressive, with the remains of 940 organisms from twenty-four species that were killed by two to four shrews inhabiting the study site over two winters (approximately ten months). Insects were the dominant food source (54%), followed by mollusks (22%), and amphibians (18%), although these preferences changed as prey availability changed over winter. Common frogs (*Rana temporaria*) tended to be the sole amphibian prey item, despite the presence of toads (*Bufo bufo*), which were likely too large to hunt. Fish consumption varied widely, from 0.7% in the first winter to 11.3% in the second winter. Prey items, frogs in particular, were thoroughly cleaned relative to other autumn foraging behaviors and reflected an increased feeding intensity. There was also evidence that, despite proportional differences in diet, shrews were catching any available prey item and possibly storing frogs in early winter for consumption later in the season. As with all species facing dramatic temperature and environmental changes in winter, shrews have a keen focus on balancing energetic needs with energetic outputs, which requires a multi-faceted strategy for success.

Temporal Niche Partitioning

Comparable to spatial partitioning because of species interactions, species also distribute themselves temporally for a variety of reasons. Night and day create special environments that link to seasonal changes, such as winter months in northern latitudes. The natural availability of light, either through day length or moon phase, might restrict an animal's activity depending on the physical limitations of their eyes. Overnight

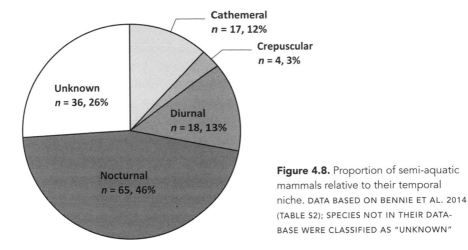

Figure 4.8. Proportion of semi-aquatic mammals relative to their temporal niche. DATA BASED ON BENNIE ET AL. 2014 (TABLE S2); SPECIES NOT IN THEIR DATABASE WERE CLASSIFIED AS "UNKNOWN"

or daytime maximum temperatures can also limit nighttime or midday activities because of thermoregulatory challenges. For those species with high metabolism, such as shrews, energetic requirements necessitate regular foraging bouts regardless of light condition. As well, predation and competition can result in a temporal separation of species (Bennie et al. 2014). In an extensive study of 5,276 species of mammals, Johnathan Bennie and his fellow researchers at the University of Exeter conducted a global assessment of time portioning in nonmarine and fossorial mammals. They also identified distinct latitudinal and temperature influences related to the timing of activity for many mammals (Bennie et al. 2014).

Semi-aquatic mammals, as a distinct group, follow similar trends as the mammals assessed during the University of Exeter study. Most are **nocturnal** (active at night), with **diurnal** (active during the day) species comprising fewer than half the species, followed by **cathemeral** (active at any time of the day or night) and then **crepuscular** (active at twilight) species (Fig. 4.8; Table 4.1). Of note, the temporal ecology of several semi-aquatic mammals is unknown. Beavers present an interesting test of these temporal categories. Often they are described as crepuscular species; however, they are consistently active well into the early morning hours and sometime during the day. They seem to be able to alter their activity times but still are most active at night. A study on Eurasian beavers determined that even when predators were not present, beavers did not relax their nocturnal activity patterns, something the researchers

Table 4.1. Temporal niches of semi-aquatic mammals

Nocturnal	Nocturnal (cont.)	Diurnal	Cathemeral	Crepuscular
Afrosoricida	Lagomorpha	Artiodactyla	Artiodactyla	Artiodactyla
Limnogale mergulus	Sylvilagus spp.	Babyrousa spp.	Tragelaphus spekii	Hydropotes inermis
Micropotamogale spp.		Blastocerus dichotomus	Eulipotyphla	Monotremata
Potamogale velox	Rodentia	Elaphurus davidianus	Chimarrogale spp.	Ornithorhynchus
	Anotomys leander	Kobus leche	Condylura cristata	anatinus
Artiodactyla	Castor spp.	Kobus megaceros	Neomys spp.	Rodentia
Choeropsis liberiensis	Colomys goslingi		Sorex spp.	Deomys ferrugineus
Hippopotamus amphibius	Dasymys spp.	Carnivora		Otomys lacustris
Hyemoschus aquaticus	Delanymys brooksi	Aonyx cinereus	Rodentia	
	Holochilus spp.	Hydrictis maculicollis	Arvicola spp.	
Carnivora	Hydromys chrysogaster	Lontra felina	Neotomys ebriosus	
Aonyx spp.	Ichthyomys spp.	Lontra longicaudis	Rattus lutreolus	
Atilax paludinosus	Lundomys molitor	Lutrogale perspicillata	Rheomys raptor	
Cynogale bennettii	Microtus richardsoni	Pteronura brasiliensis	Synaptomys spp.	
Genetta piscivora	Myocastor coypus			
Lontra canadensis	Nectomys spp.	Eulipotyphla		
Lontra provocax	Neofiber alleni	Chimarrogale platycephalus		
Lutra lutra	Neusticomys spp.	Nectogale elegans		
Lutra sumatrana	Ondatra zibethicus			
Mustela lutreola	Oryzomys spp.	Rodentia		
Neovison vison	Oryzomys palustris	Crossomys moncktoni		
Prionailurus spp.	Pelomys fallax	Hydrochoerus spp.		
Procyon cancrivorus	Scapteromys spp.			
	Sigmodontomys alfari			
Didelphimorphia	Thryonomys swinderianus			
Chironectes minimus	Xeromys myoides			
Lutreolina crassicaudata				
Eulipotyphla				
Crocidura mariquensis				
Desmana moschata				
Galemys pyrenaicus				

Note: Activity times for 36 species (26%) are unknown.

called "the ghosts of predators past" (Swinnen, Hughes, and Leirs 2015). Temporal niches, as with many other aspects of a species' relationship with its environment, have a complicated connection to evolutionary influences that endure even in changing environments.

The role an organism plays in the environment is an expression of its morphology, physiology, and behavior that has evolved over time. The ecological community in which it lives performs a critical role in how a species interacts with its environment in spatial and temporal dimensions. The majority of semi-aquatic species that are active during the darkness of night encounter growing challenges as the world becomes increasingly illuminated with anthropogenic light pollution. Others face a warming world that presents thermoregulatory challenges, as well as advantages. The niche is a characteristic of a species or population, not the resources; however, all dimensions of the environment must be in place to support the unique roles these species play in the broader ecosystem.

PART II
PHYSICAL ADAPTATIONS

5

Morphology

The scientific felt inclined to class this rare production of nature with eastern mermaids and other works of art; but these conjectures were immediately disproved by an appeal to anatomy.

Robert Knox, 1823, *Observations on the Anatomy of the Duck-Billed Animal of New South Wales*

Morphological Adaptations

As humans, all of us have likely experienced our body's response as we move from land into water. Our skin immediately reacts to changes in temperature, while our limbs and muscles work harder to push against the resistance posed by the density and viscosity of water. In moving water, these challenges increase. Until we learn to move efficiently through water, we struggle to stay afloat and to move forward once we can no longer walk along the bottom of the waterbody. Adaptation to a semi-aquatic lifestyle generally requires specialized external and internal structures to accommodate movement, foraging, and thermoregulation in water and on land. Water and air are both considered to be fluids, but the unique properties of water—such as higher viscosity and density, lower oxygen levels, high surface tension, and high specific heat capacity—present specific challenges for semi-aquatic organisms not encountered by their fully terrestrial counterparts. But just as movement through water requires morphological adaptations to counter the added energy demands of aquatic living, the ability to function on land requires adaptations that are important for semi-aquatic mammals. Some species, such as the American mink (*Neovison vison*), are very limited in their structural adaptations to aquatic environments, despite spending much of their time in water. Other species, such as the giant otter (*Pteronura brasiliensis*), are highly adapted to aquatic habitats and are morphologically distinct from their terrestrial counterparts in several ways.

In 1982, an article by Stephen Jay Gould and Elisabeth S. Vrba presented a new term to differentiate between a currently used feature derived from natural selection (**adaptation**) and a feature used for a function that differs from the one for which it originally evolved: **exaptation**. Although there has been lively discussion around their article since 1982, biologists including A. Brazier Howell (1930) and James Estes (1989) have long noted modifications providing distinct advantages to a semi-aquatic mammal while also being present (and useful) in terrestrial species. Such is the case with two talpids (the Russian desman, *Desmana moschata*, and the Pyrenean desman, *Galemys pyrenaicus*), both of which benefit while submersed in water from a lack of external ears (**pinnae**).

At first glance, a tempting comparison would be to equate the lack of pinnae to a similar characteristic in a fully aquatic mammal, such as a whale, until one notices that many fossorial talpids also have the same feature, perhaps to prevent soil from entering their ears while digging extensive networks of tunnels. The broadly webbed feet of river otters, on the other hand, are adaptations derived through natural selection for aquatic locomotion and provide no advantage on land. If anything, extensively webbed feet are a terrestrial liability both energetically and functionally, and they are not found in terrestrial mustelids. In exploring various physical characteristics of semi-aquatic mammals, it is important to note that some modifications indicate a true progressive adaptation for aquatic life, while others are shared terrestrial features, easily adapted for aquatic activities.

As seen in previous chapters, semi-aquatic mammals are taxonomically and geographically diverse, and their physical specializations for aquatic habitats are as varied as the freshwater systems they call home. Within that variation, however, there are some common morphological characteristics shared by many of these species, specifically streamlining of the body and enlargement of specific appendages used for propulsion (Dunstone and Gorman 1998). Underwood's water mouse (*Rheomys underwoodi*) provides an excellent example of semi-aquatic adaptation. In 1970, biologists Andrew Starrett and George Fisler described it as being "among the most highly hydrodynamically specialized mammals" because of its streamlined body and suite of other physical characteristics (Starrett and Fisler 1970: 7). Its forelimbs are dramatically reduced, as are its pinnae and tiny eyes. Underwood's water mouse has nostrils that have a posterior opening behind flap-like valves and, as with many other semi-aquatic mammals, webbed feet and a flared tail that provides stability in water. These characteristics in varied combinations are found in many semi-aquatic mammals. Others bring their own combination of morphological adaptations that help bridge the ecological challenges of living in land and water.

Body Mass and Shape

One of the first descriptors we often use for mammals relates to body mass. All facets of an animal's biology are directly influenced by its size and weight, in particular its physiology, ecology, and behavior. Generally, semi-aquatic mammals tend to be larger than their fully terrestrial relatives to accommodate the rigors of an aquatic environment (Wolff and Guthrie 1985). As such, with a weight reaching 79 kg, the capybara (*Hydrochoerus hydrochaeris*) is the largest rodent in the world. The common hippopotamus (*Hippopotamus amphibius*) weighs from 1,000 to 4,500 kg (Fig. 5.1) and is the largest of the even-toed ungulates (Artiodactyla) and the third-largest extant land mammal after elephants and some species of rhinoceroses.

Across several mammalian taxa, the trend of larger semi-aquatic mammals tends to hold true. For example, within the subfamily Cricetidae (water rats, swamp rats, and fish-eating rats), the mean body length of freshwater-dependent species is over 72 mm longer and 190 g heavier than their terrestrial counterparts (Wolff and Guthrie 1985). Prior to their extinction in the early 1900s, the Antillean giant rice rat (*Megalomys desmarestii*) and the Santa Lucian giant rice rat (*Megalomys luciae*), formerly of the islands of Martinique and St. Lucia, respectively, were the largest of the Cricetinae at approximately 360 mm long and

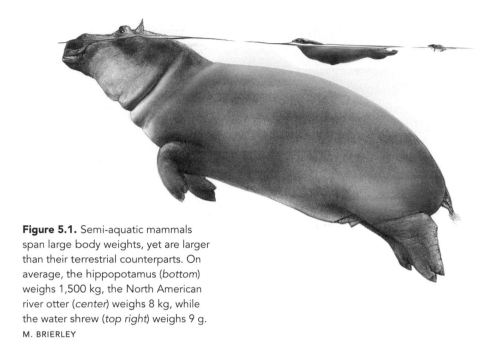

Figure 5.1. Semi-aquatic mammals span large body weights, yet are larger than their terrestrial counterparts. On average, the hippopotamus (*bottom*) weighs 1,500 kg, the North American river otter (*center*) weighs 8 kg, while the water shrew (*top right*) weighs 9 g.
M. BRIERLEY

1,025 g in weight. Given the impact of introduced mongoose on these two species and their subsequent extinction, very little is known about their ecology. The size difference is especially apparent in the subfamily Arvicolinae (voles, lemmings, and muskrats), where arvicolines from aquatic habitats average almost 82 mm longer and 32 g heavier than terrestrial arvicolines (Wolf and Guthrie 1985). Even when muskrat (*Ondatra zibethicus*) is removed from the analysis, arvicolines from aquatic habitats are 78% heavier, on average, than their terrestrial relatives.

The same trends hold for murids, shrews, marsupials, the European and American mink (second in size only to the polecat, *Mustela putorius*), and the sea mink (*Mustela macrodon*), which was hunted to extinction sometime between 1860 and 1902 (Nowak and Paradiso 1983). Oral descriptions by fur traders and indigenous people describe the sea mink as the largest of the mink, likely because of its enriched coastal diet. Of the closely related otters, the giant river otter (*Pteronura brasiliensis*) is more than twice as large as the largest of the terrestrial mustelids (Nowak and Paradiso 1983; Gittleman 1985; Estes 1989). Large males can reach 32 kg in weight and over 2 m in length.

Three explanations exist for the evolution of larger body sizes in aquatic habitats. The most common one relates to thermoregulation and the smaller surface-area-to-body-mass ratios in larger mammals. Having a larger body results in a slower rate of heat loss, an advantage when living in water where heat is easily conducted away from the body. Larger and more streamlined bodies reduce surface-to-volume ratios, which aids in heat conservation and movement through water. Along with an increase in body size of freshwater mammals, extremities tend to shorten and increase in surface area. The second explanation likely relates to predation, whereby it is more difficult to conceal a larger body on land; however, being larger in freshwater systems would be advantageous in an environment where fish are the primary aquatic predators and tend toward smaller-sized prey items (Wolff and Guthrie 1985). This hypothesis suggests that being a larger mammal would reduce predation risk and provide an evolutionary advantage. Lastly, larger body sizes could result in an animal being a more successful competitor in freshwater systems where high quality food is more patchy and harder to obtain and defend. Being larger would facilitate intraspecific competition and thus result in larger body sizes for semi-aquatic species (Wolff and Guthrie 1985). However, pinpointing one explanation as to why semi-aquatic mammals tend to be larger than their terrestrial counterparts is still difficult.

Given that larger species can experience greater efficiency of movement in water than when on land, as seen with a common hippopotamus moving along the bottom of an aquarium at the zoo versus walking on land, energetics and movement also come into play. To address this multifaceted evolutionary conundrum, Wolff and Guthrie suggest in their 1985 paper "Why Are Aquatic Small Mammals so Large?" that "first predation by fish and establishment of dominance select for large size and that high quality resources, reduced competition, and the water medium 'allow' this to occur" (Wolff and Guthrie 1985: 372). They are less convinced that temperature stress plays a large role given the broad geographical distribution of semi-aquatic mammals, with some of the largest (e.g., common hippopotamus and capybara) in the warmest waters at equatorial latitudes.

Along with increased body size, many semi-aquatic mammals have streamlined bodies to reduce drag when swimming and diving. Body shape strongly influences how water flows around the body and the

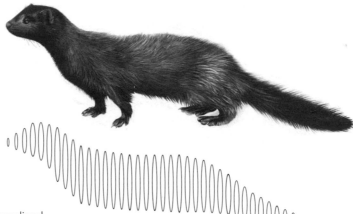

Figure 5.2. The streamlined, fusiform shape of the mink reduces drag when the mink is swimming and diving. M. BRIERLEY

efficiency with which an animal hunts, avoids predation, and travels through its habitat. A **fusiform** body, tapered at the head and at the tail, is a common body form for many semi-aquatic mammals, regardless of phylogeny, and provides a good example of convergent evolution within aquatic habitats (Fig. 5.2). As A. Brazier Howell states, "all existing aquatic mammals, excepting, if one prefer[s], the hippopotamus, are of streamline form to a greater or lesser degree dependent upon the stage of aquatic perfection which they approach in other respects" (Howell 1930: 49). Along with the hippopotamus, many of the other Artiodactyla identified as freshwater-dependent by the IUCN also lack a "stage of aquatic perfection" of the other semi-aquatic mammals. Aquatic habitats favor species that are able to counter drag and increase thrust in water. Therefore, the greater a species' dependence on aquatic habitats, the more physically adapted it becomes.

Hand in hand with a streamlined body shape, semi-aquatic mammals tend to have shorter limbs, modified feet, specialized tails, and modified vertebrae to aid streamlining and efficiency. Shorter extremities are also found in mammals from colder climates and, as with semi-aquatic mammals, this adaptation restricts heat radiation. Unlike in fully aquatic mammals, such as whales and dolphins, a reduction of limb length still must accommodate terrestrial locomotion. With an ecological link to terrestrial systems and a need to travel on land, the limbs of semi-aquatic mammals are still functionally long enough to bridge both worlds, albeit a bit clumsily for some, depending on aquatic adaptations such as webbed feet and flattened tails.

In most semi-aquatic mammals, swimming using a paddling motion is common and, in combination with their tail, requires at least two if not all four limbs to generate adequate propulsive force. Bipedal or quadrupedal padding is especially evident for mammals where swimming is primarily done at the water's surface. River otters have a broad repertoire of swimming modes (Fish 1994), including bipedal and quadrupedal paddling. As such, they vary their engagement of their webbed hind and fore limbs in concert with their powerful tail to swim. Other species, such as Eurasian and North American beavers (*Castor fiber, C. canadensis*), use only their strong hind limbs with powerful webbed feet and flat tail to swim. The combined surface area of their hind feet and tail equals 30% of the total surface area in adults, and over 50% in younger beavers (Reynolds 1993). Their unwebbed forepaws are held immobile against their chest and are not used at all for swimming, but rather they are used for handling food and moving materials.

There are three common adaptations in the feet of semi-aquatic mammals: (1) webbing between the toes, (2) stiff hairs along the margins of their feet (**fimbriation**), and (3) modification in size and shape without webbing or fimbriation. Some species, such as the American mink and the giant otter shrew (*Potamogale velox*), lack any modification to their feet relative to their terrestrial counterparts. The American mink's toes are partly webbed, but so are those of closely related, terrestrial mustelids.

For many species, such as otters, beavers, and the platypus (*Ornithorhynchus anatinus*), the presence of webbing increases the surface area of at least the hind feet to boost swimming ability. In several species, including the platypus and otters, all four feet are webbed. Given the generous web of skin on an otter's foot, surface area doubles in width once the digits are spread. On land, the platypus folds back the anterior part of the webbing on its feet for terrestrial movement and digging. Although not as extensive, the feet of the common hippopotamus and pygmy hippopotamus (*Choeropsis liberiensis*) are webbed as well, but less so for the pygmy hippopotamus.

Many other species of semi-aquatic mammals lack webbing between their toes; instead the toes are fringed with stiff hairs (fimbriation) to increase the surface area of the foot. The American water shrew (*Sorex palustris*) has stiff hairs 1 mm long on its large hind feet and a smaller fringe of hairs on its forefeet. Using a combination of webbing and air bubbles trapped in the **fibrillae** of its hind feet, this shrew can run across the surface of the water with the aid of corkscrew motions of its tail. The elegant water shrew (*Nectogale elegans*) has both webbing between its digits and a fringe of stiff white hairs along the edge of the digits of each foot that provide distinct advantages for aquatic life. As with webbing, the stiff hairs increase the surface area of the foot and enhance **thrust** while swimming. The **drag** caused by the hairs and/or webbing is reduced by drawing the toes together while moving the foot forward through the water to ready for the next stroke.

Similar to the foot, tails of semi-aquatic mammals often change to aid in swimming, either by increasing surface area or by changing shape. The tails of many semi-aquatic mammals either increase in surface area by being dorsoventrally flattened or laterally flattened. The platypus and both species of beavers are well known for their dorsally flattened tails. In the case of the platypus and beavers, their tails aid underwater propulsion and fat storage, especially in cold water or winter environments. However, the platypus's furry tail is not used to slap the water in alarm like the scaly tail of a beaver, but rather it aids the female in incubating eggs by holding the eggs against her warm body (Hawkins and Battaglia 2009). The Neotropical river otter (*Lontra longicaudis*), smooth-coated otter (*Lutrogale perspicillata*), and giant otter also have dorsoventrally flattened tails, but they lack the breadth of those of beavers and the platypus.

Muskrats, on the other hand, have laterally flattened tails that also act as rudders while swimming. Alternatively, tails of numerous species are fimbriated on the ventral part, as seen in the Russian desman, Pyrenean desman, Ruwenzori otter shrew (*Micropotamogale ruwenzorii*), and various genera of water shrews (*Sorex, Neomys, Nectogale*, and *Chimarrogale*). Along with these adaptations for streamlining the body and increasing surface area for propulsion, an animal's body size and shape is intimately associated with the structural and functional components of the body, such as the **integumentary** and musculoskeletal systems described below.

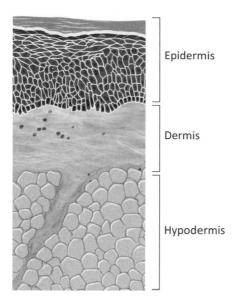

Epidermis

Dermis

Hypodermis

Figure 5.3. Generalized drawing of the skin's layers: epidermis (primarily formed of keratinocytes), dermis (fibrous tissue, blood vessels, and chromatophores), and hypodermis (connective tissue and fat). M. BRIERLEY

Integumentary System

The integumentary system, which is composed of various organs, is the first line of defense between a mammalian body and the external environment. It includes the skin, hair, scales, claws, nails, hooves, exocrine glands, and nerves, and it plays a key role in mammalian morphology and physiology. The integumentary system contains the body's largest organ, the skin, which protects muscles and internal organs from damage and provides the basis for other structures such as fur and scales (Fig. 5.3).

Skin

There are several ways in which the skin is integral to the rest of the body. It affords a physical and chemical protective barrier against pathogens, while also increasing mechanical support and strength for the entire body. The outermost layer of the skin, the epidermis, creates a waterproof barrier to the movement of water ions and forms key protective structures, such as fur, claws, nails, scales, and horns in mammals.

Just below the epidermis is the dermis, composed primarily of fibrous (collagenous) connective tissue, blood vessels, and **chromatophores** (cells containing pigments that can reflect light). In aquatic and some semi-aquatic marine mammals (e.g., seals), the arrangement of collagen fibers in the dermis presents as tightly laminated layers of cells (**stratum compactum**) that create a tough, flexible layer that supports mammals who have trunk-based aquatic locomotion (Withers et al. 2016). This type of movement is mainly found in fully aquatic mammals and would be too restrictive for species that need to move effectively on land as well as in water. As with many aspects of evolution, however, there is a gradient of characteristics depending on the extent of an animal's dependency on aquatic habitats. Studies regarding this trait in semi-aquatic mammals are scarce.

The deepest layer of the integumentary system is the hypodermis, which contains connective tissue and fat. The connective tissue secures the skin to the fascia (fibrous tissue) of the bones and muscles, while fat (adipose tissue) provides insulation and aids buoyancy. Most mammals have a combination of **white adipose tissue** ("white fat") and **brown adipose tissue** ("brown fat"). The former, along with storing lipids (fatty acids and their derivatives), is considered an endocrine organ that plays a role in metabolism and other physiological processes (Trayhurn and Beattie 2001), including the secretion of the protein leptin, which is critical for energy balance in mammals. Brown adipose tissue, on the other hand, is important for heat production (thermogenesis) unrelated to shivering. Together, the fat and blood vessels of the hypodermis provide insulation and protection to mammals, which is especially important for species that move so readily between aquatic and terrestrial habitats in all seasons. For example, imagine an American water shrew in Alaska swimming under the ice and then emerging onto a snowy shoreline in the middle of winter.

Conversely, as a species from sub-Saharan Africa, the common hippopotamus stands out for its almost hairless body and surprisingly low amounts of subcutaneous fat (Luck and Wright 1964). Its skin reaches a thickness of 35 mm and comprises approximately 18% of its body weight (Luck and Wright 1964). In comparison, human skin at its thickest point, on the heels of our feet, is 4 mm. Hippopotami need to stay submerged for much of the day to prevent dehydration, and they have specialized glands to produce "blood sweat," described later in this chapter. A closer look at their skin resembles clay that has been etched with a small knife. Their skin shows the complexity of the integumentary system and, as with other aquatic adaptations such as their partly webbed feet, represents the morphological diversity of the integumentary system of semi-aquatic mammals.

Also derived from the integumentary system is the pouch of marsupials. The water opossum (*Chironectes minimus*) is the only semi-aquatic mammal with a pouch (*marsupium* is Latin for "pouch"). Both males and females have a well-developed rear-facing pouch associated with a strong pouch muscle (*pars pudenda*) to keep the interior of the pouch watertight (Enders 1937; Edwards and Deakin 2013). This feature is especially important for females while the young are attached to the teats inside. The male water opossum and the now extinct Tasmanian tiger, the thylacine (*Thylacinus cynocephalus*), are the only marsupials where the male also has a pouch. The male water opossum pulls his scrotum into his pouch while in the water (Marshall 1978a), likely to reduce heat loss and injury. The lack of obvious external genitalia in beavers, muskrats, and water voles (*Arvicola* spp.) in the nonbreeding season likely serves a similar purpose (Dozier 1942; Baumgartner and Bellrose 1943; Osborn 1955).

Interdigital Webbing

Almost all mammalian embryos have membranes of skin between their digits (interdigital webbing), but as the embryo develops, cell death occurs in these membranes and the webbing is resorbed before birth. However, in many semi-aquatic mammals, interdigital webbing persists throughout their lives. The mechanism involved is not well known for many species, but for bats, Scott Weatherbee of the Yale School of Medicine and his colleagues (2006) determined that it was a combination of high levels of specialized cells that send signals to proteins involved in embryonic development (fibroblast growth factor signaling) and an inhibition of cells that would otherwise trigger the reabsorption of the webbing (bone morphogenetic protein, or BMP, signaling). By inhibiting BMP signaling, cell death does not occur and digital webbing remains intact. Whether this combination of events occurs in mammals other than bats is unknown; however, several species of semi-aquatic mammals have either partially or fully webbed feet to aid in swimming and diving. The extent of webbing and its association with front or hind feet varies dramatically among these species (Table 5.1), which was alluded to earlier during the discussion of the increased surface area of the appendages of semi-aquatic mammals.

For example, the webbed-footed tenrec (*Limnogale mergulus*) of Madagascar and the Rwenzori otter shrew (*Micropotamogale ruwenzorii*) of central Africa have webbing on the front and hind feet that extends beyond the part of the digits of the hand or foot closest to the metatarsals (proximal phalanges), as do both genera of rivers otters of Eurasia (*Lutra*) and the Americas (*Lontra*). Muskrats native to North America and Alfaro's rice water rat (*Sigmodontomys alfari*) of Central America and northern South America have partly webbed feet with webbing that ends before the end of the bones at the tips of the fingers and toes (distal phalanges). Other species, such as beavers and the water opossum, only have webbed hind feet, while the front feet of the platypus are fully webbed and the hind feet less so.

Webbed feet are common among various taxonomic orders across several biogeographic realms. Within the marsupial order Didelphimorphia, the water opossum has webbed feet, and within order Monotremata, the platypus is the only species with webbed feet. Order Rodentia includes several genera with at least one species possessing either fully or partially webbed feet, including *Amphinectomys*, *Baiyankamys*, *Castor*, *Crossomys*, *Holochilus*, *Hydrochoerus*, *Hydromys*, *Ichthyomys*, *Lundomys*, *Myocastor*, *Nectomys*, *Ondatra*, *Oryzomys*, *Parahydromys*, *Rheomys*, *Scapteromys* (only *S. meridionalis*), and *Sigmodontomys*. Within order Eulipotyphla, there are three monotypic species with webbed feet: the elegant water shrew, the Pyrenean desman, and the Russian desman. Webbed-footed species of order Afrosoricida include *Micropotamogale* (otter shrews) and *Limnogale* of Madagascar. Several semi-aquatic carnivores (order Carnivora) have webbed feet, including all genera of otters (*Aonyx*, *Hydrictis*, *Lontra*, *Lutra*, *Lutrogale*, and *Pteronura*), the otter civet (*Cynogale bennettii*), mink (*Mustela* and *Neovison*), and two semi-aquatic cats (*Prionailurus*). Rather than webbing, inside the hind feet of the giant otter shrew there is a flap of skin that is held snuggly against the body when pulling the foot forward through the water. Other species, including the coypus (*Myocastor coypu*), have four long, unwebbed digits on the front feet, one of which lacks a claw. The longer hind feet are webbed, except for the fifth digit, which is free and used for grooming. For semi-aquatic deer, although

Table 5.1. Characteristics of appendages of species with adaptations to their feet and/or tails

Species	Forefeet Fully webbed	Forefeet Partially webbed	Forefeet Fimbriated	Forefeet Enlarged	Hindfeet Fully webbed	Hindfeet Partially webbed	Hindfeet Fimbriated	Hindfeet Enlarged	Tail Dorsoventrally	Tail Laterally flattened	Tail Ventrally	Tail Tail end tuft	Tail Tapered	Tail Long (L), Short (S)	Tail Scaled (S), Furred (F)
Ucayali water rat, *Amphinectomys savamis*	−	−	−	−	+	−	+					+			S
Ecuador fish-eating rat, *Anotomys leander*	−	−	−	−	+	−							+		F
European water vole, *Arvicola amphibius*	−	−	−	−	−	−	−							S	F
Southern water vole, *Arvicola sapidus*	−	−	−	−	−	−	−	−					+	S	F
mountain water rat, *Baiyankamys habbema*	+	−	−	−	+	−	−	+						L	F
Shaw Mayer's water rat, *Baiyankamys shawmayeri*	+	−	−	−	+	−	−	+						L	F
marsh mongoose, *Atilax paludinosus*	−	−			−	−							+	L	F
African clawless otter, *Aonyx capensis*		−	−	−		+	−	+					+	L	F
Asian small-clawed otter, *Aonyx cinereus*		+			+		−	+					+	L	F
Hairy babirusa, *Babyrousa babyrussa*	−	−			−	−				−	−	+	−	S	Skin
Sulawesi babirusa, *Babyrousa celebensis*	−	−	−		−		−			−	−	+	−	S	Skin
Togian Islands babirusa, *Babyrousa togeanensis*	−	−	−			−	−			−	−	+	−	S	Skin
marsh deer, *Blastocerus dichotomus*	EM			+	EM			+	−	−	−	−	−	S	F

Note: Plus sign (+) indicates that feature is present; minus sign (−) indicates feature is absent; blank cell indicates feature is undocumented or not applicable; EM indicates an elastic membrane instead of webbing.

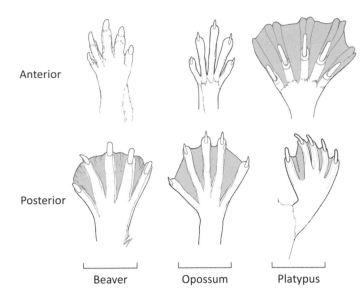

Figure 5.4. Webbed feet are common among various taxonomic orders across several biogeographic realms. Represented here are examples from orders Rodentia (beavers) and Didelphimorphia (water opossum), with webbing only on the posterior feet, and Monotremata (platypus), with webbing on both front and back feet. M. BRIERLEY

Anterior

Posterior

Beaver   Opossum   Platypus

their feet only have two digits, there is web-like material between the hooves. The marsh deer (*Blastocerus dichotomus*) has an elastic membrane between its large hooves, while Père David's deer has skin between its hooves. Whatever the presentation, interdigital webbing, or some version thereof, is an excellent example of convergent evolution as an adaptation for semi-aquatic living (Fig. 5.4).

Fat

Fat is also an important component of the integumentary system. Although the role of fat is often oversimplified as a means to store energy and provide a soft buffer of protection to an animal, the biochemical and physiological roles of these tissues are much more complex. Semi-aquatic mammals do not hibernate, thus making fat storage an important part of their thermoregulatory toolbox in colder environments. Where white adipose tissue stores energy as calories, brown adipose tissue is often called the "hibernating gland" because of its ability to burn energy and create heat (Feldhamer et al. 2007: 175). Instead of primarily relying on shivering to warm up (**shivering thermogenesis**), which is energetically demanding, some mammals benefit from brown adipose tissue because it is rich in energy-producing mitochondria and serves as a subcutaneous heat source. As a monotreme, the platypus lacks brown fat (Hayward and Lisson 1992) and likely relies on high levels of thyroid hormones for energy production (Nicol 2017). Brown adipose tissue is found in some marsupials and many eutherians (Feldhamer et al. 2015), where it lies mainly in proximity to blood vessels and vital organs (Hyvärinen 1994). It can be found in the interscapular, clavicle, groin, and axial (head, neck, chest, back, and abdomen) regions. Of the semi-aquatic mammals, the muskrat has brown adipose tissue and possibly the water opossum does (MacArthur 1979, 1989a).

Muskrats and beavers accumulate large amounts of dietary fat in peripheral adipose tissues in their tails and feet (Aleksiuk 1970; Käkelä and Hyvärinen 1996a), particularly during autumn in preparation for winter. For beavers, fatty acids are obtained exclusively from bark, leaves, and herbaceous plants, while muskrats in some areas occasionally supplement their diet with invertebrates and fish (Käkelä and Hyvärinen 1996a). Käkelä and Hyvärinen (1996a) noted that muskrats have little subcutaneous white adipose tissue, but, similar to the findings of other researchers (MacArthur 1986), identified a body of brown adipose tissue between the muskrats' scapulae and in the thoracic region. Beavers, on the other hand, have large amounts of subcutaneous white adipose tissue, while the Eurasian otter has a thinner layer of subcutaneous fat (Käkelä and Hyvärinen 1996b). As with many semi-aquatic mammals, thick waterproof fur plays an equally important role in thermoregulation.

Hair

Of the three layers of the epidermis, the stratum corneum is the topmost layer of skin; it receives dead keratinized cells from the middle layer (stratum granulosum). These cells from the stratum granulosum are derived from daughter cells (genetically identical cells) produced in the deepest epidermal layer, the stratum basale (Fig. 5.5). The presence of keratin is important because its water-insoluble properties prevent the underlying skin from drying out. Equally important, keratin is the protein providing structure for hair, scale-like structures, nails, claws, and hooves.

Hair, primarily called fur in nonhuman species, is unique to mam-

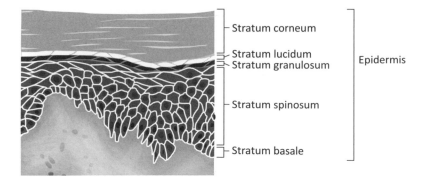

Figure 5.5. The top layer of the epidermis (stratum corneum) is replenished by the proliferation of keratinocytes. These cells are originally formed at the deepest layer of the epidermis (stratum basale). Keratinocytes migrate from the stratum basale to the stratum spinosum and the stratum granulosum, where they become known as granular cells. The stratum corneum receives dead keratinized cells from the stratum granulosum. M. BRIERLEY

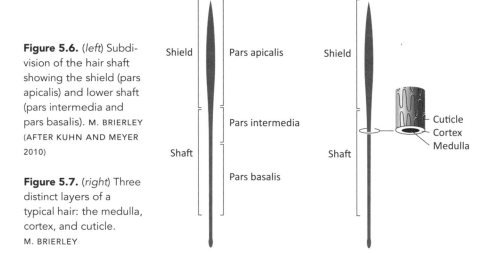

Figure 5.6. (*left*) Subdivision of the hair shaft showing the shield (pars apicalis) and lower shaft (pars intermedia and pars basalis). M. BRIERLEY (AFTER KUHN AND MEYER 2010)

Figure 5.7. (*right*) Three distinct layers of a typical hair: the medulla, cortex, and cuticle. M. BRIERLEY

mals and is directly related to **endothermy** (heat produced through internal chemical reactions of tissues in the body) and buoyancy (Fish et al. 2002). There are three distinct layers of a typical hair, the center of the shaft (medulla), the cells tightly packed around the medulla (cortex), and the surface of the hair (cuticle), all with distinct patterns that help identify individual species (Figs. 5.6, 5.7). Although some semi-aquatic mammals are virtually hairless (e.g., hippos), most have a thick pelage consisting of an outer layer of guard hairs over a thick, soft underfur. The guard hair is composed of three regions: the shield (pars apicalis) extends from the tip of the hair to the middle region (pars intermedia), which then connects to the basal region (pars basalis). Unlike various other mammal species, many semi-aquatic mammals (e.g., muskrat, beaver, and otter) only replace their fur once a year during an annual molt, likely because of their constant association with the relative uniformity of aquatic habitats (Ling 1970). Muskrats have one long molt throughout the year, the beaver molts in late summer and autumn (Thomas 1954), and the otter molts in autumn (Matthews 1952).

The form and function of hair varies in mammals. Renowned mammalogist George Feldhamer and his colleagues (2015) describe two main functional kinds of hair: pelage (guard hairs underlain by underfur) and long stiff vibrissae (whiskers). For most semi-aquatic mammals, guard hair is in the form of an awn, which tends to lie in one direction and grows to a fixed length (definitive growth), followed by shedding and regrowth. Underfur is short and dense, providing insulation and trapping air to aid positive buoyancy, thus allowing the animal to float (Table 5.2). A study of North American beaver, sea otter (*Enhydra lutris*), Australian water rat (*Hydromys chrysogaster*), river otter (*Lontra canadensis*), mink, muskrat, and platypus determined that their underfur was three to eight times denser when compared to two terrestrial species: the Virginia opossum (*Didelphis virginiana*) and the Norway rat (*Rattus norvegicus*) (Fish et al. 2002). In his 1962 study of muskrats, Kjell Johansen determined that air trapped in their underfur accounted for 21.5% of their overall volume (Johansen 1962a). As such, the structure of the dense underfur in most semi-aquatic mammals acts like a flotation device which then reduces the extra energy required to stay afloat.

To maintain important qualities of fur while in water, species such as otters and beavers waterproof their fur with oil from specialized glands (Liwanag et al. 2012). Additionally, the underfur of many semi-aquatic mammals is structured with kinks (Sokolov 1962) and interlocking cuticular scales of neighboring hairs to maintain air pockets in the fur (Williams et al. 1992; Fish et al. 2002). The fur of many semi-aquatic mammals has a silvery appearance underwater because of the presence of trapped air within the underfur (Dunstone and Gorman 1998). Two German mammalogists, Rachel Kuhn and Wilfried Meyer (2010), deter-

Table 5.2. Number of hairs per 1 cm^2 of skin in various semi-aquatic mammals

Species	Back	Abdomen
common shrew, *Sorex araneus**[*]	15,000[a]	14,000[a]
platypus, *Ornithorhynchus anatinus*	81,538[b]	
North American beaver, *Castor canadensis*	37,730.7[b]	
coypu, *Myocastor coypus*	6,000[a]	14,000[a]
muskrat, *Ondatra zibethicus*	11,000[a]	12,000[a]
muskrat, *Ondatra zibethicus*	38,615[b]	
rakali, *Hydromys chrysogaster*	45,448[b]	
Eurasian water shrew, *Neomys fodiens*	16,000[a]	15,000[a]
Russian desman, *Desmana moschata*	30,000[a]	37,000[a]
American mink, *Neovison vison*	33,845[b]	
Eurasian otter, *Lutra lutra*	35,000[a]	51,000[a]
North America river otter, *Lontra canadensis*	80,312[b]	

Note: Based on data in Sokolov 1962 (a) and Fish et al. 2002 (b).
*Terrestrial for comparison from Sokolov 1962.

mined that the guard hairs in otters are fusiform and have wide, flattened shields, which is a common trait for guard hair in semi-aquatic mammals that allows the hairs to lie flat thus providing a protective layer for the underfur (Toldt 1933). Otters living in colder waters (e.g., Eurasian otters, *Lutra lutra*) have the longest guard hairs, while those in warmer waters (e.g., smooth-coated otter and giant otter) have the shortest. Conversely, underfur of several semi-aquatic mammals is approximately twice as dense (>300 hairs/mm^2) and shorter than terrestrial species (Fish et al. 2002). This increased density aids insulation and buoyancy.

Other specialized hairs are more sensory in nature. Vibrissae (whiskers) are commonly found on all species of semi-aquatic mammals. They are usually long, straight, stiff hairs with highly innervated roots within a blood-filled sinus located deeper in the skin than other hairs. Some are even able to be moved independently. Vibrissae are very sensitive to touch and water pressure, which is translated from the root nerves to

Figure 5.8. Otter civet (*Cynogale bennettii*) with long vibrissae.
M. BRIERLEY

the sensory part of the brain (somatosensory cortex in particular). The giant otter uses its highly sensitive vibrissae to track changes in water pressure and currents.

The otter civet has a spectacular display of long vibrissae around its snout and under its ears; the vibrissae are tactile for nocturnal hunting (Fig. 5.8). Similarly, Pyrenean desmans, aquatic genets (*Genetta pisciv-ora*), and crab-eating rats (*Ichthyomys* spp.) use their distinctive vibris-sae for hunting aquatic prey. The aquatic genet uses its long, downward pointing vibrissae to touch the water to attract insectivorous fish, which it then catches and eats. Of the ichthyomyine species, the Ecuador fish-eating rat (*Anotomys leander*) distinguishes itself as the only one with one or more vibrissae above the eyes. In many cases, including, for ex-ample, the crab-eating rats of South America (*Neusticomys*), the loca-tion, length, and number of vibrissae aid in identifying different species within the same genera.

The importance of vibrissae for semi-aquatic mammals cannot be overstated, especially in relation to carnivorous species. In 1914, British zoologist Reginald Innes Pocock published a seminal article titled "On the Facial Vibrissae of Mammals," where he describes vibrissae relative to "their high development in the matter of thickness and length in piscivorous or insectivorous aquatic or semi-aquatic genera like *Chiro-nectes*, *Potamogale*, *Lutra* and *Cynogale*, and their comparative feebleness or deficiency in aquatic herbivores, like the Sirenians, Hippopotamus and Hydrochoerus" (Pocock 1914: 912).

Pocock also notes the lack of visible vibrissae in the platypus, which he associates with their highly modified jaw. Without additional adaptations, such as electroreception and electrolocation in the platypus, the role of vibrissae as a sensory aid in foraging is apparent. In one study, a Eurasian otter was challenged with finding prey in both clear and darkened water. It took four times longer to find prey in the darkened water than in clear water. Thereafter, the vibrissae of the otter were trimmed off and the researcher noted that, although the otter was able to catch fish in clear water in the same amount of time as before, in dark water it took twenty times longer to catch prey than it did with whiskers intact (Green 1977).

Eurasian otters and Australian water rats not only have increased innervation of their vibrissae compared to terrestrial species, their vibrissae are also of higher density (Hyvärinen et al. 2009). The use of vibrissae in finding prey is so important that Heikki Hyvärinen and his colleagues refer to this trait as "vibrissal sense" in their 2009 paper. In Chapter 8, "The Predators: Foraging Strategies and Niches," I discuss the use of vibrissae for hunting; there are some fascinating adaptations for aquatic specialization.

A third type of hair, as indicated earlier, aids in locomotion. Specific to the feet and tail, many semi-aquatic mammals possess stiff hairs (fibrillae) that increase surface area of the feet and serve as a keel on the underside of the tail. These hairs are much shorter than vibrissae and have no known sensory functions. As with webbing, a number of taxa have fibrillae on either the fore or hind feet. Those families with a fringe of hair on both front and hind feet but no webbing include Cricetidae (water rats/Amazonian mouse, *Neomys* spp., round-tailed muskrat *Neofiber alleni*), Muridae (false water rat, *Xeromys myoides*), and Soricidae (*Chimarrogale, Neomys, Sorex*). The American water shrew also has partly webbed hind feet combined with fibrillae. Two species from family Talpidae have webbing and fibrillae on both front and hind feet (Russian desman and Pyrenean desman), as does one from family Soricidae (elegant water shrew).

Those with fibrillae only on the hind feet include family Cricetidae (water mice, *Chibchanomys*; fish-eating rats, *Neusticomys*, except *N. ferreirai*; Goldman's and Thomas's water mice, *Rheomys raptor* and *R. thomasi*; swamp rats, *Scapteromys*) and family Muridae (Ethiopian

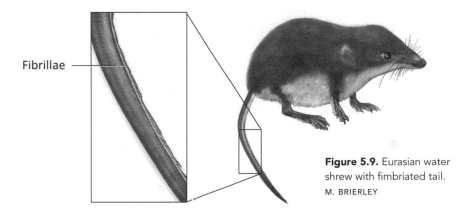

Fibrillae

Figure 5.9. Eurasian water shrew with fimbriated tail.
M. BRIERLEY

amphibious rat, *Nilopegamys plumbeus*). Several species also have partly webbed hind feet combined with fringes. These species, all from Family Cricetidae, include the muskrat, Mexican water mouse (*Rheomys mexicanus*), Ucayali water rat (*Amphinectomys savamis*), Ferreira's fish-eating rat (*Neusticomys ferreirai*), Western Amazonian nectomys (*Nectomys apicalis*), crab-eating rats (*Ichthyomys*), and Lund's amphibious rat (*Lundomys molitor*). The various combinations of these adaptive structures are uniquely associated with a species' ecological niche.

As mentioned previously, these stiff hairs are also found on the tails of many semi-aquatic mammals (Fig. 5.9). Just as the keel of a boat runs along the centerline of the hull, the fibrillae on tails of semi-aquatic mammals generally run ventrally along the centerline of the tail and provide directional control and stability. Occasionally, the fringe also runs laterally along the edges of the tail, which increases the surface area of the tail to increase propulsion through water. In the case of the elegant water shrew, white lateral fringes merge together along the underside of the tail, which gives the tail a pyramid-like appearance. In some species, such as the southern water vole (*Arvicola sapidus*), the stiff hairs are present but sparsely distributed. Several species from five widely distributed families have ventrally fimbriated tails.

As with other convergent evolutionary traits, fimbriated tails are specific adaptations to habitat characteristics rather than phylogenetic relationships. For example, fimbriated tails evolved within family Cricetidae, including the webbed-footed marsh rat (*Holochilus brasiliensis*), all Neotropical crab-eating rats (*Ichthyomys*), all South American water rats (*Nectomys*), all Central American water mice (*Rheomys*), and South American swamp rats (*Scapteromys*). Of the Muridae, the same trait exists in the African wading rat (*Colomys goslingi*) and New Guinea's earless water rat (*Crossomys moncktoni*). Within family Soricidae, all Asiatic water shrews (*Chimarrogale*), the Congo's Kahuzi swamp shrew (*Crocidura stenocephala*), and all three of the Old World water shrews (*Neomys*) have a stiff fringe along the underside of their tails, despite dramatically different geographic distributions. Similarly, both species of European desmans share the same trait. Finally, from family Tenrecidae, the webbed-footed tenrec, which independently evolved on the island of Madagascar over the past twenty-five to forty-two million years, possesses a similarly fimbriated tail as these other species living from the tropics to the Arctic.

Several semi-aquatic mammals, including the coypu and round-tailed muskrat, have a scale-like growth covering their tail formed via clumped (agglutinated) keratin fibers that develop similarly to hairs. These "scales" are often derived from stunted fur and evolved differently than scales in fish, reptiles, and birds. They likely provide a streamlining advantage while swimming and offer some protection when the animal drags its tail behind it while on land. Just as hair and its associated structures are derived from the epidermis, so are the glands associated with hair follicles that keep hairs moist and waterproof.

Exocrine Glands

Glands are small but powerful organs derived from epithelial tissue. They produce and secrete critical substances inside the body via the blood stream (endocrine glands) and onto the epidermis (exocrine glands) or into a cavity of another organ via ducts (exocrine glands). Of the exocrine glands, sebaceous glands tend to be associated with hair follicles, where their oily secretions (sebum) aid in waterproofing the pelage. Contrary to common belief, anal glands of beavers are not used to oil fur; rather, once beavers groom their pelage (e.g., clean and realign hairs), they oil the fur with **squalene** found in the film of lipids from the sebaceous glands at the base of the guard hairs (Rosell 2002). Oiling fur, combined with high fur density, the alignment of guard hairs, and surface tension of water, allows beavers and many other semi-aquatic mammals to maintain a waterproof coat of fur. Otters also use oil glands in their skin to aid in waterproofing. Another species, the sitatunga (*Tragelaphus spekii*), has a body covered in shaggy, greasy, long, thin fur that is coated with an oily, water-repellent secretion. These semi-aquatic antelopes from the swamps of sub-Saharan Africa often submerge their whole bodies in water to avoid insects and predators, only keeping their eyes and nose above the surface. Lipids play an important role in countering the effects of extensive immersion in water.

Waterproof pelage plays an essential role in heat balance. The harderian gland, located within the orbit of the eye, was first described by Johann Jacob Harder in 1694 (Payne 1994). There is some variation in products produced by this gland, but in muskrats it produces thermoregulatory lipids. These lipid secretions are spread through the fur by muskrats to maintain a water-repellent pelage. To demonstrate the importance of this gland, Henry James Harlow removed it from one group of muskrats to create a comparison of insulative value of the fur upon submergence (Harlow 1984). He also shampooed a second group of muskrats. Both groups of animals were then compared to a control group. Both the group with the gland removed and the shampooed group had significantly higher rates of heat loss via evaporation than the control group.

Unlike sebaceous glands, sweat glands are found deep in the dermis and reach the skin's surface through coiled ducts (Feldhamer et al. 2015). Evaporative cooling comes from eccrine sweat glands, which produce thin layers of sweat; while apocrine sweat glands near the hair follicles produce thick, sticky (viscous) sweat commonly found in the armpit and groin regions. Both sebaceous glands and sweat glands are involved in scent-marking and pheromone production, which play important roles in attracting mates, defending territories, and deterring predators. Scent-marking also allows for individual identification during social interactions (Sun and Müller-Schwarze 1998).

As with various other mammals, however, both the common and pygmy hippopotami lack sweat glands. Both species produce a unique substance called "blood sweat," which is secreted from subdermal glands that cover the body, especially on the back and flanks (Allbrook 2009). This slimy fluid ranges from colorless, to pink, to red, to dark brown, and it oozes over the entire body to protect from sunburns when the animal

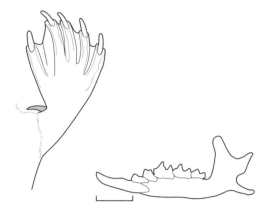

Figure 5.10. The spur of the male platypus (*left*) and grooved lower incisor of the Eurasian water shrew (*right*) both aid the flow of venom.
M. BRIERLEY

is out of the water. It also helps regulate temperature, and it aids in the healing of wounds because of its antibacterial properties (Hashimoto et al. 2007).

A third type of exocrine gland, the mammary glands, have unclear evolutionary origins. The location and number of mammary glands vary among semi-aquatic mammals, and in monotremes these glands lack nipples. For the female platypus, an accumulation of ducts releases milk directly onto the surface of the mammary area (areolae) of the mother's abdomen and the baby platypus licks milk off the wetted hair. Other glands of note include specialized salivary glands in water shrews (*Neomys anomalus* and *N. fodiens*) and the crural gland of male platypuses, both of which produce toxic venom (Fig. 5.10).

For the Mediterranean and Eurasian water shrews (*Neomys teres* and *N. fodiens*), the venom is delivered from submaxillary salivary glands and then flows along the concave inner surface of their lower incisors to immobilize prey (Rode-Margono and Nekaris 2015). The crural gland in the platypus is located in the male's thigh, and toxin is released via hollow keratinized spurs on the ankles of the hind legs during predator defense or sexual competition during mating season. The spurs are actively raised via small muscles and small articulating bones (Rode-Margono and Nekaris 2015); the male platypus then wraps its legs around its target and drives the spurs into the unfortunate recipient. For animals as large as a dog, the toxin can be fatal; for humans it can mean weeks of excruciating pain and swelling without relief. Fortunately, the venom is only active during mating season.

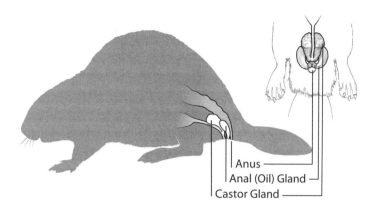

Anus ────
Anal (Oil) Gland ──
Castor Gland ──────

Figure 5.11. Anal gland and castor gland of beavers. M. BRIERLEY

Special glands between the pelvis and base of the tail in both species of beavers aid in communication (Fig. 5.11). As previously mentioned, secretions from the anal gland are not used for oiling the fur (Rosell 2002). Rather, a thick paste in the form of anal gland secretions mixes with castoreum from the castor sacs to produce unique chemical signatures for individual beavers, much like distinctive fingerprints for humans. Castor sacs are often called castor glands; however, they are really two pockets that form off the distal wall of the urethra (Svendsen 1978). The castor sacs then join the urethra and open to the urogenital pouch, which then releases castoreum. Castoreum, a prized ingredient in expensive perfumes, is mainly composed of secondary metabolites of urine (Rosell 2002).

Mustelids are also well known for scent-marking behavior through the conspicuous depositing of feces (spraint) within their territories. Spraint surveys are a common method of detecting and estimating otter populations. Otters have two pairs of glands, the anal glands and the proctodeal glands, on either side of the rectum that open into the anus via ducts. Anal glands, consisting of a storage sac and a duct, secrete proteins and monosaccharides that combine with droplets of fatty acids (Gorman, Jenkins, and Harper 1978; Chanin 1985). Once the animal defecates, it ejects anal gland secretions on top or to the side of the spraint (Gorman, Jenkins, and Harper 1978), thus using complex compounds to leave an individual chemical signature at the site.

Ungulates, such as Père David's deer (*Elaphurus davidianus*) in China and the marsh deer of South America, have preorbital glands in the corner of their eyes that aid in scent-marking (Geist 1999). The marsh deer also has nasal glands that appear to serve no purpose in a pheromonal system (Jacob and von Lehmann 1976), and although they have tarsal and pedal glands, they lack the metatarsal glands found in other closely related species of deer. The lack of metatarsal glands in both Père David's deer and the marsh deer might reflect the infectiveness of such glands in species that spend a significant amount of time in water. As with other aspects of mammalian morphology, the glandular system is extensive given the role of glands in almost every function of animal biology.

Claws, Nails, and Hooves

Another important part of the integumentary system is the specialized keratinized structures originating in the stratum corneum of the epidermis. Claws, nails, and hooves are essential to the protection and functioning of fingers and toes of mammals in almost all aspects of daily activities. Regular activities keep these ever-growing structures trimmed to an ideal length. Although claws, nails, and hooves vary dramatically in presentation, their underlying components are similar. Function, however, is much more varied and is closely associated with foraging. The only mammals lacking claws, nails, or hooves are found in the whale families and sirenians (manatees and the dugong), both of which have adopted fully aquatic marine habitats.

At their simplest, claws, nails, and hooves provide protection to the sensitive, highly innervated tips of an animal's digits. Where they differ lies in the dominance and location of the unguis and the subunguis. The unguis comprises the dorsal part of the claw or nail. In claws, the unguis is more heavily keratinized and surrounds the top and sides of the end of the digit. The subunguis, which is softer, continues from the foot pad and joins the unguis at the tip of the claw. Claws are the most common of these structures found in semi-aquatic mammals. Some species have reduced or absent claws, as seen with the clawless thumb (pollex) of the semi-aquatic Kemp's grass mouse (*Deltamys kempi*) and the clawless otters (*Aonyx*).

In felids, claws are highly developed and are retractable, which allows for greater dexterity in maneuvering objects. As with the cheetah (*Acinonyx jubatus*), both species of semi-aquatic cats (flat-headed cat, *Prionailurus planiceps*, and fishing cat, *P. viverrinus*) have retractable claws, but with reduced sheaths. Many species of the family Viverridae have retractable or semi-retractable claws. This might also apply to the otter civet and aquatic genet; however, this information is not readily documented. Claws aid individuals in manipulating objects, defending themselves, digging, and grooming if necessary. They are also a key component in hunting and capturing prey. Curved claws likely allow for more versatile grooming. The earless water rat (*Crossomys moncktoni*) has strongly curved claws and is noted as one of the most aquatic of the Muridae (Helgen 2005).

Nails offer their own advantages. They are only on the upper surface of the ends of the digits. In this case, the harder unguis dominates and the subunguis is reduced to connective tissue between the nail and the pad of the toe or finger. Evolutionarily derived from claws, nails allow for more dexterity in the hands and feet when holding and manipulating objects.

Hooves are also derived from claws, and they are also structurally unique. The unguis completely surrounds the subunguis, which forms the distal underpart of the foot, while the foot pad lies farther back on the bottom of the foot. Hooves often accompany a reduction in the number of digits in ungulates, however, both species of hippopotami have four toes on each foot.

The twelve species of otters found in freshwater systems provide an ideal example of the diversity of claws and nails in mammals relative to adaptive diversification in prey specialization. Three species of otters are named specifically for their claws, or lack thereof. As its name implies, the African clawless otter (*Aonyx capensis*) lacks claws altogether on its front digits, and instead it has rough pads on its palms and opposable thumbs to handle prey. The African clawless otter is born with claws, but it loses all except those on the hind digits 2, 3, and 4, which are used for grooming. The loss or reduction of claws allows the animal greater paw dexterity as it eats small prey items such as amphibians and invertebrates. The Congo small-clawed otter (*Aonyx congicus*) has no claws or webbing on its forepaws, but it has rudimentary claws on the second, third, and fourth digits of the hind foot (these claws are shed as the

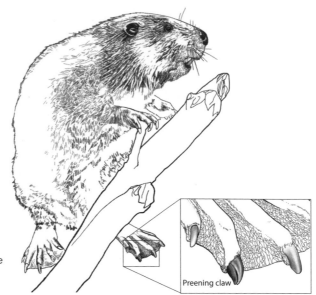

Figure 5.12. Preening claw on the second digit from the inside on the hind foot of a beaver. M. BRIERLEY

Preening claw

animal ages). As its name implies, the Asian small-clawed otter (*Aonyx cinereus*) only has little spikes on all four feet for digging in the mud and lifting rocks while foraging (Nowak 1999). They use their forepaws with almost human dexterity.

Conversely, the spotted-necked otter (*Hydrictis maculicollis*), North American river otter, southern river otter (*Lontra provocax*), Eurasian otter, hairy-nosed otter (*Lutra sumatrana*), Neotropical otter, and the marine otter (*Lontra felina*) have well-developed claws on all four fully webbed feet. The smooth-coated otter has webbed feet and strong dexterous paws that are equipped with sharp claws and smooth footpads, while the giant otter has thick webbing on its feet to the end of each digit and sharp claws on each foot.

As noted with the African clawless otter, grooming claws (sometimes called the "toilet claw") are common in many semi-aquatic mammals and are important in maintaining waterproof fur. The claws of the first and second hind toes in both species of beavers are slightly narrower and curve inward toward the body to conform to the head and chest during grooming (Fig. 5.12). Bailey provides an excellent description of these claws: the inner claw has a "long double-edged nail that clamps down over a long soft lobe, opening and closing like a duck's bill" (Bailey 1923,

78). He then described the more specialized claw as split claw. As with the inner claw, this grooming claw is supported by a soft toe pad and has a finely serrated keratinous growth between the lobe of the toe and the "real nail." The sharp-edged nail then fits over this growth, yet opens up slightly once the pressure is released. In essence, with the outer claw the beaver has a fine-toothed comb, and it has a coarser comb represented by the inner claw. Grooming claws in other semi-aquatic mammals do not appear to have a similar structural specialization.

Semi-aquatic mammals are an interesting mix of species with pedal adaptations for swimming, yet robust claws for digging. The star-nosed mole (*Condylura cristata*) is an odd combination of semi-aquatic and fossorial mammal whose long claws on its front and back feet are specialized for digging (Merritt 2010; Feldhamer et al. 2015). Other semi-aquatic mammals, including the muskrat, platypus, and coypu, are a small selection of water-dependent species that dig burrows and tunnels, and even extensive foraging channels in the case of beavers. Unlike claws, true nails are less common in mammals.

True nails are most often associated with primates, although not exclusively. Nails evolved later than claws and present evolutionary advantages relative to our sense of touch and manual dexterity. In particular, the whorls and patterns on our fingertips allow better traction when gripping fine objects. Although claws tend to dominate in semi-aquatic mammals, the African clawless otter has rudimentary nails, and the Asian small-clawed otter's claws are thin ungulae, much in the same form as other mammalian nails (Hamrick 2001). Similarly, the water opossum has thin, narrow nails that aid in dexterity when catching prey. The African wading rat has long digits, with the first digit sporting a rudimentary minute nail on its front feet. As with most of their terrestrial counterparts, fish-eating rats, crab-eating rats, and water mice, (e.g., *Anotomys leander*, *Ichthyomys*, *Nectomys*, *Neusticomys*, and *Rheomys*) have a small pollex with a nail, with all other digits endowed with strong, sharp claws (Voss 1988). Similarly, the platypus has broad nails on its forepaws, while retaining sharp claws on its hindfeet.

Of the even-toed ungulates (Artiodactyla) adapted to freshwater habitats, several have hooves that help them walk along soft, muddy substrates. Most notably, the sitatunga has elongated hooves that are widely separated at the tips and enlarged false hooves. Combined with a "peculiar flexibility" of the pasterns (joint between the proximal and middle phalanx) that rest directly on the ground (Nowak 1999: 1136), this foot morphology makes for awkward walking when the sitatunga is out of the water. The marsh deer, although not as dramatic as the sitatunga, has large feet with an elastic membrane between the hooves, while Père David's deer has skin between its hooves. The southern lechwe (*Kobus leche*) of south-central Africa also has splayed elongated hooves, while the Nile lechwe (*K. megaceros*) has long, slender hooves. Unlike many of the semi-aquatic Artiodactyla, hippopotami have nail-like hooves at the end of each of the four toes on their front and hind feet. The diversity of claws, nails, and hooves among semi-aquatic mammals reflects what an animal eats, how it catches its prey, and how it moves in aquatic habitats as efficiently as possible.

Nerves and Sensory Organs

Detecting prey and maintaining vigilance for predators are two of the most important activities for animals in the wild. For many semi-aquatic mammals, the ears, nose, and eyes are positioned higher (dorsally) on the skull, and often in-line, to allow all three sensory organs to stay above water while the animal swims. As a result, semi-aquatic mammals can detect both terrestrial and aerial threats. The eyes of flat-headed cats are positioned closer together and farther forward on the head to maximize binocular vision, thus facilitating hunting aquatic prey. To maintain amphibious activities, there is a delicate balance between the structure and function of the mammalian nervous system and the sensory organs it supports. Early studies suggested that aquatic mammalian carnivores had larger brains than their terrestrial counterparts to support cognitive and sensory requirements for an aquatic or a semi-aquatic lifestyle (Bininda-Emonds et al. 2001), but this idea has since been refuted for both fully aquatic and semi-aquatic mammals (Marino 2007; Reidenberg 2007).

The different physical properties of air and water offer special sensory challenges for species required to function in both mediums. For example, sound travels approximately four times farther in water (1,450 v_w(m/s)) than air (331 v_w(m/s)), which might seem advantageous to a semi-aquatic mammal. However, water depth, water velocity, temperature, and even the substrate of the water body influence the distance and strength of sound (Hanke and Dehnhardt 2013). Although different frequencies of sound travel at the same speed through the same medium, as accommodated by different wavelengths, if the wavelength is more than four times the water depth, the distance sound travels can be significantly reduced.

Wolf Hanke and Guido Dehnhardt (2013: 418) describe in their article "Sensory Biology of Aquatic Mammals" how a sound frequency of 120 Hz or less is "severely dampened" in a waterbody that is 30 m deep. They then describe how soft substrates, including muddy pond bottoms, exacerbate the situation by additionally lowering the cut-off frequency. Running water makes it especially difficult for semi-aquatic mammals to hear low-frequency sounds generated by their prey (Hanke and Dehnhardt 2013). Regardless of the properties of sound in water, external ears, plugged by water or closed with valve-like structures, present even greater challenges to an animal's ability to hear underwater. These are only some of the sensory hurtles for semi-aquatic mammals.

Light always travels faster than sound, although light and sound change their relative speed differently depending on the medium. Sound travels slower in air than in water, while light does just the opposite. We can witness the difference in the speed of light and sound when watching a fireworks display. First we see the bright light display, and then we move our hands to our ears. For many mammals, eyes are closed or otherwise ineffective in water. To compensate, semi-aquatic mammals tend to have more sensory nerves for vibrissae than terrestrial species of the same family (Hyvärinen et al. 2009). With aquatic foraging challenged by low light and poor visibility, one strategy of the Eurasian otter is to have four times the number of myelinated nerve fibers of deep vibrissal nerves than the closely related polecat (Hyvärinen et al. 2009).

An unusual sensory strategy in mammals is the use of electroreception. The platypus is the only semi-aquatic mammal with electroreception. It uses forty thousand electroreceptors to detect prey and other objects underwater combined with sixty thousand push-rod mechanoreceptors to assess distance (Wilkens and Hofmann 2008). Electroreceptors are located along the **parasagittal plane** of the upper and lower bill and pick up weak electrical signals from prey as the platypus moves its head from side-to-side. Both the platypus and echidnas form electroreceptors from specially adapted glands that secrete mucus in the skin (Czech-Damal et al. 2013), although the number of electroreceptors in the platypus is up to ten times more numerous than in the terrestrial echidnas.

Quality of vision relies on the amount of available light and how that light is scattered. Once light enters water, there is a reduction of light intensity (**light attenuation**) as photons are either absorbed or scattered. **Turbidity** and water depth play important roles in how much light is available in water. A greater amount of suspended soil in the water column results in less available light. As depth increases, light decreases. Refractive properties of water make it difficult to perceive distance accurately. Temperature and water pressure also affect vision. As terrestrial mammals, humans experience the effects of water on our vision first hand. While underwater, a terrestrial mammalian eye, without any morphological adaptations, is farsighted (hypermetropic), so objects seen close up appear blurry. Additionally, as seen when we put a pencil into a glass of water, refraction in water poses challenges. In air, the cornea is an important refractive structure, but it loses this function in water (Ballard, Sivak, and Howard 1988). To compensate, we often wear some sort of goggle to create a layer of air between our eyes and our environment to aid the focusing power of the cornea. Even then, our stereoscopic vision can be compromised.

To compensate for the need for vision in air and in the water, the amphibious eye requires improved focusing ability (dioptric power), which can be aided with increased lens curvature (Dunstone 1998). In otters, the sphincter iridis muscle can change the shape of the lens to aid focusing ability while underwater (Walls 1942). A comparison of the intraocular muscles in the eyes of North American river otters (carnivores) and North American beavers (herbivores) reveals significantly greater intraocular musculature in otters, in particular in the sphincter and dilator muscles (Ballard, Sivak, and Howard 1988). Although beavers can see underwater, their eyesight is primarily adapted for vision in air. However, the otter is well adapted for amphibious vision, with eye musculature that can adapt the lens to counter the refraction of light in water, thus making it effective in both air and water. This adaptation

Table 5.3. Morphology of ocular structures in various taxa of semi-aquatic mammals

Family	Species	Common name	Eye morphology
Ornithorhynchidae	*Ornithorhynchus anatinus*	platypus	Eyes tightly closed by oblique fold of skin edges of facial furrow
Felidae	*Prionailurus planiceps*	flat-headed cat	Extreme depression of skull; eyes unusually far forward and close together (better stereo vison)
Mustelidae	*Aonyx cinereus*	Asian small-clawed otter	Nictitating membranes close when underwater
	Lontra canadensis	North American otter	Nictitating membranes close when underwater
	Lontra longicaudis	Neotropical otter	Nictitating membranes close when underwater
	Lutra lutra	Eurasian otter	Nictitating membranes close when underwater
	Pteronura brasiliensis	giant otter	Nictitating membranes close when underwater
Castoridae	*Castor canadensis*	North American beaver	Nictitating membranes close when underwater; eyes aligned with nostrils and ears; greater dorsal inclination than other rodents
	Castor fiber	Eurasian beaver	Nictitating membranes close when underwater; eyes aligned with nostrils and ears; greater dorsal inclination than other rodents
Caviidae	*Hydrochoerus hydrochaeris*	capybara	Eyes near top of head
	Hydrochoerus isthmius	lesser capybara	Eyes near top of head
Myocastoridae	*Myocastor coypus*	coypu, nutria	Eyes set high on head
Hippopotamidae	*Hippopotamus amphibius*	common hippopotamus	Eyes slightly periscoped, aligned on top of head along with ears and nostrils; directed laterally and slightly forward

is ideal when hunting fish and small mammals underwater, while the herbivorous nature of beavers, who forage extensively on land, requires less visual acuity in water. Beavers are not without adaptations though, as can be seen in Table 5.3.

In 1973, Ronald J. Schusterman and Barry Barrett described how the Asian small-clawed otter uses its iris sphincter to compress and distort the curvature of the lens to avoid hypermetropia underwater. However, low light levels still reduced the otter's visual acuity. The air-water equivalence in focusing ability was only maintained in relatively bright light. For many other semi-aquatic species, including the rodents and shrews, eye size is reduced and underwater foraging is more tactile in nature than visual.

One structure of the eye, well known in beavers, and also in the platypus, otters, the common hippopotamus, cats, and mink, is a third eyelid, often called a **nictitating membrane** (also referred to as "plica semilunaris," "membrana nictitans," or "palpebra tertia"). It also occurs in many terrestrial and fossorial mammals. In 1928, E. Phillip Stibbe called it the "so-called nictitating membrane" because in mammals it does not nictate (blink) across the eye as it does in birds, but rather muscular action protrudes and creates tension in the membrane to fix it in place (but not move it across the eyeball), while the eyeball rolls inward underneath the membrane. He further describes a cartilage originating within the subconjunctival tissue that then turns the mucous membrane covering the eye (conjunctiva) back onto itself (evaginates). As such, the "intercepting cartilage," as Stibbe wished it were called, is able to protect the front of the eye, remove particles from the eye's surface, aid in underwater vision, and act as a "sort of 'squeegee' at the inner side of the orbit" (Stibbe 1928: 166). Despite his request for a new term, it often retains the term nictitating membrane in mammals.

EARS

Unlike many fully aquatic mammals, which tend to lack external ears, semi-aquatic mammals still rely on their ears while on land. Pinnae are especially important for detecting the proximity and direction of potential predators. In many semi-aquatic mammals, there is no discernible difference in the form and function of their ears compared to terrestrial counterparts, but some species have ears more dorsally situated than their fully terrestrial relatives to keep the ears above water while swimming. Other species, including all otters, have valve-like structures that close to prevent water from entering the ears while underwater. As with several species of semi-aquatic mammals, the southern water shrew has very small ears, often completely covered in thick fur. However, as noted by A. Brazier Howell (1930), many terrestrial species among the Eulipotyphla and Rodentia also have ears well hidden within their pelage, and this characteristic, although advantageous, is not a specialized adaptation for aquatic habitats. In general, most semi-aquatic mammals do not rely on hearing while underwater and cannot compensate for the technical challenges when sound propagates differently underwater.

The common hippopotamus is unique in its auditory adaptations within the semi-aquatic mammals. It hears and communicates as we do above water, but its underwater communication highlights its fascinating adaptations for aquatic communication. With its ears closed while underwater, it listens via waterborne sounds through its jaws, similar to the way dolphins do. The fat deposits around the neck and vocal cords help transfer sound first to the squamosal bone and then to the tympanic and periotic bones of the middle ear, which are suspended by ligaments. Thus, it is able to hear sounds underwater, regardless of its ears being above or below the waterline. Remarkably, if the jaw is submerged while the nose and ears are above the water, the hippopotamus can transmit sounds in air and water simultaneously (Barklow 2004). Although not as dramatic an adaptation as seen in the common hippopotamus, several semi-aquatic mammals have modifications to their pinnae and ear structure to accommodate swimming and diving (Table 5.4).

The platypus is also unique in the morphology of its middle ears. In most mammals, the left and right middle ears are independent of each other acoustically; however, in the platypus the tympanic cavities are open to the pharynx, a trait more common with reptiles and amphibians (Mason 2016). This structure, which allows the tympanic membrane to sense pressure on both sides, allows the cavities of the middle ear to intercommunicate, something University of Cambridge Physiologist Dr. Matthew Mason describes as a "pressure-difference receiver." This ear morphology could allow the semi-aquatic platypus to locate sounds at the lower frequencies common to aquatic prey (Mason 2016).

Table 5.4. Morphology of external ears in different taxa of semi-aquatic mammals

Family	Species	Common name	Pinnae morphology
Ornithorhynchidae	*Ornithorhynchus anatinus*	platypus	Lack pinnae, but musculature allows orifice of ears to be cocked forward; aural aperture at posterior end of facial furrow dilated and contracted when eyes open; ears closed by oblique fold of skin edges of facial furrow when swimming; aural aperture dilated and contracted when eyes open
Soricidae	*Chimarrogale* spp. (6)	Asian water shrews	Can be sealed with a flap of skin
	Crocidura stenocephala	Kahuzi swamp shrew	Covered with short hairs and partly covered by pelage
	Nectogale elegans	elegant water shrew	Only insectivore to lack pinnae
	Neomys anomalus	southern water shrew	Small, nearly hidden in fur
	Neomys fodiens	Eurasian water shrew	Small with white tufts, nearly hidden in fur
Talpidae	*Condylura cristata*	star-nosed mole	Barely evident externally
	Galemys pyrenaicus	Pyrenean desman	Can close off ears, lack pinnae
	Desmana moschata	Russian desman	Can close off ears, lack pinnae
Herpestidae	*Atilax paludinosus*	marsh mongoose	Fur thick, completely covers pinnae while swimming
Viverridae	*Cynogale bennettii*	otter civet	Flaps to close underwater
Mustelidae	*Aonyx cinereus*	Asian small-clawed otter	Ears have valve-like structure
	Lontra canadensis	North American otter	Small valvular ears
	Lontra longicaudis	Neotropical otter	Ears close when swimming
	Lutra lutra	Eurasian otter	Valvular nose and ears
	Pteronura brasiliensis	giant otter	Valvular ears, small rounded
Castoridae	*Castor canadensis*	North American beaver	Ears are valvular
	Castor fiber	Eurasian beaver	Ears are valvular

Table 5.4. *continued*

Family	Species	Common name	Pinnae morphology
Cricetidae	*Anotomys leander*	Ecuador fish-eating rat	Not visible; slit-like ear can be closed by muscle
	Chibchanomys orcesi	Las Cajas water mouse	Ears buried in fur of head
	Chibchanomys trichotis	Chibchan water mouse	Ears buried in fur of head
	Nectomys apicalis	Western Amazonian nectomys	Small well furred
	Neofiber alleni	round-tailed muskrat	Small, nearly concealed in fur
	Neotomys ebriosus	Andean swamp rat	Fur covered
	Ondatra zibethicus	muskrat	Inconspicuous valvular ears
	Oryzomys palustris	marsh rice rat	Short and hairy
	Scapteromys meridionalis	Waterhouse's swamp rat	Rounded, covered in dense fur
Muridae	*Otomys lacustris*	Tanzanian vlei rat	Furry ears
	Pelomys hopkinsi	Hopkins's groove-toothed swamp rat	Small, haired ears
	Pelomys minor	least groove-toothed swamp rat	Hair on ears
	Rattus lutreolus	Australian swamp rat	Small and concealed by fur
	Rhcomys mexicanus	Mexican water mouse	Inconspicuous ears
	Rheomys raptor	Goldman's water mouse	Small ears
	Rheomys thomasi	Thomas's water mouse	Small, visible, and covered with hair
	Rheomys underwoodi	Underwood's water mouse	Small and concealed by fur
Caviidae	*Hydrochoerus hydrochaeris*	capybara	Ears near top of head
	Hydrochoerus isthmius	lesser capybara	Ears near top of head
Myocastoridae	*Myocastor coypus*	coypu, nutria	Valvular ears set high on head

Table 5.4. *continued*

Family	Species	Common name	Pinnae morphology
Leporidae	*Sylvilagus aquaticus*	swamp rabbit	Smaller than other cottontails
	Sylvilagus palustris	marsh rabbit	Rounded, smaller than other cottontails
Hippopotami-dae	*Hippopotamus amphibius*	common hippopotamus	Well formed; aligned on top of head along with ears and eyes; fold into recessed area and use of valvular plug to close underwater

NOSE AND OLFACTORY STRUCTURES

Once again, the precarious balance between the aquatic and terrestrial worlds results in some degree of sensory adaptation for semi-aquatic mammals. Several studies reveal that numerous taxa of semi-aquatic mammals have reduced olfactory bulbs; these taxa include the platypus, semi-aquatic tenrecids, soricids, talpids, viverrids, herpestids, and otters (Pihlström 2008). Many species, despite their reduced olfactory bulb, rely on chemical signatures to communicate with each other. The main olfactory bulbs in beavers are quite large. While studying North American beavers in western Canada, I have seen them sniffing at vegetation as a form of presampling and also raising their nostrils into the air to determine whether I am friend or foe. These traits are akin to fully terrestrial species. A good sense of smell is a common trait among mammals. There are other semi-aquatic mammals, however, that have unusual olfactory adaptations.

The olfactory system comprises more than just the external structures of the nose (Table 5.5). At the base of the nasal cavity in many mammals is the Jacobson's organ (vomeronasal organ), an auxiliary organ used in chemoreception of pheromones in particular. It is present in almost all water-dependent mammals, except for Sirena and many whales. An extensive multi-agency study determined that the platypus has more type 1 vomeronasal receptors (V1R) than previously recorded, which might be associated with "water-soluble, non-volatile odorants, during underwater foraging" (Warren et al. 2008: 178). The common hippopotamus uses its vomeronasal organ much like a syringe when underwater (Estes

Table 5.5. Morphology of nasal structures in various taxa of semi-aquatic mammals

Family	Species	Common name	Nasal morphology
Ornithorhynchidae	*Ornithorhynchus anatinus*	platypus	Nostrils on dorsal surface of snout; nostrils closed by oblique fold of skin edges of facial furrow when swimming; aural aperture dilated and contracted when eyes open
Tenrecidae	*Potamogale velox*	giant otter shrew	Flap of skin seals nostrils
Talpidae	*Condylura cristata*	star-nosed mole	22 pink fleshy tentacles on nose
	Galemys pyrenaicus	Pyrenean desman	Chemo/mechanosensitive Eimer's organs; can close off their nostrils
Felidae	*Desmana moschata*	Russian desman	Eimer's organs at end of long, bi-lobed nose
Viverridae	*Cynogale bennettii*	otter civet	Nostrils have flaps to close underwater
Mustelidae	*Aonyx cinereus*	Asian small-clawed otter	Valvular nostrils
	Lontra canadensis	North American otter	Valvular nostrils
	Lontra longicaudis	Neotropical otter	Nostrils close when swimming
	Lontra felina	Marine otter	Slit-like nostrils can close underwater, small nose
	Lutra lutra	Eurasian otter	Valvular nostrils
	Pteronura brasiliensis	giant otter	Valvular nose; nose completely covered in fur—only slit-like nostrils visible
Castoridae	*Castor canadensis*	North American beaver	Valvular nostrils
	Castor fiber	Eurasian beaver	Valvular nostrils
Cricetidae	*Ondatra zibethicus*	muskrat	Valvular nostrils
Muridae	*Pelomys minor*	least groove-toothed swamp rat	Hair on nasal region
	Rheomys underwoodi	Underwood's water mouse	Nostrils have flap-like valves

Table 5.5. *continued*

Family	Species	Common name	Nasal morphology
Caviidae	*Hydrochoerus hydrochaeris*	capybara	Nostrils, eyes, and ears near top of head
	Hydrochoerus isthmius	lesser capybara	Nostrils, eyes, and ears near top of head
Myocastoridae	*Myocastor coypus*	coypu, nutria	Nose and mouth valvular
Hippopotamidae	*Hippopotamus amphibius*	common hippopotamus	Aligned on top of head along with ears and eyes; nostrils close before submersion
Tragulidae	*Hyemoschus aquaticus*	water chevrotain	Slit-like nostrils

1991). Biologist Richard D. Estes describes its function being much like a bulb-syringe whereby water is drawn into the organ (but not the nasal passage) to test for indicators of reproductive status in urine voided into the water by other hippopotami.

Axial and Appendicular Skeleton

Adaptations to the skeletal anatomy of most semi-aquatic mammals are likely limited by their need to use their limbs effectively while on land (Fish 1982a). Generally, there are only slight skeletal modifications for swimming in many species (MacArthur 1989a), although some commonalities exist in several taxa. Mammalogist Frank Fish describes four specializations in the limbs of semi-aquatic mammals to aid in aquatic locomotion: (1) a short, dense humerus and femur, (2) elongated digits on at least one set of feet, (3) larger foot area facilitated by webbing and fimbriation, and (4) increased bone density to compensate for positive buoyancy resulting from the thick underfur (Stein 1989; Fish 1996). The common hippopotamus has especially dense bones and spends much of its time walking along the bottom of waterbodies. For many species, decreased femur length allows for the hind limb to be held closer to the body to aid streamlining during the recovery stroke while swimming (Stein 1988; Samuels, Meachen, and Sakai 2013). The Eurasian otter provides an excellent example of decreased femur length as an aquatic adaptation (Mori et al. 2015).

A multi-taxa study of seven semi-aquatic mammals highlights the variability of bone density. When comparing the North American beaver, coypu, water opossum, little water opossum (*Lutreolina crassicaudata*), muskrat, web-footed marsh rat, and South American water rat (*Nectomys squamipes*), only bone densities for beavers were within the range of fully aquatic mammals (Wall 1983; Stein 1989). However, once three species of terrestrial *Microtus* were removed from the analysis, the species of semi-aquatic mammals tended to exhibit higher bone densities than other terrestrial mammals in the study. The rice rat (*Oryzomys palustris*) showed no special adaptations for aquatic life, and it actually had the lowest bones densities of the terrestrial and semi-aquatic mammals combined (Stein 1989). The researcher, Dr. Barbara Stein, also determined that beavers had noticeable thickening along the bone shaft, while the other species did not (Stein 1989). Additionally, the lower leg bones (tibiae and fibulae) for all genera of the seven semi-aquatic mammals were denser than the upper leg bone (femur), and the bones of the lower arm (radius and ulna) were denser than the upper arm bone (humerus). This morphological trend in the limbs is accepted as an adaptation of fully aquatic mammals (Taylor 1914; Stein 1989; Fig. 5.13).

Increased bone density has some advantages in aquatic habitats. Denser bones help overcome buoyancy and can result in neutral to nega-

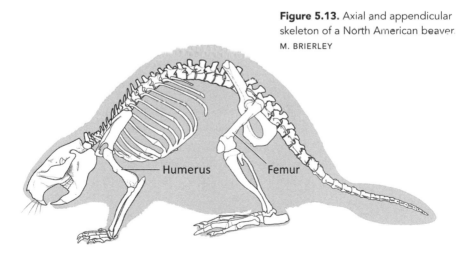

Figure 5.13. Axial and appendicular skeleton of a North American beaver. M. BRIERLEY

Humerus   Femur

tive buoyancy to facilitate diving for prey and protection from predators. Higher bone density also supports stronger muscle attachments, which facilitate forceful movements of the feet while swimming and also while walking on land. As always, the broad range of bone density of a suite of semi-aquatic mammals highlights the variability of adaptations, or lack thereof, of these species for which aquatic and terrestrial habitats are equally important (Wall 1983).

Although not consistent within semi-aquatic taxa, aquatic carnivores tend to have increased head and body lengths relative to their close terrestrial relatives (Bininda-Emonds et al. 2001), which aid in streamlining while swimming and diving. Barbara Stein notes that the "increase in length of the first metatarsal would result in the overall shape of the foot being more symmetric, thereby insuring that equal water pressure would be applied to both its borders while paddling" (Stein 1988: 508). Among Neotropical marsupials, the water opossum has an enlarged pisiform bone derived from their wrist bone that serves as an opposable sixth digit on their fore feet (Marshall 1977; Galliez et al. 2009). Although not a common aquatic adaptation, this extra "digit" could aid in seeking out and holding onto prey.

In all mammals, teeth are not specific to habitats, but rather to foraging strategies and protection. For example, the grooved teeth of the common hippopotamus provide structural support for fighting while also being effective in breaking down plant-based food (Rode-Margono and Nekaris 2015). For rodents, sharp incisors allow them to cut vegetation, while the folded grooves of their molars grind down the tough cell walls of plants. As with many **piscivores**, the sharp teeth of the aquatic genet are ideal for spearing and holding onto slippery fish. Rather than detail the range of teeth within the semi-aquatic mammals, specific descriptions of adaptations for foraging are described in Chapter 8, "The Predators: Foraging Strategies and Niches."

Muscle

Of the three types of muscles in mammals—skeletal (striated), smooth, and cardiac—the skeletal muscles allow for voluntary motion (muscle contracted at will). These muscles, joined to bones mainly via tendons, use the bones as levers to aid in locomotion and help maintain posture.

Of the skeletal muscles, the axial muscles help straighten or arch the spine, while the appendicular muscles control limb movements and play a critical role in locomotion. It is the appendicular muscles in semi-aquatic mammals that have received the most attention by researchers, albeit limited for most species. To some degree, the muscles around the eyes are also well examined in some species (e.g., otters) because of their role in underwater vision. The nonvoluntary muscles, however, are best described in their role in physiological adaptations to semi-aquatic life, as seen in Chapter 6.

The most common muscular modification in many semi-aquatic mammals is the relative size increase in some of the thigh muscles to aid increased propulsion in water (Stein 1988). However, increased size of thigh muscles, the sartorius in particular, is lacking in the semi-aquatic mustelids (Mori et al. 2015), despite some initial research to the contrary in the early 1900s (Taylor 1914). The University of Tokyo's Dr. Kent Mori and his colleagues (2015) also determined that the Eurasian otter also had smaller relative mass in several muscles (i.e., gluteus, semimembranosus, biceps femoris, semitendinosus, and tenuissimus muscles) than a seal (*Phoca* spp.) and marten (*Martes* spp.), thus refuting early findings in a similar study (Smith and Savage 1956). Overall, Mori and colleagues determined that freshwater semi-aquatic mustelids shared similar musculature with their terrestrial counterparts.

The pygmy hippopotamus and common hippopotamus, however, have retained several muscles lost to other modern hooved mammals, and they use them to counter their bulky form while moving through the water (Reidenberg 2007). Mammalogists Rebecca Fisher, Kathleen Scott, and Virginia Naples (2007) dissected two deceased pygmy hippopotami from the Zoological Park in Washington, DC, and closely examined the forelimb of each animal. Interestingly, they determined that the pectoral muscles, which support the trunk and limbs in hippos, are well developed and originate posteriorly beyond the ribs, unlike other hooved mammals. This arrangement allows for increased force when retracting the forelimb, which the researchers associate with charging forward at another animal, climbing the steep banks of water bodies, and moving forward while underwater. This last trait is important for an animal that walks on the bottom of waterbodies, rather than swimming. Hippopotami also retained a unique combination of muscles and tendons that allow each individual digit to be weight-bearing to prevent splaying while walking on muddy substrates, unlike in most hooved mammals. Also unique to hippopotami is the retention of several "primitive" muscles in the forefeet (e.g., mm. palmaris longus and flexor digitorum brevis) that help control the digits while walking underwater (Fisher et al. 2007: 692). These features also support hippopotami's close phylogenetic relationship with cetaceans and early divergence from other hooved mammals.

Morphological specialization certainly exists for many of the semi-aquatic mammals described in this book, but many species, such as the American mink and the rice rats, present very few modifications despite their dependence on aquatic habitats. Alternatively, beavers and platypus possess adaptations specific to freshwater habitat and are often presented as prime examples of semi-aquatic mammals. Given the broad range of freshwater habitats and the variety of dependence of various species on these habitats, it is difficult to find one physical feature that defines all semi-aquatic mammals. Being semi-aquatic rests on a continuum. Morphological features might not always be obvious, but behavioral ones are clear. That is the joy in ecology: exploring the whole of a species to understand how it became uniquely successful within its particular niche. Semi-aquatic mammals have survived despite the physical challenges of living a lifestyle with one foot on land and the other in water.

6

Physiological Adaptations

Anatomy is to
physiology as
geography is to
history; it describes
the theatre of events.

Jean François Fernel,
1542, *De naturali parte
medicinae*

Water as a Medium

In coining the term "physiology," French physician Jean François Fernel was able to move past the description of the body's form (anatomy) and delve into the complexity of its function (physiology) (Tubbs 2015). Species' interactions with their physical environment pose a central challenge: How does an organism remain warm and active over extended periods of time? The added complication for semi-aquatic mammals lies in how they achieve this necessary goal in two very different environments: air and water. It is what West Chester University biologist Frank Fish calls "the most energetically precarious position" (2000: 683). Four properties of water, individually and in combination, are central to the tenuous balance a mammal faces when it is resting or moving in water. Firstly, water has a thermal conductivity approximately twenty-five times higher than air; therefore, water has almost no resistance to heat flow from the warmer body of a swimming mammal (MacArthur 1989a). Secondly, water has a specific heat capacity that is roughly 3,500 times higher than air, which translates to a required input of 4,184 joules of heat to increase the temperature of 1 kg of fresh water by 1°C. It takes a relatively long time for water to warm or cool, thereby creating a fairly stable environment for aquatic organisms. Next, water is eight hundred times denser than air. On immersion, compression of the fur contributes to the loss of the insulating boundary layer of air trapped within the pelage. If the animal is swimming, or is moving quickly in water in pursuit of prey, there is an added increase in thermal conductance and corresponding heat loss (MacArthur 1989a). Lastly, water is sixty times more viscous that air, which results in more friction (**drag**) for moving animals, especially when the animal is swimming at the water's surface. The greater effort required to move through water increases the amount of energy lost—energy that could otherwise be used to warm the animal's core. All of these properties of water affect the whole of the animal, from sight to respiration, but without a stable core temperature, all other organs are compromised.

A body loses heat through four processes (Fig. 6.1). Conductive heat loss (**conduction**) occurs through the physical contact of one object with another. In the case of semi-aquatic mammals, this form of heat loss occurs immediately when the animal enters cooler water, and it increases upon full submersion. The difference in temperature causes heat to flow from the body of the animal to the surrounding water, and it will only stop if both systems are of equal temperature (**thermal equilibrium**). As soon as a semi-aquatic mammal enters the water, it also loses heat

Figure 6.1. Modes of heat loss for a North American beaver. G. A. HOOD

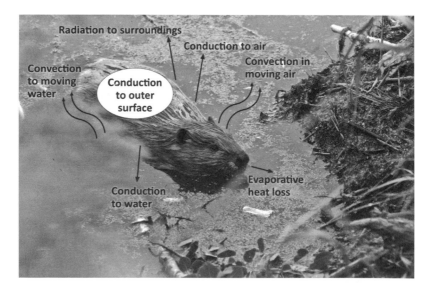

via **convection**. Convective heat loss is the major source of cooling for a swimming mammal, as water moves heat away from the body. For humans it might account for almost 97% of total heat loss in water (Bullard and Rapp 1970). On land, wind blowing across the skin, much like a fan on a hot day, also creates convective heat loss. Although convection requires the movement of a fluid (air or water), even in still water density, currents form between the body and water. Robert MacArthur called this phenomenon "free convective heat loss" during his assessment of mammals in cold conditions (1989a: 291).

While an animal is submerged, radiant heat loss is negligible. **Radiation** involves losing heat through infrared rays, rather than through direct contact. Similar to the way radiant heat from the sun provides warmth, the body gives off heat energy to the surrounding atmosphere. Once on land, radiant heat loss becomes more important for thermoregulation in semi-aquatic mammals. The same is true for **evaporation**, where the vaporization of sweat or water left on the body upon emergence creates a cooling effect as the water becomes a gas. Although sweating is not a major factor in water, evaporative heat loss increases if the pelage is not groomed, or does not dry quickly upon emergence from water. Eurasian water shrews (*Neomys fodiens*) often exit the water through narrow tunnels that help squeeze water out of their fur as quickly as possible (Köhler 1991). Several semi-aquatic mammals shake vigorously after leaving the water to remove excess moisture. There are various inherent behavioral and physiological characteristics of mammals, in general, that counter heat loss and others that are more common in several semi-aquatic mammals. These adaptations are also important relative to oxygen exchange, digestion, and sensory physiology, which can play interconnected roles.

Basic Physiology of Semi-aquatic Mammals

Central to the thermoregulatory success of mammals is **endothermy**, the ability to create and regulate heat from internal chemical processes (**metabolism**), rather than from passive external inputs from the surrounding environment (**ectothermy**). Endothermy allows mammals (and birds and some fish) to remain active over longer time periods in varying external temperatures (e.g., different seasons or times of the day) within a large range of habitats. Endothermy also plays a central role in the chemical and physical processes in the body, most of which are very sensitive to changes in temperature. Therefore, mammals must maintain a narrow internal temperature range to function fully, regardless of environmental variation. Such physiological regulation of a constant internal body temperature is also called **homeothermy**. Although broadly defined as homeothermic, some semi-aquatic mammals are also able to undergo periods of torpor, and they are able to tolerate times when there is differential exposure to the elements. Maintaining a constant core temperature requires mechanisms by which the brain can detect changes in body temperature and then create an immediate response to increase or lower heat production in one or more areas of the body (Kemp 2017). This response then creates an instantaneous physical reaction (e.g., shivering or increased blood flow) to mitigate thermal stress. Adaptive behaviors have also evolved to support physiological reactions.

There are various behavioral options that help in thermoregulation, such as alternating time spent between terrestrial and aquatic environments, favoring nocturnal or diurnal activities, huddling together, and changing body position. Prewinter weight gain is a common strategy in mammals. North American beavers (*Castor canadensis*) of all age classes increase their body mass and the size of their tails in summer, yet as adults and yearlings use up these fat stores over the winter, kits actually gain weight (Smith and Jenkins 1997). Adult beavers are more effective at conserving energy than young beavers, and the adults are more reliant on other adaptations to survive winter temperatures. Morphological adaptations include evolving a thick waterproof pelage and having a larger body size to decrease surface-area-to-volume ratio. In semi-aquatic mammals, thick waterproof fur is a more functional substitute for large stores of body fat, especially because blubber has lower insulation value than fur and also makes movement on land more difficult and energetically costly (Costa and Kooyman 1982). A mere 1 cm thickness of air provides an insulative value equivalent to 4 cm of fat (Chanin 1985). The pelage and skin of semi-aquatic mammals tend to be thickest on the ventral (anterior, front) surface of the body, which is in regular contact with water (Sokolov 1962; Dawson and Fanning 1981). In terrestrial mammals, these structures tend to be thicker on the dorsal (back) surface of the body (Sokolov 1962).

Regular grooming throughout the day is essential to maintaining in-

Figure 6.2. Beaver grooming and waterproofing its fur. M. BRIERLEY

sulation and waterproofing in the pelage of semi-aquatic mammals (Fig. 6.2). Eurasian otters (*Lutra lutra*) spend a considerable amount of time grooming after emerging from fresh water (18%), and they spend even more time grooming after being in seawater (33%), due to increased water infiltration (Kruuk and Balharry 1990). This grooming behavior can increase metabolic rates in Eurasian otters by 64% above resting levels (Kenyon 1969). Despite its importance in thermoregulation and buoyancy, "fur is an inflexible insulation" (Estes 1989), and semi-aquatic mammals must use other strategies to reduce overall energetic costs and regulate body temperature in dramatically different environments. As outlined below, various evolutionary adaptations help semi-aquatic mammals to enhance energy production and conservation by facilitating movement through water, by regulating their temperature both physiologically and behaviorally, and by maintaining a thick waterproof pelage that enhances buoyancy and thermoregulation (Fish 2000).

Temperature Regulation

Maintaining thermal **homeostasis** in water presents several challenges. Just as water changes in temperature, density, and velocity, an animal's behavior while in water varies as well. Hunting for fish or foraging for submerged vegetation involves increased movements that exacerbate heat loss through convection and conduction. Diving presents specific challenges that interrupt the delicate balance between maximal oxygen uptake during intense exercise (**VO2 max**), muscular performance, and ongoing uptake of oxygen by the blood. There is also internal conductance of heat from the body's core to the surface, which is directly linked to heat transfer from the body to water. Additionally, there is convective heat transfer via the veins and arteries as blood flows throughout the body from areas of differing temperature. Heat storage within the body and body configuration itself play an important role in thermoregulatory strategies (Withers et al. 2016).

Body Size

As previously mentioned, semi-aquatic mammals are generally larger than their terrestrial counterparts, yet smaller than fully aquatic species (Wolff and Guthrie 1985). Larger animals have a selective advantage relative to thermoregulation because of a lower surface-area-to-volume ratio. In an early immersion study using rats, dogs, and people (within limits) it became obvious that as body mass increased, the rate of cooling decreased in water (Spealman 1946). At ~13 g, the American water shrew (*Sorex palustris*) is the second-smallest mammal of the thirteen water shrews covered in this book. Of all semi-aquatic mammals, it is the third-smallest semi-aquatic mammal for which body mass was available (*n* = 115 species; Fig. 6.3). When actively swimming in 10°C to 12°C water, the American water shrew's average rate of decrease in body temperature is 2.05°C per minute (Calder 1969)—the fastest known cooling rate for aquatic mammals while swimming (MacArthur 1989a). Diving increases the cooling rate to 1.43°C per thirty seconds as air from the pelage is lost due to increased density of water with depth.

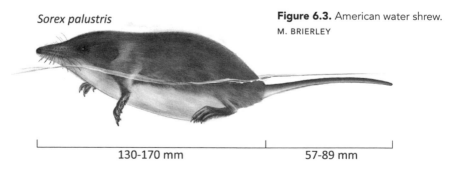

Sorex palustris

Figure 6.3. American water shrew.
M. BRIERLEY

130-170 mm   57-89 mm

Almost all mammals experience a decrease in body temperature while swimming; however, the platypus (*Ornithorhynchus anatinus*; ~1.4 kg) can withstand long periods of immersion without becoming hypothermic. The platypus has a resting metabolic rate that is 37% to 54% lower than eutherians; it can remain homeothermic for up to three hours in 5°C water (Grant and Dawson 1978b). Its large flattened guard hairs combine with a densely kinked underfur to result in fur with a higher insulative value than that of many other semi-aquatic mammals (Grant and Dawson 1978b; Fish et al. 2002). Apart from the platypus, very few species weighing less than 2 kg can sustain a constant temperature in cold water (MacArthur 1989a). There are, of course, many semi-aquatic mammals that weigh substantially less than 2 kg. The median weight of the semi-aquatic mammals identified in this book is 168 g.

More typically, a small body mass results in rapid cooling in water. Although one of the largest shrews in North America, the American water shrew weighs an average of only 8 to 18 g, and Delany's swamp mouse (*Delanymys brooksi*) weighs even less at approximately 5 g. The substantially larger muskrat (*Ondatra zibethicus*) weighs up to 1.8 kg. To make up for their smaller size, these species are especially reliant on thick fur, adaptive behaviors, and an ability to selectively cool different parts of their body. Fewer than 40 of the 140 species described in this book weigh more than 2 kg. For many species, a fusiform body shape creates an effective surface area for reduced heat loss, which compensates for increased surface area associated with webbed feet (e.g., in beavers, otters, and platypuses) (MacArthur 1989a). This body shape scales down to even the smallest of the semi-aquatic mammals. For mustelids, their elongated shape seems an exception to this strategy, but they compensate with a **basal metabolic rate** (BMR) that is 20% higher than terrestrial mammals of a similar size (Iversen 1972). They also have shorter limbs. Some of the smaller species, including muskrats and the rakali (*Hydromys chrysogaster*), have a tolerance to shallow hypothermia, which MacArthur calls "permissive cooling" (1989a: 320). As discussed later in this chapter, foraging strategies and modifying the length of time in water also play an important role in thermoregulation.

There continues to be vigorous debate as to the *mechanism* behind the inverse relationship between body mass and BMR. However, the relationship itself is clear; in 2008, Brian McNab determined that body mass alone accounted for ~95% of the variability in BMR across a suite of mammalian species. The remaining variability might be explained by diet, climate, habitat characteristics, or behavioral strategies (Withers et al. 2016). Yet the mechanisms still remain somewhat of a mystery. In the 2016 book *Ecological and Environmental Physiology of Mammals*, Philip Withers and his coauthors refer to almost a dozen explanations within the literature; however, regardless of the mechanism (and there may be several depending on species-environment relationships), this pervasive allometric relationship remains an important consideration when addressing physiological responses of species that regularly move between land and water.

Metabolism

An organism's metabolic rate is a measure of how quickly energy is produced by chemical processes in its cells; this production of energy involves muscle contractions, internal bodily functions, large amounts of oxygen, and processing of sensory signals sent to the brain (Feldhamer et al. 2015). A high metabolic rate, found in some semi-aquatic mammals, provides an internal heat source produced by chemical reactions in the body. Core body temperatures are sometimes higher as well. River otters (*Lontra canadensis*) have a BMR approximately 50% higher than expected (Chanin 1985). The platypus has a BMR that is 64% lower than eutherians of the same size but more than twice that of other monotremes (Grant and Dawson 1978a). The high BMR in the platypus, as compared to other monotremes, is supported by elevated levels of thyroid hormones (Nicol 2017), which allow it to sustain long foraging bouts. Platypuses can perform uninterrupted foraging trips lasting up to 29.8 hours, with the longest ones in winter to accommodate increased metabolic costs and diminished food resources (Bethge et al. 2009).

A high metabolic rate is not universal among semi-aquatic mammals, and species such as the water opossum (*Chironectes minimus*) of South America and the web-footed tenrec (*Limnogale mergulus*) of Madagascar have comparable metabolic rates to most eutherians (Thompson 1988; Stephenson 1994). Another study, however, determined that the BMR of the water opossum was 1.4 times higher than those predicted for other marsupials (McNab 1978). The BMR of the Eurasian water shrew was not

Table 6.1. Ratio of measured to predicted metabolic rates for semi-aquatic mammals at rest in water

Species	Measured metabolic rate/predicted metabolic rate	Reference
muskrat, *Ondatra zibethicus*	1.9 (summer)	Fish 1979
American mink, *Neovison vison*	1.8	Williams 1986
rakali, *Hydromys chrysogaster*	2.3 (summer) 4.2 (winter)	Dawson and Fanning 1981
Eurasian river otter, *Lutra lutra*	1.7 (summer) 4.5 (winter)	Kruuk, Balharry, and Taylor 1994
North American beaver, *Castor canadensis*	1.2 (summer) 1.8 (winter)	MacArthur and Dyck 1990; Williams 1989

Source: Adapted from Williams (1998: table 2.1).

Note: References indicate the source studies for the metabolic data.

as high as two shrews (*Sorex minutus* and *S. araneus*) jointly occurring in the same habitat (**syntopic**), yet its minimal thermal conductance in air was lower (Sparti 1992). The North American beaver has an equivalent BMR to similarly sized terrestrial eutherians (MacArthur 1989b). Despite appearing regularly in the scientific literature (see MacArthur 1989a: 309), an elevated BMR cannot be generalized to all semi-aquatic mammals, though it is common to some.

Terrie Williams, a comparative ecophysiologist at the University of California, Santa Cruz, has conducted extensive research on mammalian locomotion and physiology. In writing about metabolic rates of aquatic mammals, she states that the difference between resting metabolic rates and those predicted from **allometric regressions** for mammals resting in water were highest for semi-aquatic mammals in conditions similar to winter temperatures (Williams 1998). Values were derived by dividing the measured metabolic rate by the predicted metabolic rate. A ratio of 1.0 indicates that the predicted rate and the measured rate were equal; a higher value indicates that the actual metabolic rate was much higher than expected. All semi-aquatic mammals had higher measured metabolic rates than expected (Table 6.1), and many were higher than the three marine mammals in Williams's study: the California sea lion

(*Zalophus califonianus*), the harp seal (*Phoca groenlandica*), and the bottlenose dolphin (*Tursiops truncates*). As indicated earlier, it is not always a consistent trend in semi-aquatic mammals, yet higher metabolic rates are present in some semi-aquatic mammals. One reason might be that an elevated BMR might make it difficult to stay cool while active on land.

Little is known about how semi-aquatic mammals regulate temperature and metabolism in high temperatures. To conserve energy, it is important to stay within a set **thermoneutral zone** that defines the lower and upper boundaries in which the rate of basal heat production equals the rate of heat lost to the environment. Beyond those limits, increased metabolism is required to either warm or cool the body. It is rare for water to provide an environment in which water temperatures are the same as or warmer than a mammal's basal rate of heat production. However, air temperatures can exceed thermal limits regularly in summer. For many semi-aquatic mammals, the **lower critical temperature**, or LCT (minimum body temperature tolerated by an organism before generating heat to counter heat loss), is substantially lower in air than it is in water. Platypuses have an LCT of 20°C to 25°C in air, yet their LCT in water is >30°C (Grant and Dawson 1978a), which is a similar finding for the rakali (LCT$_{air}$ = 25°C; Dawson and Fanning 1981). Although LCT in water for muskrats is also >30°C, they have a lower LCT in air of 10°C (MacArthur 1984). The American mink (*Neovison vison*) has an LCT in air of 21°C to 22°C, and an LCT in water of >22°C (Williams 1983, 1986). Overall, there appears to be a high tolerance of water temperatures exceeding 30°C in the platypus, muskrat, and rakali. Upper critical water temperatures have not been established for freshwater mammals.

The dense waterproof pelage of a semi-aquatic mammal, which is so important in water, can hinder heat loss as air temperatures increase. To counter excess environmental temperatures, a mammal could expend energy to flatten hairs to remove air from its pelage, increase blood flow to the periphery, or sweat or pant to increase evaporative cooling. Thinly insulated appendages, especially plantar surfaces and tails, combined with specialized vascular arrangements, provide a "thermal window" that allows for heat loss in warm environments and heat conservation in colder ones (Williams 1998).

Regional Heterothermy

Endotherms are able to regulate their basal temperature (T_b), with a T_b of ~31°C to 32°C in monotremes, 34°C to 37°C in marsupials, and 34°C to 39°C in eutherians (Lovegrove 2012). Environmental temperatures can influence a differential regulation of a constant T_b in different body parts of many endotherms, particularly in the periphery. This ability to maintain different temperatures in different zones of the body is called **regional heterothermy**. The broad webbed feet and scaly tails of many semi-aquatic mammals are subject to higher heat loss than parts of their body that are covered with thick fur. Through "permissive cooling of the peripheral surfaces" (MacArthur 1989a: 305), mammals are able to either cool or warm specific parts of their body as needed (in particular, poorly insulated appendages). Basal temperatures in the feet and tails of several semi-aquatic mammals (e.g., muskrats, rakali, beavers, and mink) can endure sustained cooling in water temperature (T_w) ranging from 0°C to 5°C (MacArthur 1989a) without any adverse effects on core body temperatures. Conversely, these same areas can also receive excess heat from the body during warmer ambient temperatures and then release it to the environment.

The Eurasian beaver (*Castor fiber*) provides an excellent example of how semi-aquatic mammals can cool or warm peripheral tissues to regulate heat in the core. In 1965, two researchers from the University of Oslo attached temperature sensors on the tip of the tail and into the rectum of a beaver and then altered the ambient temperature (T_a) in the room (Steen and Steen 1965). At 16°C T_a, the rectal temperate was normal (~37°C), yet when the room temperature increased to 25°C, there was a notable increase in rectal temperature (by 2°C), which then stabilized at the new temperature. Skin temperature on the tip of the beaver's tail correspondingly increased from 16°C to 35°C. Yet, when the animal put its tail into the water, there was no evidence of hyperthermia. Once the room temperature was reduced to 20°C, the researchers detected momentary waves of increased temperature in the tail.

At the lowest room temperature (16°C), heat loss from the tail was <0.1 kcal per hour and rectal temperature was normal. Once air temperature increased, heat loss from the tail increased to ~1 to 2 kcal per hour, and rectal temperature was normal (Steen and Steen 1965). As a

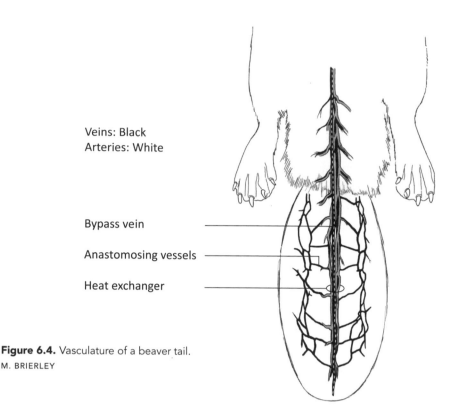

Veins: Black
Arteries: White

Bypass vein

Anastomosing vessels

Heat exchanger

Figure 6.4. Vasculature of a beaver tail.
M. BRIERLEY

species with distributions extending to the Arctic Ocean, beavers use this differential cooling to conserve heat in the winter and avoid hyperthermia in summer. Beavers, among many other mammals, employ an elaborate array of anastomosing arterioles and venules (retia) where heat transfers between the arterial and venous bloodstreams (Withers et al. 2016). Warren Cutright and Tom McKean (1979), of the University of Idaho, identified an extensive arrangement of blood vessels associated with the common iliac artery and veins in the tail of the North American beaver. They found two countercurrent arrangements: one in which two or three veins surround a central artery, and one composed of retia (finer vessels in close contact). The first form was found in the beaver's upper muscles of the thigh and in superficial chest muscles, while the second arrangement was found in the tail (Fig. 6.4). A similar vascular arrangement is found in the rakali, but the lack of a retia arrangement compromises their ability to mitigate heat loss effectively (Fanning and Dawson 1980).

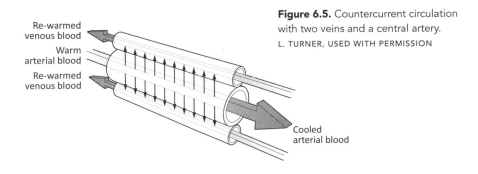

Figure 6.5. Countercurrent circulation with two veins and a central artery. L. TURNER, USED WITH PERMISSION

Re-warmed venous blood

Warm arterial blood

Re-warmed venous blood

Cooled arterial blood

Countercurrent heat exchange succeeds at regulating peripheral heat because of the proximity and connections of the blood vessels, via **anastomosing vessels**, combined with the blood vessels' opposite direction of flow (Fig. 6.5). Arteries bring warm oxygen-rich blood from the heart to the rest of the body, and veins return cooler blood back to the heart for reoxygenation. The flow of blood creates the countercurrent, while the proximity of the blood vessels allows arterial heat to transfer to the venous blood so heat is not lost to the periphery. When it is too warm, blood can bypass the rete to allow heat to be lost from the bloodstream before it returns to the heart.

Several species of semi-aquatic mammals use countercurrent heat exchange to regulate body temperature. Kjell Johansen (1962b) conducted an experiment with muskrats in which the animals were subjected to ambient temperatures >25°C. At these temperatures, blood flow through the tail was 100 to 180 times faster than when T_a was 15°C. Any experimental blocking of blood flow to the tail resulted in rectal temperatures 2°C to 3°C above the normal T_b of 37°C to 39°C. As with beavers, platypuses have a countercurrent arrangement in their hind limbs and tail (Grant and Dawson 1978a). Vasoconstriction of peripheral blood vessels is also used to cool the skin under the fur (e.g., mink, muskrats, rakali), but oxygen and nutrient exchange with the affected tissues is also reduced. Vasoconstriction is often accompanied by **piloerection**, which is the involuntary constriction of cutaneous muscles that causes hairs to stand upright and skin to tighten (IJzerman et al. 2015). Countercurrent heat exchange provides an elegant solution for mammals regularly exposed to dramatically different temperatures during their daily activities.

Thermogenesis

Two means of living in cold climates are the use of countercurrent heat exchange and avoiding the cold through increased body size, metabolic adaptations, and increased insulation. Many animals also build or live in protective structures, such as nests, burrows, and dens. Social or semi-social species huddle together for warmth, a behavior known as social thermoregulation (IJzerman et al. 2015); other species can reduce their activity levels, and therefore energy expenditure, through torpor or hibernation. Although modest, huddling in groups can lower metabolic costs. Huddling muskrats exposed to temperatures comparable to the lowest recorded temperatures in their winter lodges (from −10°C to 0°C) had a resting metabolic rate 11% to 14% below that of single animals, and thermal conductance was 8% to 10% lower (Bazin and Mac-Arthur 1992). The benefit of huddling and communal grooming did not translate to a significant increase in rewarming after emerging from cool water into a communal nest following a winter foraging bout (Mac-Arthur, Humphries, and Jeske 1997). The lack of significantly higher rewarming occurred despite almost immediate grooming of a wet individual by dry cohabitants.

Another approach to living in cold climates is through actively resisting cold temperatures by intensifying heat production, including increasing metabolic rates as previously discussed under "Metabolism." Once an endotherm experiences ambient temperatures that fall below the lower critical temperature, it must draw upon stored heat to maintain its core temperature. Exercise produces heat; however, exercise relies on stored energy that would be rapidly depleted without additional food. Two additional means to produce heat are through **shivering thermogenesis** and **nonshivering thermogenesis**.

Shivering involves involuntary muscle contractions to free heat energy. The nervous system triggers activity in antagonistic skeletal muscles, which expand and contract in opposition to each other to generate heat energy. There are few studies that document shivering in semi-aquatic mammals. Robert MacArthur observed it in the masseter muscles of wild muskrats during his 1977 PhD research, and Dawson and Fanning (1981) observed shivering during controlled thermoregulatory experiments on the rakali. In a later study, MacArthur noted that muskrats that were acclimatized to winter temperatures shivered less than those caught in summer and that any heat gains from shivering were likely lost to convective cooling (MacArthur 1986). In water temperatures <20°C, MacArthur and Dyck (1990) noted shivering in North American beaver kits when they were removed from the water, and very occasionally in older adults following cold-water immersion. In many mammals, the use of **brown adipose tissue** (BAT), or brown fat, allows for more efficient heat production through nonshivering thermogenesis.

In nonshivering thermogenesis, BAT is located in strategically advantageous areas of the body, such as interscapular areas, where it is metabolized with the aid of fat-metabolizing enzyme systems. Rather than being reduced to fatty acids that first must enter the circulatory system to be taken up by other tissues and oxidized to make heat, oxidation of BAT occurs within the fat cells themselves through a complex process involving uncoupling proteins (Randall, Burggren, and French 1997). Consequently, BAT is able to produce heat quickly when an animal is exposed to cold. BAT is common in cold-adapted small mammals, including rodents and shrews, but not in monotremes (Feldhamer et al. 2015). However, there has been limited research in this area, and more research specific to freshwater semi-aquatic mammals is required. BAT is present in some marsupials, but it appears to have a minimal role in thermoregulation (Withers et al. 2016).

Fanning and Dawson (1980) identified interscapular pads of BAT in the rakali, and Aleksiuk and Frohlinger (1971) described its presence in muskrats. MacArthur later determined that the role of BAT was to slow hypothermia in muskrats while swimming and during periods of rewarming between repetitive dives (MacArthur 1986). He did not find any evidence that it was used during diving itself. Despite the presence of interscapular BAT in the rakali, Fanning and Dawson noted that rakali were poorer at thermoregulation in water than the platypus, which lacks BAT. The rakali "displayed a general lack of homeothermy in cold water" (Fanning and Dawson 1980: 235).

Oxygen Exchange and Energetics

Large amounts of oxygen (O_2) are required to maintain high metabolic rates; as such, respiration is intimately linked to thermoregulation. Beyond thermoregulation, however, the respiratory system in mammals has evolved to afford very high rates of gas exchange. Respiration is critical to all biological functions of an organism: communication, olfaction, regulation of O_2 and carbon dioxide (CO_2), cardiovascular processes, and nutrient exchange, to name a few. It is supported by key structures: nose, mouth, pharynx, larynx, trachea, bronchi, and lungs. As the rib cage expands outwards and upwards, and the diaphragm moves toward the abdomen, the lungs fill with air. Any interruption of breathing will cause a corresponding change in all the processes supported by the respiratory system. The mammalian heart, with its complete separation of oxygenated and deoxygenated blood, systematically circulates oxygenated blood to the tissues and returns oxygen-poor blood back to the lungs and heart to restore oxygen reserves and begin the process anew. It is so much a part of mammalian physiology that unless it is stressed in some way, it often functions seamlessly as the background to all other biological functions. However, the unique oxygen demands of diving and swimming challenge many semi-aquatic mammals to effect various behavioral and physiological adaptations while underwater and also during recovery times upon emergence from the water (Mortola 2015).

SWIMMING AND O_2

Surface swimming, interspersed with short bouts of diving, is common in most semi-aquatic mammals (Fish 1982a; Williams 1983). Although diving has the most obvious impact on oxygen exchange, swimming has its own aerobic costs, primarily because of inherent energetic inefficiencies when swimming at the surface. Swimming efficiency, often measured by computing the metabolic cost of moving a unit of mass a given distance (**cost of transport**), varies relative to body morphology, mode of swimming (e.g., paddling versus undulating), velocity, and position (surface versus submerged). There are two main forces involved in swimming. Resistive force (**drag** force) occurs as a swimmer is pushing through the water and incurring a higher cost of transport, and consequently experiencing higher oxygen and energy demands. The increase of drag forces while swimming at the surface can increase energetic costs fivefold (Hertel 1966). Drag force results from the friction between the water and the animal's body, the low-pressure wake behind the animal, and wave formation as the animal pushes forward through the water. For animals such as otters that swim using undulating motions (instead of paddling), lift forces are more efficient and have an 80% propulsive efficiency, versus a 33% efficiency in drag-based swimming (Fish 1996). Fully aquatic mammals tend to have lower aerobic efficiencies than semi-aquatic mammals because they require less oxygen to burn more calories (Fish 1996). Chapter 7, "Locomotion and Buoyancy," provides greater detail on swimming and diving, but energetics, as it relates to oxygen demand, discussed here, offers important insights into physiology.

The cost of transport for semi-aquatic mammals can be three to four times higher than in terrestrial mammals and 2.4 to 5.1 times higher than marine mammals (Williams 1989, 1999). Frank Fish (2000) suggests this increased cost of transport reflects the compromise in semi-aquatic mammals of maintaining the ability to move on land and in water. Mink and muskrats use quadrupedal and pelvic (bipedal) paddling, respectively, both of which are less efficient than swimming modes used by fully aquatic mammals. American mink have the highest minimum cost of transport for any mammalian swimmer because of their use of quadrupedal paddling (Fish 2000). The American mink uses more muscle mass and causes more friction (drag) as the limbs move against the water. Fortunately, the mink's reduced limb length and use of gliding underwater to reduce swimming effort might reduce drag somewhat (Williams et al. 2000, 2002). Bipedal paddling is somewhat more efficient than quadrupedal paddling; however, the cost of transport for muskrats is 13.5 times greater than fish of a similar size (Fish 1982a), and for drag-based swimmers in general, it is 10 to 25 times higher than in fish (Fish 1996). Conversely, beavers and platypuses are well within the range of highly adapted marine mammals, with costs of transport of 3.4 and 2.4 times higher than fish, respectively (Fish 2000). The platypus is unusual in that its metabolic range does not change with swimming velocity, unlike most other semi-aquatic mammals (Fish et al. 1997). In fact, the cost of transport for the platypus decreases in a curvilinear fashion as swimming velocity increases. For other swimming animals, there are small increases in metabolic rate at even low swimming velocities (Baudinette and Gill 1985).

Underwater, beavers undulate their broad flat tails, which increases **thrust**, while the platypus uses its front webbed feet in a rowing fashion, thereby producing lift-based oscillations by generating high thrust with its forefeet (Fish et al. 1997). Both the pygmy hippopotamus (*Choeropsis liberiensis*) and common hippopotamus (*Hippopotamus amphibius*) are unique in underwater locomotion. Their large size and short legs make movement on land awkward but, although not elegant underwater, increased buoyancy in water allows for an extended gallop (Coughlin and Fish 2009). For all semi-aquatic mammals, the largest contribution to the total cost of transport is high maintenance costs (the energetic costs of keeping an organism in a healthy living state). Living in two distinctly different habitats simply takes more energy than being either a fully aquatic or a fully terrestrial specialist.

Swimming underwater reduces drag substantially, especially if an animal swims at a depth that is more than three times the maximum diameter of its body (Hertel 1966). A mink swimming at the surface is energetically inefficient, but it reduces energy costs by seven to ten times when swimming underwater (Williams 1993). Williams suggests that swimming underwater, rather than at the surface, is likely an important adaptation for decreasing energetic costs for mammals. Interestingly, underwater swimming produces a slow heart rate (**bradycardia**) in many semi-aquatic mammals, including beavers and platypuses (Castellini 1988; Fish 2000). A lower heart rate has an associated decrease in metabolism and, therefore, maintenance costs. Bradycardia is not universal in semi-aquatic mammals, thus highlighting the breadth of physiological adaptations relative to the degree of aquatic dependence among these mammals.

To swim underwater, an animal must first break through the surface of the water and its inherently high surface tension, which creates a potent resistant force. Added to the issue of surface tension is the buoyancy of thick waterproof fur. Much like a life jacket keeps humans afloat, a thick pelage combined with trapped air between the hair fibers provides positive buoyancy for semi-aquatic mammals. An average volume of air in muskrat fur is ~21.5% of their total dry volume, which translates to an average specific gravity of 0.79 kN/m^3 versus 9.807 kN/m^3 for water at

4°C (Johansen 1962a). At the surface, the pelage helps reduce the energy required to avoid sinking, but when diving to obtain food or when swimming underwater, an animal can expend at least 95% of its total mechanical energy to break through the surface of the water and counter positive buoyancy (Stephenson et al. 1989). Additional buoyancy is provided by the large volumes of air in the animal's lungs. Buoyancy plays a central role in swimming and diving energetics of semi-aquatic mammals. Mink fur provides some of the lowest buoyancy of the semi-aquatic mammal furs, which might contribute to the mink's poor energetic efficiency while swimming at the surface. However, its fur still helps conserve energy during the mink's frequent bouts of diving.

The high density of water has several effects on the body of a diving mammal. There is increased hydrostatic pressure in the lungs as air compresses when water density increases with depth (Dunstone 1998). For mink, the associated decrease in buoyancy, reduces the energy required to remain underwater at a particular depth. There is no evidence that lung volumes of most semi-aquatic mammals are proportionally larger than similar terrestrial species, although semi-aquatic mammals do tend to dive with lungs full of air (Dunstone 1998). Because shallow dives are more common with semi-aquatic mammals, the ability to store oxygen in the lungs and chemically bind it to hemoglobin in the blood and myoglobin in the muscles, as found in fully aquatic mammals, is unlikely. For example, Eurasian water shrews tend to make multiple short dives in water no deeper than 30 cm (Vogel et al. 1998). Platypuses also dive frequently and have a corresponding reduction in metabolic costs (Fish et al. 1997).

During longer dives, the blood consumes more of the lungs' stored oxygen. One strategy to avoid exhausting oxygen reserves is to dive for shorter periods of time to prevent switching to anaerobic metabolism, although even mink perform long-duration, shallow dives as a form of "travelling dive" (Harrington et al. 2012). Apnea associated with diving leads to oxygen deficits that create chemical changes in the body, while deeper dives expose the body to varied pressure changes. During apnea, the heart rate is reduced and oxygenated blood diverts to the heart and brain. Energy for swimming and tissue maintenance is derived from the anaerobic metabolism of glucose, which can also lead to lactate build up during long dives. One species with a unique approach to underwater swimming is the Russian desman (*Desmana moschata*). It often swims with its elongated proboscis above the water, much like a snorkel (Hickman 1984). It has developed a means to swim underwater while still maintaining a constant oxygen supply.

Foraging strategies can influence dive lengths and frequency. Those species that can eat underwater or are multiple-prey loaders (i.e., carry more than one prey item at a time) might dive until oxygen stores are exhausted; however, most semi-aquatic mammals surface well before they exhaust their oxygen stores (Butler 1982). Semi-aquatic feeding occurs when some behaviors, such as prey capture, occur underwater and other behaviors, like processing the prey, occur above the surface (Hocking et al. 2017). Semi-aquatic feeding behavior is found in North American river otters and some South Africa otters (Rowe-Rowe 1977b). If there is an extensive underwater search or pursuit phase, the animal risks exhausting oxygen supplies. Restoring oxygen reserves requires a recovery period at the surface. Generally, as depths increase, surface times will increase accordingly (Dunstone 1998). Surface times also increase as dive duration increases, although larger mammals are able to dive longer (Table 6.2).

Dive duration varies relative to physiological and behavioral stimuli. For Eurasian otters feeding in marine habitats, Butler and Jones (1997) found that mean dive duration was shorter when prey were caught (13.3 seconds) than when they were not (22.7 seconds). Diving experiments with muskrats in Manitoba, Canada, determined that there was an increase in average and cumulative dive times in both fall and winter, and that blood oxygen storage capacity, and consequently body oxygen reserves, were also highest at that time (MacArthur et al. 2001). Muskrats by the age of three weeks are accomplished divers and dive better than they swim (Errington 1939). The ratios of blood oxygen stores to metabolic demands are approximately the same in beavers and muskrats, at 1.6 and 1.9, respectively. Gregory Snyder (1983) suggests that both species should deplete their muscle oxygen stores within two to four minutes. Depletion of oxygen from muscles likely occurs because average diving times are generally within existing oxygen stores, and temperature changes while in water reduce muscle oxygen demand (Snyder 1983). There is a noted decrease in blood oxygen stores in Eurasian beavers within four minutes (Clausen and Ersland 1970). Beavers

Table 6.2. Dive duration of semi-aquatic mammals

Species	Max. dive time	Average dive time	Reference
platypus, *Ornithorhynchus anatinus*	11 min	31.3 sec	Evans et al. 1994; Bethge et al. 2009
Nimba otter shrew, *Micropotamogale lamottei*	15 min*	~3 min	Vogel 1983
North American beaver, *Castor canadensis*	15 min*		Irving and Orr 1935
Eurasian beaver, *Castor fiber*	6.7 min	23 sec	Graf et al. 2018
muskrat, *Ondatra zibethicus*	12 min*	19 sec	Irving 1939 (unpublished data); MacArthur 1992; MacArthur et al. 2001
rakali, *Hydromys chrysogaster*	36 sec	11 sec	Petzold, Trillmich, and Dehnhardt 1997
southern water shrew, *Neomys anomalus*	2.3 sec[†]	2 sec[†]	Mendes-Soares and Rychlik 2009
Eurasian water shrew, *Neomys fodiens*	24 sec	3–10 sec	Vogel et al. 1998
American water shrew, *Sorex palustris*	37.9 sec*	5.7 sec	Calder 1969; McIntyre 2000
star-nosed mole, *Condylura cristata*	58.8 sec	13.2 sec	McIntyre et al. 2002
American mink, *Neovison vison*	57.9 sec	9.9 sec	Dunstone 1993; Harrington et al. 2012
Eurasian otter, *Lutra lutra*	1.6 min		Kruuk 2006
spotted-necked otter, *Hydrictis maculicollis*	21 sec		Kruuk 2006
North American river otter, *Lontra canadensis*	1.5 min		Ben-David, Williams, and Ormseth 2000
marine otter, *Lontra felina*	1.1 min		Kruuk 2006
Cape clawless otter, *Aonyx capensis*	26$_{fw}$–48$_m$ sec[‡]	17–21 sec	Rowe-Rowe 1977b; Somers 2000
common hippopotamus, *Hippopotamus amphibius*	30 min		Feldhamer et al. 2015

* Provoked/forced by researcher.
[†] Aquarium only.
[‡] fw = freshwater, m = marine.

primarily forage on land, although semi-aquatic vegetation is important for them at certain times of the year. Ecologist Patricia Graf and her colleagues (2018) determined that the Eurasian beavers in their study area in Norway spent only 2.8% of their time diving, given that much of their foraging was not contingent on diving.

Diving is influenced by many variables, over and above physiological constraints, especially for semi-aquatic mammals. Environmental factors such as light levels (especially relative to depth), prey availability, and prey behavior influence behaviors in diving mammals (Butler and Jones 1997). Both diving and swimming involve a balance of metabolic capabilities, oxygen regulation, and dietary inputs. They are all interconnected.

Energy Intake and Digestion

To use energy, an animal must acquire and assimilate energy. The taxonomic diversity of semi-aquatic mammals is reflected in their diet and digestive physiology. Regardless of these differences, food energy is directed at a variety of common biological goals—reproduction, movement, metabolism, and the production and repair of tissues. All semi-aquatic mammals are heterotrophic, that is, they derive their energy from plants or other animals, all of which have varying energy content. The feeding and digestive structures used to free that energy are not unique to semi-aquatic mammals, but they help to support the physiological challenges of their variable environmental conditions. Mammalian radiations, beginning in the Mesozoic and accelerating in the Cenozoic, resulted in morphological changes that broadened mammalian feeding modes from herbivorous and insectivorous to include carnivorous and omnivorous specializations. In finer detail, these categories can be broken into diet category, including piscivores and squid-eaters, herbivore/browsers, and herbivore/grazers, to name a few (Withers et al. 2016: Table 1.4).

Philip Withers and his colleagues present a matrix of sixteen possible dietary niche combinations relative to diet category and habitat (Withers et al. 2016: Table 1.5). Following this matrix, semi-aquatic mammals occupy eight categories with a ninth category (**carnivore**) listed as a possible, but unfilled, niche. A closer diet analysis, as presented in Chapter 8, "The Predators," identifies at least seven species for which that category is the best fit (e.g., Congo clawless otter, *Aonyx congicus*, and the fishing and flat-headed cats, *Prionailurus* spp.).

The semi-aquatic monotreme, the platypus, is listed as an **insectivore/omnivore** based on its consumption of aquatic invertebrates. The water opossum and little water opossum (*Lutreolina crassicaudata*), both semi-aquatic marsupials, are also classified under insectivore/omnivore; the inclusion of invertebrates, small fish, and amphibians in their diets addresses the omnivorous component of this category.

Eutherian semi-aquatic mammals have the largest diet breadth. Philip Withers and his colleagues include them in the following categories: **crustacivore/clam-eater, frugivore/granivore, frugivore/omnivore, herbivore/browser, herbivore/frugivore, herbivore/grazer,** and **piscivore/squid-eater.** Although not categorized as such in the matrix, many semi-aquatic eutherians are also insectivore/omnivores. Chapter 8 gives more detailed assessment of specific prey items and foraging strategies, and it reassesses and expands on some of the classifications of semi-aquatic mammals in the matrix presented by Philip Withers and his colleagues. For example, several semi-aquatic mammalian carnivores are identified, and various eutherians are included in the insectivore/omnivore category (e.g., shrews). Regardless of categorization schemes, however, the extent of dietary niches in semi-aquatic mammals requires diverse physiological and morphological adaptations for acquiring and processing food to create usable energy.

Two main foraging methods are common in semi-aquatic mammals: seizing of mobile prey and browsing or grazing of nonmoving plants. Using the former method, many small semi-aquatic mammals (e.g., water shrews, web-footed tenrec, and desmans) eat arthropods and small fish; this diet provides them with high calorie food that in turn provides the added calories required to maintain a high metabolism (Stephenson 1994). Animal-based food is quickly assimilated due to lower processing times. Nigel Dunstone (1998) notes that several intermediate-sized semi-aquatic mammals, including otters, almost always eat fish, while large semi-aquatic mammals (e.g., hippopotami and the capybara, *Hydrochoerus hydrochaeris*) are herbivores. Their larger size results in a reduced metabolic rate, despite the loss of assimilation efficiency from their food. Despite being more abundant than animal-based food, plants have lower energy content and require more processing time. There are exceptions to the generalization of size and diet, with some small semi-aquatic rodents being strictly herbivorous (e.g., Delany's swamp mouse, *Delanymys brooksi*, and Tanzanian vlei rat, *Otomys lacustris*), but most small semi-aquatic mammals supplement their diet with invertebrates or small fish.

Digestion begins with the teeth with most semi-aquatic mammals, or keratinized pads in the jaw in the case of the adult platypus. There are four types of teeth, each with a special function. The incisors are located at the front of the jaw and are used to pick up food items and for grooming. Posterior to the incisors are the canine teeth, with one on each side of both the upper and lower jaw. They are used for disabling living prey, scratching the ground, fighting, and defense. The males of all three species of babirusas (*Babyrousa* spp.) have long, upwardly curving tusks formed from upper canine teeth that are used in defense. For meat-eaters, canines tear the food into smaller pieces to aid in swallowing food prior to digestion. The premolars, posterior to the canines, are used to crush and break food. The even more powerful molars at the back of the jaw cut and reduce the size of food even further. Molars of herbivores have especially tough enamel that resist wear from abrasive plant material, and so have high crowns and folded ridges to enhance grinding action. When chewing, ruminants move their jaws from side-to-side to increase contact between the food and teeth. Not all mammals have the same combination or number of teeth; there are variations depending on their diet and evolutionary history (Fig. 6.6). For example, rodents, rabbits, and deer lack canine teeth and instead have a large gap (**diastema**) between the incisors and premolars. Deer have also lost their upper incisors. Like rodents, rabbits have four incisors (two on the top and two on the bottom). However, rabbits also have a second set of auxiliary incisors (peg teeth) immediately behind the top set of incisors. Many insectivorous mammals (e.g., shrews and moles) and piscivorous mammals have numerous sharp, cone-shaped teeth for shearing and crushing their invertebrate and aquatic prey. The biting force supporting the teeth is provided by the temporalis and masseter muscles (biting muscles), except in the platypus, which has different musculature.

Once food is initially processed by the mouth, it enters the **alimentary system** (gastrointestinal tract), which begins the process of the digestion and absorption of nutrients. The platypus has a gastrointestinal

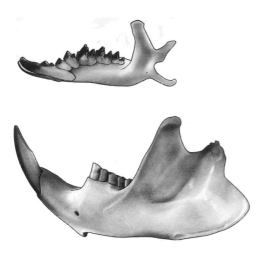

Figure 6.6. Lower jaw and teeth of American water shrew (*top*) and North American beaver (*bottom*). M. BRIERLEY

tract unlike other mammals in that the stomach is small (1.75 cm by 0.75 cm) and resembles a dilation of the esophagus rather than a distinct stomach (Krause 1974). It also lacks gastric glands. Instead, food is stored in cheek pouches until it is processed. There is no subdivision of the intestines into a small and large intestine, and there are no finger-like villi, but it does have a structure similar to a cecum, as found in many other mammals (Krause 1974).

For other semi-aquatic mammals, there are essentially four basic morphologies for the digestive system (Feldhamer et al. 2015; Fig. 6.7). Insectivores have fairly simple digestive systems composed of a stomach, a short intestine, and an anus. Most insectivores lack a cecum. Carnivores have a slightly more complicated digestive system, with a distinct esophagus leading directly into the stomach, a short intestine with a small cecum, and an anus. The permeability of the cell membranes of animal cells is much easier to break down than the cell wall of plants, so nutrients are absorbed quicker when digesting meat. Food residence time is short; therefore, consumption is often higher in carnivores than herbivores. Bile is readily released to emulsify fat, while proteases (special enzymes) help digest protein, and other enzymes help break down proteins into amino acids (Withers et al. 2016).

As with insectivorous and carnivorous mammals, omnivorous mammals have a fairly simple digestive tract. Because of a mixed diet of animal and plant material, their relatively simple stomach leads into a

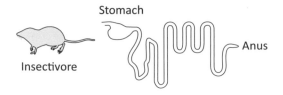

Stomach

Anus

Insectivore

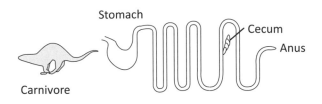

Stomach

Cecum

Anus

Carnivore

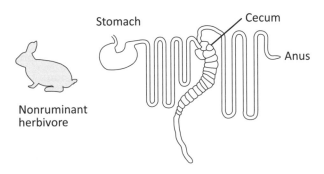

Stomach

Cecum

Anus

Nonruminant
herbivore

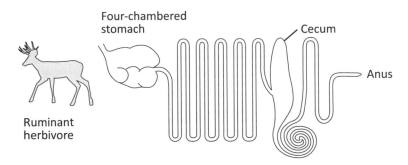

Four-chambered
stomach

Cecum

Anus

Ruminant
herbivore

Figure 6.7. Digestive systems of
mammals. M. BRIERLEY

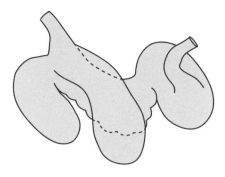

Figure 6.8. Stomach of a common hippopotamus, a nonruminating, foregut fermenter. M. BRIERLEY (AFTER LANGER 1975)

long small intestine, followed by a colon with several folds to increase processing time. In many omnivores, the cecum is just large enough to process some plant material. For herbivores, the alimentary tract is much more complex to accommodate increased residence time for food to break down and for nutrients to be assimilated. Given the broad terrestrial and aquatic forage selection available to semi-aquatic mammals, the diversity of plant food is extensive. One of the biggest challenges is the digestion of cellulose, which requires cellulolytic enzymes to break down it down. Mammals lack this enzyme, and require symbiotic microorganisms to break down and metabolize cellulose to release fatty acids and sugars through fermentation (Feldhamer et al. 2015). Consequently, there are two systems in herbivores: foregut fermentation (digastric digestion) and hindgut fermentation (monogastric digestion) (Fig. 6.8).

All artiodactyls use foregut fermentation to process their food. Ruminate foregut fermenters include the water chevrotain (*Hyemoschus aquaticus*), marsh deer (*Blastocerus dichotomus*), Père David's deer (*Elaphurus davidianus*), water deer (*Hydropotes inermis*), lechwes (*Kobus* spp.), and sitatunga (*Tragelaphus spekii*). Not all foregut fermenters are ruminants. The hippopotami and the babirusas are classified as nonruminant foregut fermenters.

Ruminants first ferment plant material in the rumen, where it is moistened and mixed with fermenting microorganisms. Larger food particles at the top of the mixture pass to the reticulum, where the mixture is softened into a mass and additional fermentation occurs to create short-chain fatty acids (Feldhamer et al. 2015). As the animal rests, it regurgitates the softened mass (cud) and chews it into even finer material, and then mixes it with its saliva. From there the plant material enters the omasum, where it undergoes muscular processing. From the omasum, material enters the abomasum (true stomach) and then progresses to the small intestine where nutrients are absorbed. The cecum provides additional fermentation and absorption. Rather than having a four-chambered stomach, nonruminant foregut fermenters have gut modifications that allow plant material to ferment before it reaches the true stomach. Both of these forms of digestion are more efficient than hindgut fermentation found in lagomorphs and rodents.

As with other herbivores, digestion in hindgut fermentation begins with mastication and the introduction of salivary enzymes to break down plant material. From there, digestion continues in a simple stomach and then quickly enters the small intestine where nutrients are absorbed. From the small intestine, small food particles enter the cecum, which acts as a fermentation chamber where microorganisms break down plant cells. One advantage is that food can be consumed at a faster rate and is only limited by how much material is in the stomach. Also, for beavers and rabbits, which both eat plants high in silica and resins, these chemicals can pass through the gut and bypass the cecum and be eliminated (Feldhamer et al. 2015).

An additional feature of some hindgut fermenters, such as beavers and rabbits, is coprophagy. Two types of fecal pellets are produced. The first is a soft, mucus-coated pellet from the cecum, which is reingested directly from the anus. The cecum in beavers is remarkably large and exceeds the size of the animal's stomach in some cases. Coprophagy allows for additional nutrient absorption from undigested materials in the cecal pellets, in turn improving nutrition and growth. The second type is a harder fecal pellet that represents the end of the digestive process and is left behind. Each digestive strategy in semi-aquatic mammals allows them to maintain metabolism and daily functions, including behaviors that maximize energy intake.

Neurobiology

The mammalian nervous system consists of the central nervous system and the peripheral nervous system, with the latter relaying messages to the brain and spinal cord. These signals translate into sensory and motor functions, including olfaction and locomotion. As animals move through the environment, there is a two-way flow of an immeasurable amount of sensory information that is processed and integrated into everything from the heartbeat to prey detection. Sensory cranial nerves allow mammals to smell, see, taste, hear, touch, breathe, digest, and maintain equilibrium, among other essential functions. Nerves in the spinal cord enable muscle movements, balance, and limb position. Driving it all is the brain, which tends to be larger in mammals than in other vertebrates, although there is some overlap with birds. The mammalian brain has three major sections: the brainstem ("midbrain": breathing, swallowing, heart rate), cerebellum ("hindbrain": coordination), and cerebrum ("forebrain": touch, vision, hearing, speech, reasoning, emotions). These examples of brain function represent a small sample of other motor, autonomic, and sensory functions performed by the brain. At the base of the forebrain, the hypothalamus controls several autonomic functions, including thermoregulation, as described earlier. Some semi-aquatic mammals exhibit fascinating sensory functions that are either unique or unusual in other mammals.

In the forebrain of the platypus, the sensory and motor area (**sensorimotor cortex**) is central to sensory perception in its bill ("muzzle"). As discussed in Chapter 5, the platypus is the only semi-aquatic mammal with electroreception, and it uses it to detect electrical fields in the water (Czech-Damal et al. 2013). The bill has two types of electroreceptors, mucus glands and serous glands, both of which are innervated by a very large trigeminal nerve (Withers et al. 2016). The electroreceptors in the bill are sensitive to high-frequencies ($\sim$50 μV per cm) and can even detect muscle activity of aquatic prey (Scheich et al. 1986).

Other semi-aquatic animals with unique sensory abilities include the star-nosed mole and desmans, both of which have Eimer's organs. The Eimer's organs on the snout of the star-nosed mole are composed of about twenty-five thousand touch receptors that are five times more sensitive than the human hand (Withers et al. 2016). Together they make twenty-two tentacle-like appendages. At their base there is a blood-filled sinus that rests on a dense network of sensory nerves (Feldhamer et al. 2015). As a mole touches an object, these appendages move on the fluid-filled sinus, thereby activating the nerve endings and ultimately initiating a signal to the central nervous system, where these signals are "decoded" to determine the nature of the object.

The Russian desman uses a different approach by employing "bubble-based" sniffing (Ivlev, Rutovskaya, and Luchkina 2013: 281). After analyzing hours of videography, Yu Ivlev and colleagues noted that as the desman explored the bottom of the experimental tank, it would bend the tip of its flexible nose counter to its forward movement and then squeeze out air bubbles from its nostrils. It would wait for the air bubbles to touch an object and then immediately inhale them back into its nose. Over 91% (392 occurrences) of these behaviors were observed when the animal was at the bottom of the tank examining the substrate. This behavior alternated between short active phases and rest phases that were three times as long in duration.

Similarly, American water shrews locate submerged prey using multiple stimuli, including motion, tactile clues, and underwater sniffing (Catania, Hare, and Campbell 2008). Shrews are very quick predators and within <25 ms of prey coming into contact with the shrew's vibrissae, an attack was initiated. Along with more common forms of prey detection mentioned above, shrews expired and inhaled air bubbles through their nostrils while submerged. In particular, they used underwater sniffing after encountering a novel object with their microvibrissae, although they also sniffed edible objects, but with fewer sniffs. They were also observed to sniff in open water, but at lower frequencies than when in contact with an object. Shrews are an excellent example of a highly adapted semi-aquatic predator that uses several sensory signals (visual, tactile, and olfactory cues) to locate and capture prey. The speed at which they can attack and capture their prey speaks to the intricate neural connections that make mammals such effective predators.

The physiology of semi-aquatic mammals is as complex and diverse as the environments in which they live. The properties of water present energetic, thermal, and sensory challenges that are not encountered in terrestrial habitats. Conversely, the necessary connection of semi-aquatic mammals to resources on land also requires energetic trade-offs. Yet, semi-aquatic mammals occupy particular niches in which they succeed, and in some cases highlight morphological and physiological adaptations that make them unique among the mammals.

7

Locomotion and Buoyancy

On land the progress
of the platypus is not
nearly so happy.

Harry Burrell, 1927,
The Platypus

Moving through Water

For semi-aquatic mammals, locomotion involves a compromise between function and efficiency. Balancing the challenges of moving through water with those required for quadrupedal movement on land has resulted in a wide range of biomechanical and physiological adaptations in semi-aquatic mammals, many of which have been covered in earlier chapters. As a transitionary phase between fully terrestrial and fully aquatic mammals, semi-aquatic mammals are not completely specialized for movement in either environment. Although semi-aquatic mammals have retained morphological, physiological, and behavioral features useful for terrestrial life, they have also developed a variety of adaptations that help them live successfully in water. Generally, the morphological and biomechanical adaptations related to locomotion in water differentiate semi-aquatic mammals most from their terrestrial counterparts. The degree of locomotive adaptation to aquatic life varies among semi-aquatic mammals and has resulted in some interesting differences among species.

As a medium, water creates more resistance to movement than when similar activities are performed on land; this resistance is due to water's higher density and viscosity. The potential energy lost for movement through water is significant, because water is eight hundred times denser and thirty times more viscous than air (Fish 2000). Some species, such as the platypus (*Ornithorhynchus anatinus*), have broad webbed forefeet and specialized rowing motions in the forelimbs that help make them as efficient swimmers as highly evolved aquatic mammals (Fish et al. 1997). Conversely, the American mink (*Neovison vison*) has relatively few structural modifications for swimming, yet it is still closely associated with aquatic habitats (Taylor 1989).

Despite the challenges of moving through two very different mediums, living an amphibious lifestyle has advantages for semi-aquatic mammals, and the physical and energetic trade-offs are not without some ecological benefits. For example, by exploiting aquatic habitats, mammals can reduce predation pressure and increase access to additional food resources in a less competitive environment.

Some species, beavers (*Castor canadensis, C. fiber*) and muskrats (*Ondatra zibethicus*) in particular, find an amphibious lifestyle advantageous when it comes to moving large branches and other vegetation. Despite the increased energetic costs of moving through water, it is actually easier for the animal to drag large objects through the water than it would be to move those same objects on land. Once in the water, the animal's efforts benefit from displacement of fluid caused by the material it is moving, which then increases the object's buoyancy (**Archimedes' principle**; Fig. 7.1). An animal's own buoyancy plays a key role in locomotion for semi-aquatic mammals; buoyancy works both to the animal's advantage and disadvantage.

Challenges faced by all semi-aquatic mammals while in water include moving within a more energetically demanding environment and being able to maintain stability in water while foraging or escaping predators. Unlike terrestrial animals, which swim across the surface of the water, semi-aquatic mammals are specialized to move through the water as a

Figure 7.1. Archimedes' principle. The buoyant force is equal to the fluid displaced by the object (e.g., 8 kg otter, 3 kg water displacement: Otter feels like it weighs 5 kg). M. BRIERLEY

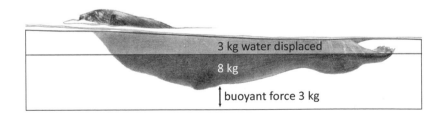

three-dimensional habitat (Taylor 1989). Effective movements while diving and swimming are aided in many of these mammals through adaptations to propulsive appendages (e.g., legs and feet). Stability requires an animal to maintain **positive buoyancy** (float) at the water's surface while countering increased energetic demands when diving below it. Ultimately, semi-aquatic mammals either developed structural adaptations to accommodate these challenges of aquatic life or adjusted their mode of swimming to accommodate both aquatic and terrestrial challenges.

General Physical Characteristics

Many freshwater environments are limited in size and depth, thus restricting the physical size to which semi-aquatic mammals could evolve (Fish 2000). In addition to physical constraints, the need to move on both land and water imposes other limitations on body size and shape in particular. In the same way that many species of semi-aquatic mammals are larger than their terrestrial counterparts (Wolff and Guthrie 1985), most semi-aquatic mammals are much smaller than their fully aquatic relatives—apart from the common hippopotamus, *Hippopotamus amphibius*, which is larger than some of its fully aquatic relatives (e.g., dolphins), but much smaller than others (e.g., many whales). Similarly, otters have shorter appendages and are somewhat more elongated than other mustelids, especially when compared to their terrestrial cousins (Estes 1989). Though some semi-aquatic mammals tend to have shorter extremities and cervical vertebrae relative to their terrestrial counterparts (from a physiological perspective, shorter appendages likely contribute to streamlining and heat conservation), their appendages are more flexible than those of their fully aquatic counterparts. Appendages of fully aquatic mammals (e.g., flippers or flukes) have evolved to serve as hydrofoils that produce lift-based oscillations, increase thrust, and maximize lift while swimming (Fish 2000).

The appendages of semi-aquatic mammals can flex to create paddling or rowing motions through drag-based oscillations. Limb action in water can also be augmented by tail movements and undulation of the body in some species. For example, giant otter shrews (*Potamogale velox*) generate horizontal undulations of their muscular tails to produce efficient movements while swimming (Kingdon 1974). Unlike fully aquatic mam-

Figure 7.2. North American river otter running on land. M. BRIERLEY

mals, semi-aquatic mammals use their limbs for both swimming and walking or running on land—movements that demand a flexible range of motion. In some cases, overland movements can be significant. In 2009, a Cree elder from northern Canada told me he had seen a river otter 10 km away from any water source. Although Tarasoff and his colleagues (1972) note that river otters are generally less mobile than most terrestrial mustelids, Taylor (1989) notes that a river otter can arch its back while moving on land, which allows it to "run" with a gait sequence that is similar to a very fast walk (Fig. 7.2). The insulative fur of semi-aquatic mammals, instead of the layer of blubber found in many fully aquatic mammals, provides increased flexibility in limb movements while on land.

As highlighted throughout this book, the role of fur in the success of semi-aquatic mammals cannot be understated. It allows them to retain heat in terrestrial habitats while reducing heat loss and increasing buoyancy in water. Semi-aquatic mammals have developed **hydrophobic** ("non-wettable") fur with higher hair density than their terrestrial counterparts (Fish et al. 2002). Despite an apparent lack of other specialized features for aquatic environments, mink have an underfur that has a higher density than the fur of any of its terrestrial relatives (Fish et al. 2002). The dense underfur found in almost all semi-aquatic mammals traps air, which prevents the underfur from becoming saturated with water and, when combined with other physical adaptations, allows these animals to function above and below the water's surface (Fig. 7.3). For many semi-aquatic mammals, dense waterproof fur plays an important role in efficient swimming.

Freshwater mammals swim both at and below the surface of the water, but metabolic costs incurred while swimming underwater can be

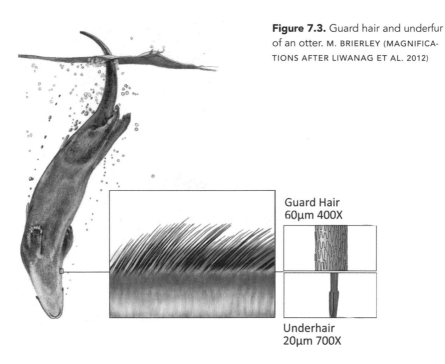

Figure 7.3. Guard hair and underfur of an otter. M. BRIERLEY (MAGNIFICATIONS AFTER LIWANAG ET AL. 2012)

Guard Hair
60μm 400X

Underhair
20μm 700X

five to ten times less than when swimming at the surface (Fish 2000; Williams 2001). As discussed in Chapter 6, this metabolic advantage is especially apparent at a depth greater than three times the maximum diameter of the animal's body (Hertcl 1966). Yet despite this advantage, semi-aquatic mammals tend to swim at the water's surface most often and make only shallow dives (Howell 1930; Fish 2000). Perhaps for this reason, body shape is not as important a factor as expected in the locomotion of many semi-aquatic mammals (Howell 1930), although some species, including otters, are noticeably streamlined and use that body shape to their advantage while swimming. A river otter (*Lontra canadensis*) swims with all four feet during surface swimming, but it switches to an undulatory mode supported by its webbed feet and **carangiform** movements of its tail once submerged (Estes 1989).

For many semi-aquatic mammals, however, body shape has not played an important role in their transition to an aquatic environment. For example, the American mink has a fusiform (torpedo-shaped) body form, which in theory would minimize drag while swimming; however, its shape is no different than that of its close terrestrial relatives. Many

Locomotion and Buoyancy 217

mustelids make good use of that shape to enter burrows to hunt small mammals (Fish 1996).

Modifications to body shape and a range of swimming styles and limb structures balance the energetic costs of both aquatic and terrestrial locomotion in many species.

Limb and Tail Modifications

Adaptations for aquatic life are reflected most in the feet and tail structures of semi-aquatic mammals (Estes 1989). As seen in Chapter 5, these structures vary tremendously in shape and function to provide specific locomotory advantages for aquatic habitats. To increase thrust during the power phase while swimming, many species have evolved feet with a larger surface area through the development of longer digits, webbing or fringe hairs between their digits, or a combination of one or more of these adaptations (Howell 1930). Biomechanically, during movement of the limb away from the center of the body, the digits spread out (**abduction**) to displace more water for increased propulsion. To reduce drag, the digits are drawn together (**adduction**) during the recovery phase when the limb is repositioned to its original location. These movements are similar to those seen in waterfowl.

Such adaptations vary based on the degree to which the semi-aquatic mammal depends on terrestrial habitats. Despite being a semi-aquatic mammal, the American mink has only partially webbed toes and shows no pedal variation from its close terrestrial relatives (Estes 1989). As previously described, the hind feet of semi-aquatic mammals can vary greatly, including having extensive webbing that reaches to the tip of elongated digits (e.g., beavers, platypuses, and water opossums), fringed feet and toes (**fimbriation**), a combination of webbing and fimbriation, or a complete lack of webbing, as seen on the relatively small feet found on swamp and marsh rabbits in the eastern United States (*Sylvilagus aquaticus* and *S. palustris*, respectively). Despite spending much of its day in and under the water, the common hippopotamus lacks any dramatic foot modifications.

Although it lacks the obvious physical limb and tail adaptations of other semi-aquatic mammals, the common hippopotamus is an adept swimmer and diver (Coughlin and Fish 2009). It remains in the water for much of the day, often with just its nostrils rising to the surface to breathe. Its abilities in water seem paradoxical, given its lack of stream-lining and foot modifications. Rather than swimming, the common hip-popotamus mainly walks underwater (Eltringham 2002); it is able to do this through a combination of high bone density and its ability to control the specific gravity of its body (Wall 1983). Mammalian physi-ologists Brittany Coughlin and Frank Fish describe the movement of the common hippopotamus underwater as an unstable gait "similar to a gallop with extended unsupported intervals" (Coughlin and Fish 2002: 675). There is also a notable dominant use of the forefeet when moving underwater. In the evenings, hippos use their short legs to venture onto land to eat, and despite their massive size, hippos can move quickly on land when necessary (Biewener 1989).

For several semi-aquatic mammals that lack webbed feet, a fringe of bristles increases the surface area of the foot to aid propulsion while swimming (Fig. 7.4). As with interdigital webbing, these hairs then fold down when the foot is drawn forward through the water during the re-covery phase of the stroke (Howell 1930). With American water shrews (*Sorex palustris*), air bubbles trapped between the bristles on their feet allow them to run across the surface of the water for three to five seconds (Pattie 1973). It is not uncommon to see an American water shrew dart out from riparian vegetation and run on the water surface briefly before it dives. These same bristles allow shrews to swim effectively underwater by increasing the surface area of the feet (Catania 2013). Closely related Russian desman (*Desmana moschata*) and Pyrenean desman (*Galemys pyrenaicus*) represent semi-aquatic mammals that use a combination of both webbing and fimbriation (Howell 1930). They have extensively webbed hind feet that are also heavily fringed along the border of the foot to further enhance surface area. Fimbriation can also occur along the tail in some semi-aquatic mammals to help prevent side slipping while swimming, much like the keel on a sailboat.

As with other physical adaptations for aquatic habitats, the tail of semi-aquatic mammals can show considerable variation, ranging from complete lateral or dorsal compression to limited or no modification (Fig. 7.5). Similar to modifications to the feet, there is no singular trend in tail structures for semi-aquatic mammals. The capybara (*Hydrochoerus hydrochaeris*) only has a vestigial tail, while the giant water shrew (*Pota-*

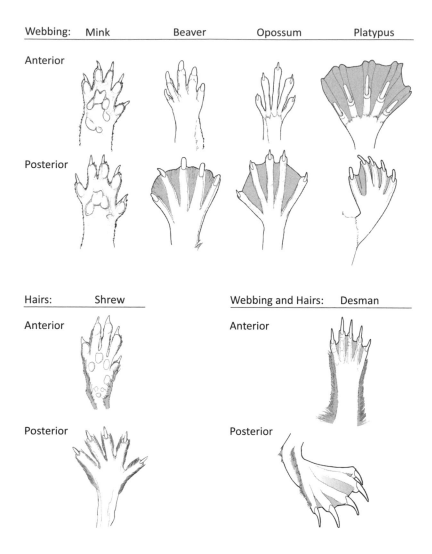

Figure 7.4. Foot modifications of semi-aquatic mammals show adaptations to increase surface areas to aid swimming and diving. M. BRIERLEY

Figure 7.5. Laterally flattened tails: river otter (a) and muskrat (b). Broad, dorsally flattened tails: beaver (c), platypus (d), and giant otter (e). No tail: capybara (f). M. BRIERLEY

mogale velox) of central Africa has a laterally flattened tail that provides its sole means of propulsion in water (Howell 1930). The web-footed tenrec (*Limnogale mergulus*) of eastern Madagascar has a ventrally fringed tail that is laterally flattened at the distal end (Howell 1930).

Beavers and the giant otter (*Pteronura brasiliensis*) of South America have broad dorsally flattened tails ("spatulate") that produce large amounts of thrust when swimming and diving (Fish 1994). Their tails can also act as a sculling oar to aid in directional movements and tracking while in the water, and their tails are sometimes the sole propulsive force for the animal. As a beaver brings food back to its lodge or a feeding platform for processing (central-place foraging), or as it tows branches through the water, its tail provides horizontal stability. Other species, including muskrats and river otters, have laterally flattened tails that are used extensively while swimming, although energetic gains can be minimal for some species (Fish 1982b). Though the tail of a muskrat contributes <2% in additional thrust, it does play an important role in counteracting yaw ("twisting or oscillation along the vertical axis") while swimming (Fish 1982b). Curiously, tail modifications have received much less attention than have other biomechanical aspects of locomotion of semi-aquatic mammals, yet the range of variation in morphology suggests evolutionary advantages in both locomotion and stability. The different combinations of limb and tail morphology represent an evolutionarily effective solution to each species' ecological niche.

Buoyancy and Stability

Closely associated with locomotion is stability control, which is directly linked to buoyancy. On land, animals must work against gravity to move vertically. In water, the effect of gravity is lessened by the effect of buoyancy, yet semi-aquatic mammals must still work to maintain positive buoyancy, to breathe, and to access resources at the air-water interface. Although swimming can aid buoyancy, it is energetically draining in the long run. Increasing buoyancy helps reduce the effort to stay afloat, but it comes at an energetic cost. To dive or to move underwater effectively requires physical effort to overcome buoyancy and surface tension of the water itself. As an added challenge, semi-aquatic mammals have increased buoyancy as a result of their fur, which they require as insulation to protect from heat loss in water. As with other morphological characteristics, semi-aquatic mammals show some common specializations in balancing buoyancy and stability control in freshwater habitats.

Unlike most marine species that have replaced the insulative value of fur with a thick layer of blubber, almost all freshwater mammals have developed a dense layer of waterproof fur that aids thermoregulation and buoyancy. Some exceptions include the capybara and both species of hippopotami. From an energetic perspective, thermoregulation is especially important because of high conductive heat loss to water. For mammals that move readily from land to water, fur is advantageous because it has a greater insulative value than blubber (Costa and Kooyman 1982; Dunstone and Gorman 1998). Air trapped in fur also reduces energy required to tread water rather than float.

In northern semi-aquatic mammals, fur comprises two types of hair:

guard hairs and dense matted underfur that traps air next to the skin (Dunstone and Gorman 1998; see Fig. 7.3). Several studies indicate that buoyancy is positively associated with hair density, which is consistently higher in semi-aquatic mammals than similar terrestrial species (see Fish et al. 2002). As discussed in Chapter 5, the underfur of semi-aquatic mammals is composed of a large number of kinks and interlocking scales among the hairs (Sokolov 1962; Grant and Dawson 1978a; Williams et al. 1992). Dense non-wettable fur that is specialized to hold air is thus a considerable advantage. Dunstone and Gorman (1998) note that, when underwater, the pelage of semi-aquatic mammals is silvery in appearance because of trapped air. Yet with increased depth, air is forced out of the fur, leaving the animal unprotected from the cold. Compression of the fur at depth is a minor issue for most semi-aquatic mammals, though, because many are shallow divers due to their association with shallow-water habitats (e.g., marshes, rivers, streams).

To counteract buoyancy when diving or swimming below the surface, semi-aquatic mammals can increase thrust with simultaneous strokes of paired hind limbs or with tail propulsion. The trade-off is that less energy is used while floating at the surface, while more energy is used to stay submerged when foraging or when accelerating from the surface into a dive. In some cases, at least 95% of an animal's total mechanical energy can be used to counter buoyant forces when submerging below the surface (Stephenson et al. 1989). This cost balances against the metabolic benefits of dense fur as insulation to reduce conductive heat loss while in water, thus highlighting the energetic value of this trade-off. Energy budgets for semi aquatic mammals are complicated to say the least.

Maintaining buoyancy properties of dense waterproof pelage can require a large portion of a semi-aquatic mammal's daily time and energy budget. Grooming is necessary to maintain the condition of the air layer within the pelage and to maintain its waterproof qualities. The air layer in the pelage of muskrats comprises approximately 21% of the body volume (Johansen 1962a) and requires regular grooming to maintain its buoyancy and waterproof characteristics. It is not uncommon to see a semi-aquatic mammal give a violent shake to fluff up the pelage and restore its insulative characteristics when emerging from water. The Eurasian water shrew (*Neomys fodiens*) squeezes water out of its fur by entering a constricting tunnel when emerging from the water (Köhler

1991). This action renews the electrostatic energy in the awn of the underfur, which helps to maintain the integrity of the air layer within the fur. The importance of pelage in the ecological success of semi-aquatic mammals cannot be understated, and thus semi-aquatic mammals must constantly maintain the integrity of their fur. Along with physical adaptations, there is a range of biomechanical strategies that play a central role in swimming both at the surface and under the water.

Swimming Modes

How an animal moves through the water should maximize energy efficiency by reducing drag and maximizing thrust (Fish 1996). Having a streamlined body and specially adapted limbs is helpful, but the very movement of the limbs also plays an important role. Semi-aquatic mammals have often retained the gait of their terrestrial counterparts and modified it to varying degrees to allow both aquatic and terrestrial travel (Howell 1930). Symmetrical gaits, such as walking or trotting, resulted in the least efficient mode of swimming, which in semi-aquatic mammals involves paddling or rowing with two or four limbs (Fish 1996). Known as **drag-based propulsion**, it consists of two phases: a power phase, when thrust ("propulsive force") is produced (e.g., limb pushes caudally), and a recovery phase, when the limb is repositioned for the next stroke and drag (resistive force from friction) is created. Asymmetrical gaits (galloping and bounding) resulted in the more efficient undulatory mode of propulsion (**lift-based propulsion**) that is seen in only a couple of semi-aquatic mammals in marine environments (e.g., seals and walruses) and all fully aquatic mammals, which tend to have limbs that are truncated and fused into flippers and fins (Feldhamer et al. 2007). This mode of propulsion produces thrust on both the upstroke and the downstroke, thereby eliminating any need for a recovery phase.

Although drag-based propulsion is inefficient, it allows semi-aquatic mammals to retain their ability to move in relative comfort in terrestrial habitats (Fish 1996). These mammals vary in the manner in which they employ drag-based propulsion. Quadrupedal paddling (four limbs) is most often used by terrestrial mammals when swimming, although it is also used by the capybara (Fig. 7.6), and occasionally by semi-aquatic mustelids (Fish 1982a; Santori et al. 2008). It is commonly described as

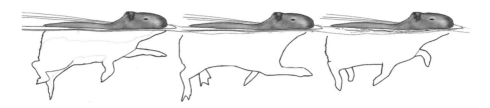

Figure 7.6. Quadrupedal swimming mode of capybara. M. BRIERLEY

the "dog paddle," with all four limbs equally engaged. Bipedal paddling (two limbs) is much more common in semi-aquatic mammals (Table 7.1). It tends to be more efficient (Howell 1930), and it reduces interference between the front and hind limbs (Fish 1996).

Bipedal paddling involves swimming with either just the front limbs ("pectoral") or just the rear limbs ("pelvic") in alternate or synchronous movements (Fig. 7.7). Alternate pectoral paddling is rare in semi-aquatic mammals and is only found in semi-aquatic mustelids and the polar bear (*Ursus maritimus*). Beavers, water opossums, and muskrats often swim by alternating movements of their hind limbs (**alternate pelvic paddling**); although once below the surface, muskrats orient their hind limbs to the side (horizontal plane) and row instead of paddle (**alternate pelvic rowing**). The only semi-aquatic mammal that rows with its front limbs (**pectoral rowing**) is the platypus (Fish 1996, 2000). This form of swimming is more like the way sea lions swim than the way other freshwater semi-aquatic mammals swim, and it results in a more efficient stroke with lower metabolic costs (Fish et al. 1997). Also similar to sea lions is the combined surface area of the platypus's forefeet (13.6%), which is larger than the webbed hindfeet of semi-aquatic rodents (Grant and Dawson 1978a; Fish et al. 1997). Pectoral rowing, combined with large webbed forefeet, helps the platypus maximize the production of thrust.

The river otter paddles with its hind feet simultaneously (Fish 1996), although it is also able to push its forelimbs out sideways to force its body into a turn (Chanin 1985). River otters and giant otter shrews can also involve their bodies in an undulating motion when accelerating underwater to gain additional thrust. Similarly, giant otters can provide a significant amount of thrust and propulsion by holding their hind limbs

Table 7.1. Swimming modes of freshwater semi-aquatic mammals

Order	Species	Bipedal		Quadrupedal	Rowing		Undulating	Reference
		Pectoral	Pelvic		Pectoral	Pelvic		
Monotremata	platypus, *Ornithorhynchus anatinus*				X			Fish 1996
Didelphimorphia	water opossum, *Chironectes minimus*		X					Fish 1996
Afrosoricida	Nimba otter shrew, *Micropotamogale lamottei*			S				Fish 1996
Afrosoricida	giant otter shrew, *Potamogale velox*		X				L	Fish 1996
Lagomorpha	swamp rabbit, *Sylvilagus aquaticus*			S				Lowe 1958
Lagomorpha	marsh rabbit, *Sylvilagus palustris*			S				Chapman and Willner 1981
Rodentia	beaver, *Castor* spp.		XA					Fish 1996
Rodentia	capybara, *Hydrochoerus* spp.			X				Santori et al. 2008
Rodentia	South American water rat, *Nectomys squamipes*		S	SB				Santori et al. 2008
Rodentia	common water rat, *Nectomys rattus*		S	SBU				Santori et al. 2008
Rodentia	muskrat, *Ondatra zibethicus*		S			U		Fish 1996
Rodentia	water rats, *Hydromys* spp.		XA					Fish 1996

Table 7.1. *continued*

Order	Species	Bipedal		Quadrupedal	Rowing		Undulating	Reference
		Pectoral	Pelvic		Pectoral	Pelvic		
Eulipotyphla	water shrews, Crocidura, Chimarrogale, Nectogale, Neomys, Sorex		X	X				Howell 1930; Hickman 1984; Fish 1993; Mendes-Soares and Rychlik 2009
Eulipotyphla	star-nosed mole, Condylura cristata			X		U		Dagg and Windsor 1972
Eulipotyphla	Russian desman, Desmana moschata			X				Ivlev, Rutovskaya, and Luchkina 2013
Carnivora	American mink, Neovison vison			XA				Fish 1996
Carnivora	otters, Lutra spp., Lontra spp.	XA	X	SA			UD	Fish 1992, 1996
Artiodactyla	common hippopotamus, Hippopotamus amphibius	UA2						Fish 1993

Note: S = surface, U = underwater, X = both surface and underwater, A = alternate strokes, UA2 = underwater alternating bipedal walking gait, B = swimming bound, L = lateral, D = dorsoventral.

to the side of their tails with the soles of their feet held upward (Fig. 7.8). To increase streamlining, they hold their forelimbs to their chest unless they are turning (Chanin 1985). This combination of body positions can produce impressive speeds. Nicole Duplaix (1980) observed a young male giant otter swim 100 m in just twenty-six seconds after being startled. Given the energy required, this form of swimming can only be sustained for short period of time. Such undulatory movements in semi-aquatic mammals are considered a transitional phase between drag-based and lift-based swimming (Fish 1996).

Unlike other semi-aquatic mammals, the common water rat (*Nectomys rattus*) and South American water rat (*N. squamipes*) have developed a "swimming bound" while swimming at the surface of the water. It is an asymmetrical gait that resembles the bounding seen in terrestrial movements of small mammals. When using the swimming bound, the animals are able to swim at faster speeds but, similar to otters using undulatory movements, cannot sustain these movements for long periods of time (Santori et al. 2008). Ecologist Ricardo Santori and his colleagues describe the swimming bound as follows.

> In the swimming bound each pair of limbs showed power and recovery phases, hence the gait cycle had 2 power and recovery phases, 1 for each pair of limbs. . . . Movement started with the power phase of the forelimbs, which were extended rotating backward while the hips were flexed and the hind limbs were protracted under the trunk. As forelimbs were accelerated posteriorly, the hips, knees, and ankles were extended and the hind limbs were retracted, producing the 2nd power phase. The movement of hind limbs raised approximately one-third of the anterior region of the body above waterline, pushing the rodent forward and upward, following maximal extension of limbs. . . . In some individuals we could see an aerial phase, due to the force generated by the hind limbs. The power phase of the hind limbs was still in progress when the recovery phase of the forelimbs started, which repositioned the forelimbs under the neck. . . . The gait cycle finished when hind limbs were flexed beginning their recovery phase, and forelimbs were again extended to a new power phase. (Santori et al. 2008: 1556)

Another less-studied form of locomotion, often interpreted as a form of play, is sliding (or "tobogganing") behavior in North American river

Bipedal: Pelvic

Bipedal: Pectoral

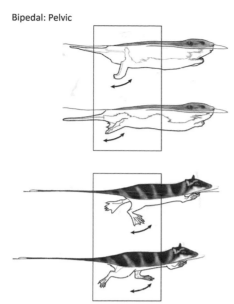

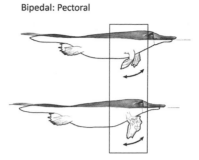

Figure 7.7. Pelvic and pectoral bipedal swimming modes of beaver (*top left*), water opossum (*left*), and platypus (*above*). M. BRIERLEY

Figure 7.8. River otter modes of swimming: quadrupedal paddling (*top*) at the surface and undulating (*center*) and pushing off (*bottom*) underwater. M. BRIERLEY

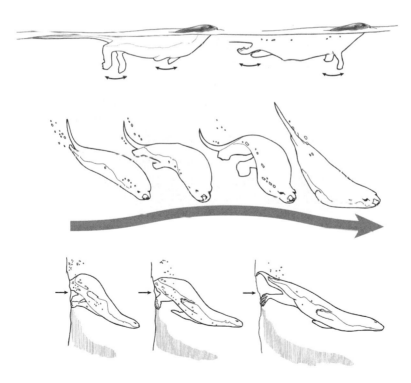

otters (Stevens and Serfass 2005). In particular, this form of travel is often observed on snow (Chanin 1985; Melquist, Polechla, and Toweill 2003), although otter slides made of mud often can be seen along some rivers inhabited by otters. Following an otter reintroduction on the Youghiogheny River in Pennsylvania, Sadie Stevens and Thomas Serfass (2005) used trail and video cameras to monitor river otter behavior at a number of active latrine sites. Within a 5.82-minute video, they observed fifty-three seconds of sliding behavior by otters at the site. While sliding, the otters held their limbs back so that the front limbs were parallel with the body and the hind limbs parallel with the tail. Given the repeated slides, interspersed with wrestling behavior, Stevens and Serfass concluded that sliding is not only a form of locomotion for otters but also a form of play. As a form of locomotion, otters used natural ice and snow-covered slopes to extend the length of their slides (Beckel-Kratz 1977). Such behavior reduces energetic expenditures and facilitates movement through snow, given the relatively short length of an otter's legs.

The need to function both on land and in water has resulted in energetic compromises in semi-aquatic mammals, as demonstrated by a variety of morphological characteristics and locomotor behaviors. As an intermediate stage between terrestrial and fully aquatic mammals, semi-aquatic mammals need to balance speed, acceleration, and endurance with thermoregulatory requirements and energetic inefficiencies during locomotion. Effective locomotion and energetics are thoroughly connected. Unlike fully aquatic animals, freshwater semi-aquatic mammals depend on drag-based swimming modes that are less efficient in both energetic savings and overall performance. Some, including otters, are able to increase their efficiency by incorporating caudal undulations into their movements, but only for short bouts. Despite these inefficiencies, however, semi-aquatic mammals have evolved physical and behavioral mechanisms that allow them to succeed in energetically challenging habitats.

PART III
FEEDING ECOLOGY

8

The Predators: Foraging Strategies and Niches

For one species to mourn the death of another is a new thing under the sun.

Aldo Leopold, 1949, *A Sand County Almanac*

Types of Predation

All semi-aquatic mammals are predators in one form or another, regardless of whether their prey items consist of large fish, benthic invertebrates, or various forms of vegetation. In each case, the predator benefits, while the prey is killed and most often consumed. Predation includes carnivory (eating meat), herbivory (eating plants), and omnivory (eating from more than one trophic level). There are various foraging strategies predators use to maximize success relative to prey type, seasonal availability, and habitat conditions. A predator's skill represents a combination of morphological adaptations that function within a broader behavioral context. For some species, foraging is immediately instinctive, as evidenced by **precocial** young, born in an advanced state and able to fend for themselves quickly after birth. Others are **altricial** and are more dependent on social learning to become effective hunters. For example, Eurasian otters (*Lutra lutra*) are not self-sufficient hunters until they are thirteen months or older (Watt 1993). The majority of semi-aquatic mammals are solitary foragers (unless they are young animals with their parents), even when their individual foraging efforts return food back to a family group, as with winter food-caching behavior of both species of beavers (*Castor* spp.). Less common is cooperative foraging, where a group hunts together to herd prey and improve individual catch rates. These various behavioral strategies are also adapted to prey type, and thus can be grouped into strategies specific to **carnivores, herbivores,** and **omnivores.**

Semi-aquatic mammals cover a range of predators with prey items most often being substantially smaller than the predators, such as aquatic invertebrate larvae consumed by the Ecuadoran fish-eating rat (*Anotomys leander*). For fewer species, prey items can also be substantially larger, as with a towering trembling aspen (*Populus tremuloides*) felled by a beaver (Hood and Bayley 2008a). Similarly, the false water rat (*Xeromys myoides*) can attack and consume crabs much larger than itself (Nowak 1999). Foraging involves searching for prey, catching prey, processing prey, and then digesting prey. Each component of foraging requires an intricate balance between the forager and its environment, which results in a foraging strategy specifically adapted to that context.

Foraging Strategies

Solitary and Group Foraging

Ultimately, foraging is a balance between energetic gains and energetic losses. To survive, the losses cannot outweigh the gains. Search behaviors, including underwater sniffing by American water shrews (Catania, Hare, and Campbell 2008) and "fish drives" by some species of otters, can be optimized by either foraging alone or by foraging in groups (Hershkovitz 1969).

The advantages of solitary foraging are that any resources that are found can be completely consumed by the individual (unless brought back to a communal area), dominance interactions are reduced, and the forager might also be less conspicuous to predators. Solitary foraging is especially effective when resource availability is high.

Group foraging also has several benefits. A group can be much more effective at defending resources, especially with territorial species, as found in several species of otters. There is also an increased efficiency in finding food and catching it and, because of social cues, intraspecific competition can also be reduced. Larger prey items can be secured with more foragers working cooperatively. However, competition with other group members is ever present and needs to be mediated through social cues, maximum group size, and interspecific behaviors. The energetic advantages offered by different foraging behaviors have been summarized in theoretical terms for both solitary and group foraging strategies.

Optimal foraging theory works well in describing solitary foraging (MacArthur and Pianka 1966), although it has inspired some debate (Pierce and Ollason 1987; Parker and Smith 1990). At its core, it presents a theoretical framework to help predict or explain functional decisions made by individual foragers in the context of environmental constraints. As with all foraging models, optimal foraging is based on rationalization of costs and benefits of specific choices a forager makes. For example, is it more energetically efficient to sit and wait for a prey to come to you or to actively seek out prey? When does the prey item become too small to afford energetic returns, or so large that capturing and processing it requires an excess of available energy stores? And when do specific foraging choices increase the risk of predation of the predator itself? All of these considerations are bound by species-specific morphological, physiological, and behavioral constraints.

Even though a beaver can cut down a 25 m tall aspen tree, it cannot bring the whole tree back to the pond in one trip, thus making its repeated trips back to the tree more predictable to terrestrial predators. With that said, beavers do cut down very large trees, which reflects a net energetic gain over the long run, within additional constraints (e.g., distance, handling time). Water shrews, on the other hand, capture very small prey items but need to maximize foraging time to maintain a very high metabolism and prevent starvation. For herbivores, an additional constraint is processing time. Many ruminants, for instance the marsh deer (*Blastocerus dichotomus*), not only need to find an adequate quality and quantity of food, they also need additional time to ruminate the food in their four-chambered stomachs before they can obtain more. Of course, competition with other foragers also influences diet choices and foraging strategies, including diet breadth. Within optimal foraging theory, there are several models that help describe many of the foraging choices of semi-aquatic mammals. Within optimal foraging theory, in particular, are additional theories that explore the energetic trade-offs applicable to specific environmental conditions.

The optimal foraging model provides a framework for assessing prey choice relative to prey availability and profitability (time required for capture and processing, and the energetic content of the prey). In a feeding experiment with a captive flat-headed cat (*Prionailurus planiceps*), the individual cat captured and consumed live frogs, but it ignored two sparrows that were placed in its cage (Muul and Lim 1970). Similarly, in a Brazilian study, fish comprised over 97% of the Neotropical otter's diet, with ~90% of them from a single species of *Geophagus*, which was more abundant and easier to catch (Helder and Andrade 1997). As more profitable prey populations decline, consumers are predicted to then consume the next most profitable prey item. Even though the Neotropical otter appears to specialize on fish, it also eats crustaceans, amphibians, mammals, insects, and birds, depending on availability and ease of capture.

Patch selection theory also explores optimal foraging strategy of solitary foragers, but in situations where prey is typically found in patches that are varying distances apart. The question here, then, is how long is it energetically beneficial to stay in a patch before expending energy to travel to the next one? Within this context, the **marginal value theorem** (Charnov 1976) predicts foragers will stay in a patch until prey density (and therefore search time) reaches a threshold below which it is too energetically costly to remain in the patch. It also predicts that a patch that is resource-rich will support a longer residence time than those that are resource-poor; however, distances between patches also influence how long an individual remains in a patch (Sinclair, Fryxell, and Caughley 2006).

This theorem implies that an animal understands the spatial distribution of patches in advance, which might be a stretch for some species, but not for those with well-defended territories that are well known to an individual. On the north-west coast of Shetland, UK, 10% of the entire foraging area for Eurasian otters was distinct "fishing patches" where the otters performed a significantly higher number of consecutive dives than elsewhere within their foraging area (Kruuk, Wansink, and Moorhouse 1990). Interestingly, capture success did not differ between feeding patches and neighboring sections where they also foraged. The key difference was that the sea bottom in the feeding patches was much more accessible than elsewhere, where open space in the marine vegetation was less common. In this case, otters could perform more dives in feeding patches with less resistance from vegetation than in other areas. Fishing in patches can present other difficulties for semi-aquatic mammals, though. Because of limited oxygen reserves, semi-aquatic mammals must resurface, which not only reduces search and pursuit times but also gives mobile prey a chance to disperse (Dunstone 1998). Additionally, coming to the surface could disorient predators and make it difficult for them to relocate prey on a subsequent dive (Dunstone 1978). To counter this difficulty, American mink (*Neovison vison*) tend to peer into the water from shore to locate prey prior to diving (Poole and Dunstone 1976).

A second foraging model based on patch selection theory is central-place foraging (Orians and Pearson 1979). This theoretical model also incorporates energy maximization as a forager travels through a patch from a central place (e.g., den or lodge); however, these individuals travel to distinct foraging locations and then return to their central place with their prey. As with other foraging theories, there is a balance between search and handling costs. Because prey items are brought back to a central place for processing, the energetic value of the prey should balance the energetic costs of obtaining it and consuming it. The weight of the prey item and handling time are additional considerations, as is predation risk.

North American beavers (*Castor canadensis*) present a model species for central-place foraging by semi-aquatic mammals. Their ponds act as a central place, and in fall the food caches in front of their lodges are where all the cut stems are stored for winter. In one study, Dr. Suzanne Bayley and I determined that beavers would forage 100 m or more away from the pond edge, but the bulk of foraging was within 30 m of the pond (Hood and Bayley 2008a). The farther away beavers foraged from the pond, the more energy was required to bring back stems from the woody plants they cut, along with an increase in vulnerability to predators. However, if preferred species were at greater distances, it was often more energetically profitable to go the extra distance to cut an aspen tree than to go to a nearby patch of nutritionally poor beaked hazel (*Corylus cornuta*). Of interest was that, depending on the area, energy-maximization strategies differed within a central-place foraging behavior. In some areas, beavers cut smaller stems as distance increased (Donkor and Fryxell 1999), while in other areas stem sizes increased with foraging distances (Gallant et al. 2004). Not surprisingly then, in other areas there was no relationship between stem size and foraging distance (Barnes and Dibble 1988). Different combinations of ecological conditions result in different energy-maximization strategies.

Nigel Dunstone (1998) suggests that diving mammals are central-place foragers, with the surface acting as the central place. Almost all semi-aquatic mammals bring each prey item to the surface to consume it and to empty water out of their mouths prior to eating. Dive depths and foraging times both have metabolic costs that must be replenished at the surface before another foraging dive can begin. To maximize benefits over costs, it is best to surface well before oxygen is depleted and recovery times are increased (Butler 1982). In all foraging strategies, when energetic gains per unit time begin to decrease, the forager reaches a **giving-up time**, which represents the length of time from when the animal last fed to the time it leaves the patch for another one. This same strategy applies to **giving-up density**, which represents the prey density in a patch at which the animal moves on to other patches. Solitary foragers "give-up" patches that are declining in prey density sooner than group foragers, likely because the benefits of foraging elsewhere outweigh the costs of staying (Somers and Nel 2004; Carthey and Banks 2015).

Group foraging is uncommon in semi-aquatic mammals, but it is found to some degree in some species of otters (Kruuk 2006). In this situation, an individual's foraging success is not only dependent on its own behaviors but also on the behaviors of others.

Foraging in groups can take various forms. Capybaras (*Hydrochoerus hydrochaeris*) have group home ranges and restrict their activities to specific locations within these ranges (Herrera and Macdonald 1989). They live in stable groups of up to forty individuals of varying ages. Adults actively fend off intruders and can remain in the group for up to two years. The group itself is fairly static in its distribution, with only small seasonal shifts (300 m) in foraging areas (Herrera and Macdonald 1989). Typical of group foraging in other species, the group of capybaras concentrates its foraging on the best grazing habitat within its home range, and individuals within the group feed for similar amounts of time regardless of overall habitat availability for the group as a whole. There are two benefits from this behavior. Foraging in groups reduces predation risk to an individual, while grazing stimulates plant growth and increases primary production of preferred grasses (Ojasti 1978). Of course, any gains are tempered by maintaining an appropriate group size to reduce density-dependent declines in habitat quality. If habitat suitability declines, then the next best alternative habitat will be used. The resulting distribution of the group is described by a theory called **ideal free distribution** (Fretwell and Lucas 1970).

As seen with capybaras, each individual is free to choose the most ideal option available for itself. The ideal pattern of distribution holds when the density of the forager population is low, and individuals would shift to less favorable patches when it becomes high. Conversely, a higher availability of resources would result in a higher number of individuals. A somewhat different group foraging method that still fits within this theoretical framework is the use of prey drives by otters. In particular, smooth-coated otter (*Lutrogale perspicillata*) are often reported to hunt in small groups (Duplaix and Savage 2018). In Thailand, otter specialist Hans Kruuk observed two otters swimming side by side while driving fish into the reeds where the otters then caught them (Kruuk 2006). Giant otters (*Pteronura brasiliensis*), which live in family groups, are also noted to drive schools of fish into shallow waters where they are then caught and consumed. Hershkovitz (1969) observed eight giant otters swimming abreast in what appeared to be a group drive, although it is not the dominant form of fishing for this species (Duplaix 1980).

One interesting departure from group foraging behaviors is seen in the common hippopotamus (*Hippopotamus amphibius*). Hippopotami forage in groups similar to the way capybaras do; however, modeling of their foraging revealed behavior more accurately described as central-place foraging (Lewison and Carter 2004). As with other species, forage quality was important, but distance to the hippopotamus's daytime waterbody was also an important determinant in foraging decisions. Hippopotami often entered the water at the same location they exited for their nighttime foraging forays. Additionally, animals increased their forage intake as distance from water increased, as determined by increased bite rates and step rates. Although there is no immediate return to a central place once vegetation was obtained, many other aspects of hippo foraging behavior appear to be best represented by a central-place foraging model (Lewison and Carter 2004). All foraging models are tools to help predict behaviors relative to energy maximization within an ecological context. Their utility lies in helping us understand how animals interact with critical components of their environment.

Caching and Hoarding

Just as group living can help secure more prey, storing food can ensure a ready food supply while minimizing contact with environmental hazards (e.g., predators, cold weather, low prey densities). Another benefit of hoarding food is to isolate it from competitors; this is a behavior seen with some mustelids and felids (Vander Wall 1990). The most common group to cache food is the rodents, as seen with winter food-caching behavior of beavers. Although North America water voles (*Microtus richardsoni*) do not store food, the European water vole (*Arvicola amphibius*) is well known for storing grasses in its burrow to help it through the winter. Similarly, Bunn's short-tailed bandicoot rat (*Nesokia bunnii*) also stores food in its burrow on occasion (Al-Robaae 1977), and the rakali (*Hydromys chrysogaster*) is known to store dead cane toads in its nest site (Cabrera-Guzmán et al. 2015). The Transcaucasian water shrew (*Neomys teres*) stores living prey that it has immobilized with venom injected from its teeth (Kryštufek and Bukhnikashvili 2016), and the Pacific water shrew (*Sorex bendirii*) also caches excess food (Cassola 2016). Food storage for these species is especially important during winter when energetic demands increase.

Foraging Niches

While foraging theories help explain the energetic decisions behind foraging behavior, foraging niches describe the morphological, and sometimes behavioral, adaptations to specific prey items. Philip Withers and his colleagues (2016) adapted a classification scheme developed by mammalogist John F. Eisenberg (1981) to create subcategories within the broader categories of predators (carnivores, omnivores, and herbivores) that represent more specific ecological niches. As noted in Chapter 6, there are nine categories that apply to the diets of the semi-aquatic mammals covered in this book.

1. Carnivore
2. Crustacivore/Clam-Eater
3. Insectivore/Omnivore
4. Piscivore/Squid-Eater
5. Herbivore/Browser
6. Herbivore/Grazer
7. Herbivore/Frugivore
8. Frugivore/Granivore
9. Frugivore/Omnivore

Each category applies to eutherian mammals; however, semi-aquatic monotremes and marsupials are restricted to the insectivore/omnivore category. Using the best information available, I have assigned all of the species described in this book to one of these nine categories, where possible.

Diet studies are rare for some species, and their classifications are based on the most dominant food items in their diets. At times, species with broad dietary niches (**euryphagous**), depending on the source, might seemed suited to more than one category, but the nuances of their diet help establish the most suitable category. For species with more selective diets (**stenophagous**), a specific category might appear too broad, yet the species' foraging choices are still encompassed within its assigned category. Still, the foraging niches presented above provide a helpful means to differentiate the breadth of foraging strategies among semi-aquatic mammals. With more research, some species assignments might change slightly. Much depends on the location and seasonality of available studies. These niches are described in greater detail below.

Carnivorous Mammals

As a mode of feeding, carnivores consume other animals. Carnivorous mammals are most often associated with Order Carnivora, yet there are some semi-aquatic mammals within Order Carnivora that eat a substantial amount of plant material, and there are mammals from other orders that prey exclusively on other animals. Despite eating invertebrates, crustaceans, fish, and **herptiles** (e.g., frogs and small turtles), the crab-eating raccoon (*Procyon cancrivorus*, Procyonidae: Carnivora) includes a good portion of fruit, nuts, and vegetables in its diet (Feldhamer, Thompson, and Chapman 2003). A good understanding of a species' dominant prey items helps define the most accurate foraging category for the species as a whole. There will be within-species exceptions. The ability of semi-aquatic mammals to occupy both terrestrial and aquatic habitats also allows for a range of prey items that might be more representative of the habitat than diet specialization. In addition, taxonomic classifications (e.g., Carnivora) do not necessarily reflect diet breath of members of a specific taxon.

Animal matter comprises the bulk of the diet of carnivorous mammals. Occasional or situational consumption of meat by otherwise herbivorous mammals should not be seen as a dominant shift to a carnivorous diet. As an example, in Alaska, North American beavers have been observed eating the carcasses of Chinook salmon (*Oncorhynchus tshawytscha*) during the annual spawning runs up coastal rivers (Gleason, Hoffman, and Wendland 2005). Beavers were seen to feed on fresh carcasses discarded by anglers on three occasions, with chewing observed on one occasion. They avoided the older carcasses that are commonly found on spawning streams. Despite this unusual example, beavers are well known to be herbivorous, specializing on woody vegetation.

Cannibalism and carnivory are examples of occasional dietary switches within semi-aquatic mammals that are linked to specific environmental conditions. These behaviors have been noted in both the common hippopotamus and the muskrat (*Ondatra zibethicus*), which are both typically considered herbivores. For the hippopotamus, both cannibalism (Dorward 2015) and carnivory (Dudley et al. 2016) are uncommon, yet these behaviors have been documented with some regularity in eastern and southern Africa. Carnivorous behavior in the hippopotamus, an obligate herbivore, is likely driven by a nutrient deficiency (Dorward 2015); however, it could result in the spread of anthrax throughout the population once contaminated meat is consumed (Dudley et al. 2016). As with carnivory, cannibalism in hippopotami is likely linked to nutritional stress (Dorward 2015). With muskrats, cannibalism is associated with high population densities and increased competition for resources (Errington, Siglin, and Clark 1963). Kevin Campbell and Robert MacArthur (1996) determined that muskrats can digest high amounts of animal tissue. Even the consumption of even small amounts of animal tissue (i.e., 2% to 3.5% of their diet) provides their daily requirement of nitrogen.

Semi-aquatic mammals that are considered carnivores predominantly consume animal tissue with only occasional use of plant material. Carnivores eat a broad range of animal matter, some of which requires specialized morphological adaptations that allow the predator to hunt efficiently. The digits and blunt claws of the Congo clawless otter (*Aonyx congicus*) are specially adapted for overturning stones and sifting through muddy substrate of streams and wetlands in search of crabs, mollusks, and small vertebrates. As with Asian small-clawed otters (*A. cinereus*) and African clawless otters (*A. capensis*), Congo clawless otters capture prey with their hands rather than with their mouths. Their broad, robust teeth are ideal for crushing shells (Nowak 1999). Similar to all semi-aquatic mammals, their morphology often represents dietary specialization within the broader categories of carnivory, omnivory, and herbivory.

Withers et al. (2016) identified four foraging categories in which animal matter comprises the majority of food items: carnivores, crustacivores/clam-eaters, insectivores/omnivores, and piscivores/squid-eaters. Using a combination of sources (e.g., Nowak 1999; IUCN 2019), I assigned applicable species of carnivorous semi-aquatic mammals to one of these categories, with insectivore/omnivore being the most speciose of the four.

CARNIVORES

Of the 140 mammals described in this book, six have an exclusively carnivorous diet. These species hunt and consume a range of both terrestrial and aquatic vertebrates, and they can consume invertebrates to a lesser degree. Carnivores fit two trophic levels: **secondary consumers**, which eat herbivores (**primary consumers**), and **tertiary consumers**, which consume both primary consumers and secondary consumers.

Carnivores have a range of hunting strategies and diets. The otter civet (*Cynogale bennettii*) is a pursuit predator, but it can also submerge itself in the water with only the tip of its nose exposed above water as it waits for unsuspecting birds and mammals to come for a drink (Nowak 1999). Similar to the otter civet, European mink (*Mustela lutreola*) have a broad diet consisting of mammals, birds, and frogs. Philip Youngman

(1990) describes a 1976 Russian study by Danilov and Tumanov that determined that muskrats, voles, shrews, and moles comprised a third of the European mink's winter diet, while consumption of birds, frogs, fishes, and insects increased substantially in the summer. Despite being less aquatically adapted than the North American mink, most of the European mink's prey are aquatic animals (Maran et al. 1998). As with the European mink, there is seasonal variation in the diet of American mink (*Neovison vison*). Summer prey include crayfish, frogs, shrews, rabbits, mice, fish, waterfowl, and muskrats; however, in winter small mammals make up the bulk of its prey (Banfield 1974). Of course any prey item is subject to its availability in an area. Crayfish are not widely distributed throughout the entire range of mink but small mammals are.

While conducting research in Wood Buffalo National Park, I was able to obtain fur trapping records dating to the mid-twentieth century and noticed an interesting trend in muskrat and mink trapping returns (Fig. 8.1). As muskrat numbers increased, there was a slightly delayed, yet mirrored, increase in mink numbers. The same pattern occurred as muskrat numbers declined, especially following the declines in water coverage in the Peace Athabasca Delta after the installation of a large hydroelectric dam on the Peace River in British Columbia. There was a clear predator-prey cycle reflected in the trapping returns, linking mink density to muskrat (prey) density, which was later confirmed to me by Cree elders who were born and raised on the Delta within the park.

The Congo clawless otter is more terrestrially adapted than the other two otters in its genus (Nowak 1999). It has a broad carnivorous diet consisting of frogs, crabs, fish, mollusks, invertebrates, and small mammals (Jacques, Duplaix, and Chapron 2004). Its placement in the carnivore category is not a perfect fit; yet, unlike the African clawless otter (a crustacivore/clam-eater), it does not appear to eat crabs to any great degree. A rare study focused on the Gabon and the Congo found the Congo clawless otter eats softer prey items than the African clawless otter. In particular, it consumes worms that it digs from muddy riverbanks (Jacques, Duplaix, and Chapron 2004). Its molars are smaller than the African clawless otter, which is reflected in its choice of softer prey. Despite its consumption of worms in this particular study, it is not an insectivore given the breadth of other animal species in its diet (Somers and Nel 2003).

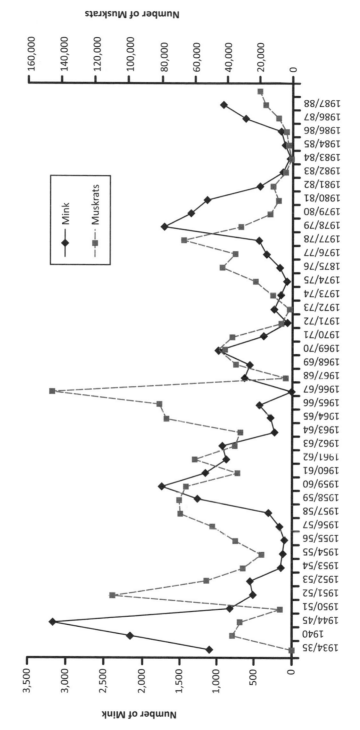

Figure 8.1. Mink-muskrat trapping returns from Wood Buffalo National Park, Canada, 1935 to 1988. G. A. HOOD

Figure 8.2. Fishing cat in the San Diego Zoo. D. PATRIQUIN, USED WITH PERMISSION

Felids are highly specialized carnivores. Flat-headed cats primarily prey on fish, shrimp, birds, and small rodents (Nowell and Jackson 1996). In 1970, there was a rare opportunity to examine the stomach contents of a flat-headed cat that was killed on a river bank in Selangor, Malaysia. It had consumed the flesh and vertebrae of fish (Muul and Lim 1970). As indicated earlier, flat-headed cats also eat frogs. A captive kitten immediately took to a basin of water and fully submerged its head to a depth of 12 cm to seize pieces of fish (Muul and Lim 1970). It also appeared to "wash" objects in the water. Interestingly, when the kitten captured its food, it would pounce on it with a snarl and then carry it at least 2 m from where it was obtained.

The closely related fishing cat, *P. viverrinus* (Fig. 8.2), primarily eats fish, but it also consumes shrimp, birds, herptiles, insects, and small rodents (Haque and Vijayan 1993). When fishing, it moves its paws above the water to attract aquatic prey. Although it is primarily a wetland species, the fishing cat also consumes terrestrial prey, including civets, small mammals, and even livestock and other domestic animals. Annual rodent consumption (e.g., black rat, *Rattus rattus*, and the Indian mole-rat, *Bandicota bengalensis*) ranges from 365 to at least 730 rodents per animal (Adhya 2015), thus helping moderate rodent populations. In his 1929 book *Sterndale's Mammalia of India*, Frank Finn describes the fishing cat's diet as follows: "It presumably kills birds, and must kill wild mammals also, for it takes tame ones—dogs, sheep, and even calves—while it is also a child-eater on occasion, for it will take off native infants even up to four months old" (Finn 1929: 89). There have been no other reports of it "taking off" infants since 1929.

Another category of carnivorous mammals includes the crab and clam eating specialists (crustacivore/clam-eater), of which eleven are semi-aquatic mammals. Four of those species are within genus *Ichthyomys*, the aptly named crab-eating rats. Morphological and behavioral adaptations help crustacivores break open the shells and avoid injury from pincers. Nimba otter shrews (*Micropotamogale lamottei*) specialize on soft-shelled crabs, which they bring to shore and then attack from behind to evade being pinched (Vogel 1983). Crabs make up the bulk of their diet, although they also eat catfish, insects, and tadpoles (Kuhn 1964). They do not eat small mammals (Vogel 1983). In captivity, Nimba otter shrews eat as much as 40 g of fish per day. As with many semi-aquatic mammals, they use highly developed facial vibrissae to detect aquatic prey (Guth, Heim de Balsac, and Lamotte 1959). In a 1983 study of captive Nimba otter shrews, biologist Peter Vogel observed an animal holding its head over the water and contacting the surface with its vibrissae prior to entering the water, potentially to detect underwater prey.

Giant otter shrews (*Potamogale velox*) also specialize in catching and eating crabs, although they will eat frogs, fish, mollusks, and prawns as well (Nicoll 1985). They often use their forefeet to hold their prey down and to flip crabs over to expose their more vulnerable ventral side. Because they are nocturnal predators, giant otter shrews rely on scent and touch to find their prey. Not much is known about the crab-eating rats (*Ichthyomys* spp.) of South America, except that, as their name suggests, they primarily eat crabs and supplement their diet with aquatic invertebrates (Nowak 1999). The rakali, a member of family Muridae, has been studied to a much greater extent. Crustaceans, mollusks, and fish comprise the bulk of its diet, although it also eats aquatic insects, frogs, mice, waterfowl eggs, poultry, turtles, and even the occasional bat (Fanning and Dawson 1980; Watts and Aslin 1981). They use their strong incisors to open mussels. They are also one of the few mammals capable of eating poisonous cane toads (Cabrera-Guzmán et al. 2015).

Crustaceans also comprise a substantial proportion of the diet of two species of otters. All species of *Aonyx* are considered to be adapted for hunting and consuming crabs much more than fish, the African clawless otter (Fig. 8.3) and the Asian small-clawed otter in particular. As noted previously, what little is known about the Congo clawless otter indicates

Figure 8.3. African clawless otter in San Diego Zoo. D. PATRIQUIN, USED WITH PERMISSION

it has a broader diet of animal matter, thus explaining its inclusion in the carnivore category above. African clawless otters are well suited for a diet of crabs (*Potamonautes*), supplemented with fish, frogs, insects, and mammals. As expected, there are seasonal differences, with crab consumption increasing from autumn to summer and frog consumption being negligible in summer (Somers and Nel 2003). Occasionally, they also consume some plant material (Somers and Nel 2003; Nel and Somers 2007).

Asian small-clawed otters, as with other species of *Aonyx*, use their highly dexterous forepaws to find and catch prey. Like Congo clawless otters, Asian small-clawed otters dig in the muddy riverbanks and turn over stones in search of clams, mussels, and crabs. They also use their front paws to catch small vertebrates and invertebrates in shallow water (Hussain, Gupta, and de Silva 2011). Crabs can comprise over 80% of their prey, although Asian small-clawed otters also eat a variety of fish, herptiles, and invertebrates (Foster-Turley 1992). In Thailand, crabs comprised over 90% of their diet, with fish, amphibians, and rodents as less dominant prey items (Kruuk, Balharry, and Taylor 1994). In a rare case, crabs as large as 44 cm were consumed by this small otter (~5 kg). Though they have broad molars that are ideal for breaking open shells, Asian small-clawed otters also dig up shellfish and lay them in the sun so the heat will cause the shellfish to open on their own (Timmins 1971). In West Sumatra, they also consume the invasive golden apple snails

(*Pomacea canaliculata*) in rice fields. The snail is an introduced aquatic pest of global concern (Aadrean, Novarino, and Jabang 2011).

A third category of carnivorous mammals are the insectivores/omnivores. Although they may also consume plant material, their diet reflects the dominance of insects and other invertebrates (including aquatic invertebrates) as a prey base. Of the 140 semi-aquatic species, seventy-two are in this category, including the only semi-aquatic monotreme (platypus), both marsupials (water opossums), species in all five genera of semi-aquatic shrews (*Crocidura, Chimarrogale, Nectogale, Neomys, Sorex*), both desmans, and numerous rodents, among others (Table 8.1). These species primarily eat invertebrates (e.g., insects, arthropods, earthworms, and mollusks), along with some vertebrates and occasional plant material. Several species have fairly predictable diets of invertebrates, mollusks, small fish, and often some plant matter. Others are known to eat primarily insects, but much more research is required to understand their diets more fully. Of these species, there are several with foraging behaviors and diets that represent the essence of the insectivore/omnivore category.

Platypuses are classic insectivores, primarily eating benthic aquatic invertebrates, although they also consume tadpoles, eggs, and small shrimp (Grant and Temple-Smith 1998). As noted before, they are one of only two mammals to use electroreception to location prey. The water opossum (*Chironectes minimus*) holds its front paws out front while swimming to help it detect aquatic prey, although its long vibrissae are also fundamental in sensing prey (Marshall 1978a). It consumes crayfish, shrimp, and fish, as well as aquatic plants and fruit (Hunsaker 1977). The less-aquatic little water opossum (*Lutreolina crassicaudata*) is known to be "savage in temperament but can be tamed" (Nowak 1999: 35). Typically, it eats crabs and beetles, but it will also stalk and kill many species of invertebrates, small mammals, birds, reptiles, and fish.

Foraging on the bottom of water bodies can lead to injury, so the nose pad of the Rwenzori otter shrew (*Micropotamogale ruwenzorii*) is thick and leathery. Its vibrissae can identify worms, insects, crabs, fish, and small frogs, which it then consumes out of the water (Kingdon 1974). Little is known of the critically endangered Ethiopian amphibious rat

Table 8.1. Insectivores/omnivores

Species	Species
platypus, *Ornithorhynchus anatinus*	marsh/shaggy rats, *Dasymys* spp.
water opossum, *Chironectes minimus*	Congo forest mouse, *Deomys ferrugineus*
little water opossum, *Lutreolina crassicaudata*	New Britain water rat, *Hydromys neobritannicus*
web-footed tenrec, *Limnogale mergulus*	Ziegler's water rat, *Hydromys ziegleri*
Rwenzori otter shrew, *Micropotamogale ruwenzorii*	Ethiopian amphibious rat, *Nilopegamys plumbeus*
Cabrera's hutia, *Mesocapromys angelcabrerai*	waterside rat, *Parahydromys asper*
Ecuador fish-eating rat, *Anotomys leander*	Sulawesi water rat, *Waiomys mamasae*
water mice, *Chibchanomys* spp.	false water rat, *Xeromys myoides*
Kemp's grass mouse, *Deltamys kempi*	swamp shrews, *Crocidura* spp.
water rats, *Nectomys* spp.	Asian water shrews, *Chimarrogale* spp.
fish-eating rats, *Neusticomys* spp.	elegant water shrew, *Nectogale elegans*
rice rats, *Oryzomys* spp.	Eurasian water shrews, *Neomys* spp.
water mice, *Rheomys* spp.	American water shrews, *Sorex* spp.
swamp rats, *Scapteromys* spp.	star-nosed mole, *Condylura cristata*
water rats, *Baiyankamys* spp.	Russian desman, *Desmana moschata*
African wading rat, *Colomys goslingi*	Pyrenean desman, *Galemys pyrenaicus*
earless water rat, *Crossomys moncktoni*	marsh mongoose, *Atilax paludinosus*

Source: Withers et al. 2016.

(*Nilopegamys plumbeus*), but its extensive vibrissae likely play a critical role in detecting aquatic invertebrates. As noted earlier, the false water rat defies its relatively small size by taking crabs larger than itself, although it also consumes smaller invertebrates (Nowak 1999). Not much is known about the specific prey species of many of these mammals (e.g., *Nectomys, Neusticomys, Rheomys, Scapteromys, Dasymys*), let alone the degree of dependence on those species. This lack of knowledge makes it difficult to identify key prey species and foraging habitats used by poorly understood species. One strategy is to take a broad-based management

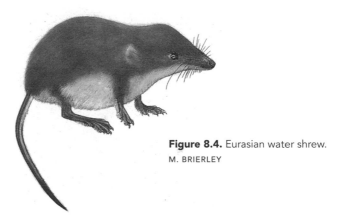

Figure 8.4. Eurasian water shrew.
M. BRIERLEY

approach to maintain their existing habitats and the ecological communities they support.

For some of these species, hunting behavior is driven by specific physiological or metabolic needs. Shrews are solitary foragers with high energy requirements, which are needed to support a very high metabolism. The Eurasian water shrew (*Neomys fodiens*; Fig. 8.4), weighing in at approximately 14 g, can consume 1.2 to 1.7 g of prey within a four-hour period, which means a shrew consumes almost 73% of its body weight every day (Rychlik and Jancewicz 2002). To do so, they must eat every few hours or they will starve. Some species, such as the Himalayan water shrew (*Chimarrogale himalayica*), consume only insects and spiders (He et al. 2010), while the Sumatran water shrew (*C. sumatrana*) also eats crustaceans and small fish.

Some shrews add to their success as predators by using venom. Three water shrews, the southern water shrew (*N. anomalus*), the Eurasian water shrew, and the Transcaucasian water shrew (*N. teres*), weaken their prey by secreting venom from their submaxillary gland (Kowalski et al. 2017). Eurasian water shrews obtain 50% to 95% of their prey in water (Vogel et al. 1998). As with the Old World shrews, they capture prey primarily in water (Fig. 8.5) but also hunt on land. The use of venom to subdue their prey may help these species gain sufficient food to meet their high metabolic demands. No shrews within the genus *Sorex* have venom. Captive American water shrews (*Sorex palustris*) eat almost every ten minutes and require food within three hours (Nagorsen 1996).

All semi-aquatic talpids are also insectivores/omnivores. The star-nosed mole uses the twenty-two tentacles of its Eimer's organ to detect its

Figure 8.5. American water shrew hunting prey in water. M. BRIERLEY

prey. Almost half its diet comprises annelid worms, with aquatic leaches dominating at ~80% (Hamilton 1931). Insects are also important prey items. Both the Russian desman (*Desmana moschata*) and Pyrenean desman (*Galemys pyrenaicus*) eat various aquatic organisms including fish, mollusks, amphibians, crustaceans, and insects (Nowak 1999). Similar to shrews, desmans need to eat at least 66% of their body weight each day. As a result, they have developed adaptations that maximize their hunting efficiency. The Pyrenean desman not only swims very quickly in swift-flowing streams, it also uses its sensitive prehensile nose and echolocation to navigate through water and find prey (Richard 1973).

In order Carnivora, the marsh mongoose (*Atilax paludinosus*) eats a variety of foods including insects, larvae, crabs, mollusks, reptiles, birds, frogs, and various types of fruit (Nowak 1999). It sometimes breaks open shells by throwing them onto hard surfaces. Despite being wetland dependent, it is the only mongoose without webbed feet. Lack of webbing allows greater dexterity for detecting aquatic prey as the marsh mongoose searches the mud and under stones of marshes, streambeds, and tidal estuaries (Nowak 1999).

PISCIVORE/SQUID-EATER

The fourth and final category of carnivorous semi-aquatic mammals are the piscivores/squid-eaters (Withers et al. 2016). Of the ten semi-aquatic species in this category, one is the aquatic genet (*Genetta pi-*

scivora), which is endemic to the Democratic Republic of the Congo, and nine are otters. As with other specialized carnivorous species, they combine a preferred diet of fish with other prey items. The aquatic genet touches the surface of the water with its long vibrissae to attract insect-eating fish, which it then attacks (Van Rompaey and Colyn 2013). It also combines its specialized slow gliding walk with sensory signals from its vibrissae to detect, and perhaps attract, prey. It catches its prey with rapid movements of its open mouth (Van Rompaey and Colyn 2013). As one of the rarest carnivores in Africa, much of what is known of the aquatic genet comes from personal accounts of local hunters. Aquatic genets have also been seen eating cassava tubers; however, their sharp teeth are specialized to eat meat.

Ecologists tend to group otters based on diet: those otters that specialize on fish and those that specialize on invertebrates, including crabs (Chanin 1985). A depth of research on these species highlights the various behavioral and morphological adaptations of semi-aquatic species, relative to their preferred prey. The piscivorous otters include river otters (*Lutra* and *Lontra*), the spotted-necked otter (*Hydrictis maculicollis*), the smooth-coated otter, and the giant otter. Despite their preference and specialization for catching and consuming fish, none are exclusive to one prey type.

There has been an extensive amount of research about the Eurasian otter's foraging ecology since the 1960s (Chanin 1985; Kruuk 2006). There is a broad size range of fish taken by Eurasian otters, from a Harris pike weighing at least 9 kg to fish smaller than 6 cm, including sticklebacks (Weir and Bannister 1973; Webb 1975). As with many other semi-aquatic mammals, Eurasian otters are highly dependent on their vibrissae to detect food underwater. They also exhibit location specific and seasonal food preferences. Fishing is almost always associated with dives, often repeated in habitat patches (Kruuk and Moorhouse 1991). Occasionally, they use "swim-fishing" (Kruuk 2006: 142), where they swim along the surface, dive, and then resurface a bit farther down from their original direction of travel. They depend on their sensitive vibrissae to detect prey, with eyesight being secondarily important. Along with fish, Eurasian otters eat crustaceans, clams, small mammals, invertebrates, amphibians, birds, eggs, and even vegetation. The adults eat smaller prey in the water and take larger fish to land.

Given its rarity, very little is known about the hair-nosed otter (*Lutra sumatrana*), one of the world's least-studied otters (Wright, Olsson, and Kanchanasaka 2008). They often forage during the day in small groups of two to five otters (Nguyen 2005). Over 80% of their diet consists of fish, with crabs and water snakes eaten to a lesser degree (Kanchanasaka 2007). They also supplement their diet with frogs, lizards, insects, birds, and small mammals (Wright, Olsson, and Kanchanasaka 2008). With human overhunting of other large predators (e.g., leopards and tigers), the hairy-nosed otter is likely one of the largest predators remaining in Vietnam and Cambodia (Wright, Olsson, and Kanchanasaka 2008).

The more crepuscular spotted-necked otters are social, but they are solitary when foraging (Kruuk and Goudswaard 1990). These African otters forage relatively close to shore (~2 to 10 m) and search for small fish (~10 cm long) between rocks and the branches of overhanging shrubs. They are very agile and pursue fish with some speed. The smooth-coated otter of southern Asia, another social species, frequently hunts in groups. As with many other otters, it is piscivorous, but it also consumes insects, snakes, frogs, and small mammals (e.g., rats). In coastal habitats, it also eats crabs (Haque and Vijayan 1995). It differs notably from the Eurasian otter because of its massive head and robust teeth (Hwang and Larivière 2005). Where their ranges overlap, smooth-coated otters eat larger fish than Eurasian otters and fewer amphibians (Kruuk et al. 1994).

New World otters, despite geographic separation, share many similarities with the piscivorous Old World otters. North America river otters (*Lontra canadensis*) detect prey, in part, with their vibrissae and catch it with their mouths. In an extensive study in west-central Idaho, USA, Wayne Melquist and Maurice Hornocker (1983) determined that fish preyed upon by these otters were quite diverse. Fish from seven different families occurred in 93% to 100% of the 1,902 scats analyzed in their study. Prey sizes ranged from 2 to 50 cm, with kokanee salmon (*Oncorhynchus nerka*), mountain whitefish (*Prosopium williamsoni*), and largescale suckers (*Catostomus macrocheilus*) being the top three prey items. North America river otters also eat invertebrates, mussels, waterfowl, and small mammals, including muskrats. Less commonly, they will eat beavers (Greer 1955). As with many other otters, they also hunt in coastal marine environments, but they need freshwater nearby. Although they are generally solitary foragers, North American river otters forage jointly at times. In northeastern Pennsylvania, Thomas Serfass (1995) watched four otters creating a disturbance at the center of a large pool of water; they then separated into pairs and began swimming in a zig-zag pattern to herd fish to a more restricted area where the fish were then readily caught.

All four otters of the Neotropical realm specialize on catching fish, although, similar to other otters, they will supplement their diets with other prey items. The marine otter (*Lontra felina*), the smallest otter in its genus, is almost exclusively found in coastal marine habitats. It hunts within 100 m of the coast (Larivière 1998) and only occasionally ventures into coastal freshwater environments. It consumes fish, but it is also an adept predator of crabs, shrimp, sea urchins, and invertebrates. Occasionally it will consume birds and small mammals (Ostfeld et al. 1989). Of the piscivorous otters, it most often offsets its diet with other prey.

The more extensively studied Neotropical otter (*Lontra longicaudis*) primarily eats fish, which was present in 97.2% of all scats that were examined during a key study in Brazil (Helder and Andrade 1997). As with other otters, it consumes crustaceans and mollusks, and it occasionally consumes insects, reptiles, birds, and small mammals. When foraging in coastal waters, ~67% of its diet is fish and ~28% is crabs (Alarcon and Simões-Lopes 2004). In contrast, the endangered southern river otter (*L. provocax*) includes crabs in its diet in an almost equal or even greater proportion to fish, depending on habitat types (Medina 1998; Medina-Vogel and Gonzalez-Lagos 2008). In a study of the southern river otter in Chile, crayfish were the dominant prey (85%), while fish (9%), eels (4%), and small crabs (3%) were also consumed. Amphibians and birds made up an additional 2% of its diet (Medina Vogel et al. 2013). It presents an excellent example of a carnivorous species that has a diet breadth that spans more than one predefined dietary niche.

Nicole Duplaix (1980) studied giant otters extensively during her PhD research in Suriname, South America. Unlike the southern river otter, giant otters have fish as a dominant prey item. Giant otters exhibit a broad range of hunting behaviors; they hunt alone, in pairs, and in family groups (Duplaix 1980). They are very active while hunting and combine a sequence of active swimming and diving as they pursue fish on the bottom and under floating vegetation. As Duplaix explains, "The prey is rushed from above or below, the otter swiveling or turning at the last instant to seize it in its jaws, usually close behind the head dorsally or ventrally" (Duplaix 1980: 544). Although they hunt in groups at times, each otter catches its own prey and consumes it almost immediately, except for larger prey, which is brought to shore. There is no food sharing or hoarding. An adult otter consumes up to 2.8 to 4.0 kg of fish per day (Duplaix 1980). Each meal is digested and passes through an otter on average in 3.13 hours (Carter et al. 1999). When fish are less available, giant otters will supplement their diets with crustaceans, snakes, and small caimans in some locations (Ribas et al. 2012).

Herbivorous Mammals

As discussed in Chapter 6, herbivores are morphologically and physiologically adapted for eating plants. Following the dietary niches in Withers et al. (2016), each of the herbivorous semi-aquatic mammals falls into one of five possible niche classifications, depending on the type or specific parts of plants consumed. They generally also combine some aspect of grazing, browsing, frugivory, granivory, and omnivory. Herbivores are primary consumers and eat mainly from one trophic level of **primary producers** (e.g., plants). Often associated with herbivorous diets, omnivores consume prey from more than one trophic level, and so they consume both plants and animals. As a group, herbivores lack or have reduced canine teeth; they have broad molars and extensive processing capabilities of their digestive tracts to aid the break-down and digestion of plant material. The male babirusas have maintained their canine teeth, but they have no function in eating and rather serve a defensive purpose (Fig. 8.6). Their functional teeth do, however, reflect a diet consisting of plant and animal matter. Similarly, water deer (*Hydropotes inermis*) and male water chevrotains (*Hyemoschus aquaticus*) have large canine teeth used for dominance and defense, which form into long, slightly curved tusks (Nowak 1999).

One physiological aspect of herbivore digestion, discussed in Chapter 4, is coprophagy, the eating of one's own feces to allow for additional digestion and nutrient gain from partially digested plant material. It is common in rodents and lagomorphs, both of which are **hindgut fermenters**. Coprophagy is commonplace in swamp rabbits (*Sylvilagus aquati-*

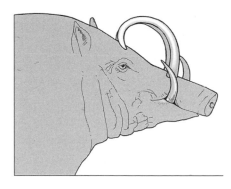

Figure 8.6. Tusks of a male babirusa.
M. BRIERLEY

cus), marsh rabbits (*S. palustris*), beavers, capybaras, and Tanzanian vlei rats (*Otomys lacustris*). In both hindgut and **foregut fermenters**, there is a mutualistic relationship between the herbivore and microorganisms within the herbivore's digestive organs. Given the number of semi-aquatic rodents, it is very likely that many others use coprophagy to maximize nutritional gains from otherwise difficult to digest plant material.

Just as with any predator-prey interactions, plant-herbivore interactions involve a fascinating diversity of plant defenses to deter herbivory (Gong and Zhang 2014). Herbivores consume more than 15% of plant biomass in temperate and tropical ecosystems annually (Gong and Zhang 2014), thus influencing the composition and structure of plant communities. Rather than being passive prey, plants have physical and/or chemical responses to hinder foraging by herbivores. Some plants have physical features (e.g., thorns or thick bark) to deter herbivores, while others use chemical resistance. Once attacked by an herbivore, some plants produce plant secondary metabolites (also called secondary compounds), of which there are more than fifty thousand recorded to date (Gong and Zhang 2014). Some of these chemicals include terpenoids and tannins, both of which can be toxic to mammals (Lindroth and Batzli 1984). To counter these chemicals, some mammals, including beavers, produce tannin-binding proteins that are specific to those produced by their preferred forage species (willow, *Salix* spp.), while more generalist herbivores produce tannin-binding proteins that bind to several different types of tannins (Hagerman and Robbins 1993). Once bound, these toxins are quickly moved through the digestive tract and are eliminated before they can cause any adverse effects. Production of secondary compounds by plants can be energetically costly and so often are not sustained over long periods of time. If herbivory is constant, targeted plant species will often decrease in abundance and be replaced by less desirable forage plants (Hood and Bayley 2009).

Alternatively, many plants have evolved tolerance mechanisms to allow the plant to compensate or even repair damage done by herbivores. One strategy is to increase leaf size to heighten photosynthetic productivity. Another is to encourage compensatory growth, which activates new stem growth in response to browsing, as seen in the stumps of poplar trees (*Populus* spp.) felled by beavers. The roots of various grass species can quickly produce new growth, which is notable in capybara grazing areas (Ojasti 1978), while other species of plants increase their seed production to send out more propagules to counter losses. Plant-herbivore interactions are often equated to an evolutionary arms race, whereby one defensive measure is countered by an evolutionary adaptive response by the herbivore. These relationships are complex, and for many herbivores, the selection for specific plant species involves a suite of genetic and physiological adaptations that result in foraging choices to produce the best energetic gains. The five different categories of herbivores capture that range of adaptation in semi-aquatic mammals.

HERBIVORE/BROWSER

Herbivore/browsers consume plant stems, twigs, buds, and leaves as dominant components of their diet (Withers et al. 2016). There are only two species of semi-aquatic mammals within this category, the North American beaver and the Eurasian beaver (*Castor fiber*). Given the highly visible effects of their foraging activities in the form of cut trees and well-worn pathways to their foraging area, their foraging behaviors have been extensively studied. In a study in Elk Island National Park in Alberta, Canada, my research team and I found a broad diet; our analysis of over 21,400 stems found thirty-seven different shrub and tree species cut by beavers at ten pond locations (Hood and Bayley 2008a). Both species of beavers are considered choosy generalists, consuming several different species that provide necessary micronutrients, yet focusing on a small number of preferred forage species (e.g., poplar and willow) (Jenkins 1975). They do not eat wood, but rather they forage on the bark, leaves, twigs, and buds of woody plants (Fig. 8.7). As described earlier, beavers

Figure 8.7. North American beaver chewing aspen tree.
J. DIXON, USED WITH PERMISSION

store woody stems in winter food caches immediately outside the main entrance of their lodge. The trees they fell can easily exceed 50 cm in diameter. To cut down a tree, they brace their upper incisors against the trunk and use a fore-and-aft movement of the lower incisors and lower jaw to cut large pieces of wood from the main stem (Kim et al. 2005). When available, beavers also eat herbaceous plants, including sedges, grasses, and submerged aquatic vegetation.

HERBIVORE/GRAZER

Herbivore/grazers primarily eat grasses and other soft-stemmed plants, but some species will augment their diet with woody browse at times. This category is by far the largest of the semi-aquatic herbivores and covers three taxonomic orders: Lagomorpha, Rodentia, and Artiodactyla (Table 8.2).

Marsh rabbits and swamp rabbits eat a variety of plant material, but they supplement their diet with bark and seedlings on occasion (Wilson and Ruff 1999; Helm and Chabreck 2006). Preferred foods of the swamp rabbit include the graminoid savannah panic grass (*Phanopyrun gymnocarpon*) and herbaceous dicots, such as false nettle (*Boehmeria cylindrica*). Marsh rabbits also include blackberries (*Rubus ursinus*), various rhizomes and tubers, and aquatic plants (e.g., cattails, *Typha latifolia*, and water hyacinths, *Eichhornia crassipes*) in their diet. Similarly, greater cane rats eat grasses and cane but also eat bark, fruits, and nuts (Nowak 1999).

Capybaras and coypus (*Myocastor coypus*) are somewhat similar in their dietary preferences and primarily consume grasses and aquatic vegetation, although bark, fruit, and roots are consumed occasionally. As an invasive species in many parts of the world, coypus can overgraze areas to the detriment of native herbivores (Carter and Leonard 2002). European water voles can also cause extensive root damage in agricultural systems. As with the southern water vole (*Arvicola sapidus*), the European water vole occasionally eats animal matter (insects, mollusks, water snails, and fish), although vegetation is a primary food source. North American water voles, however, appear to be strictly vegetarian (Ludwig 1984).

Very little is known of the marsh rats (*Holochilus* spp.), although the web-footed marsh rat (*Holochilus brasiliensis*) is known to eat tender parts of plants, and at high population levels it can cause extensive damage to cultivated crops (Nowak 1999). Even less is known about Lund's amphibious rat (*Lundomys molitor*), the Andean swamp rat (*Neotomys ebriosus*), Alfaro's rice water rat (*Sigmodontomys alfari*), and swamp rats (*Pelomys* spp.), although all are confirmed herbivores (Nowak 1999). Round-tailed muskrats (*Neofiber alleni*) are similarly vegetarian and have a diet of aquatic grasses, roots, stems, and seeds.

Muskrats are primarily grazers of riparian, emergent, and submerged aquatic vegetation, but unlike the round-tailed muskrat, they will consume animal matter as well. Their digestive system is not designed to process animal matter; however, a study by Kevin Campbell and Robert MacArthur in 1996 determined that there were notable nutritional gains when meat was added to their diet during feeding trials. Specifically, energy and protein absorption improved. Eating even 2% to 3.5% of animal tissue allowed muskrats to meet their daily nitrogen requirements, although most meat is likely obtained opportunistically (Campbell and MacArthur 1996). Southern bog lemmings (*Synaptomys cooperi*) also consume some animal matter (e.g., invertebrates), but they are primarily focused on herbaceous plants and grasses, similar to the northern bog lemming (*Synaptomys borealis*) (Banfield 1974).

All semi-aquatic ungulates are grazers, except for babirusas (*Baby-*

Table 8.2. Herbivores/grazers

Species	Species
rabbits, *Sylvilagus* spp.	Alfaro's rice water rat, *Sigmodontomys alfari*
greater cane rat, *Thryonomys swinderianus*	northern bog lemming, *Synaptomys borealis*
capybaras, *Hydrochoerus* spp.	swamp rats, *Pelomys* spp.
coypu, *Myocastor coypus*	pygmy hippopotamus, *Choeropsis liberiensis*
web-footed marsh rat, *Holochilus brasiliensis*	common hippopotamus, *Hippopotamus amphibius*
Lund's amphibious rat, *Lundomys molitor*	marsh deer, *Blastocerus dichotomus*
North American water vole, *Microtus richardsoni*	Père David's deer, *Elaphurus davidianus*
round-tailed muskrat, *Neofiber alleni*	water deer, *Hydropotes inermis*
Andean swamp rat, *Neotomys ebriosus*	lechwes, *Kobus* spp.
muskrat, *Ondatra zibethicus*	sitatunga, *Tragelaphus spekii*

Source: Withers et al. 2016.

rousa spp.) and the water chevrotain (*Hyemoschus aquaticus*), which are better suited to the frugivore/omnivore dietary niche. Both the pygmy hippopotamus (*Choeropsis liberiensis*) and common hippopotamus are pseudo-ruminants. As foregut fermenters, they have three-chambered stomachs consisting of a parietal blind sac, stomach, and glandular stomach, but unlike ruminants, they do not chew their cud (Clemens and Malioy 1982). Their digestive system is designed to optimize nutritional gain from low-energy foods, including grasses and leaves. Pygmy hippopotami are strictly vegetarian and eat herbaceous plants, grasses, aquatic plants, leaves, and fallen fruit (Nowak 1999). On average, both species of hippopotami retain food in their digestive tract longer than other ruminants and chew their food less due to their interlocking canines that limit lateral grinding motions when chewing food (Schwarm et al. 2006; Dudley et al. 2016). Hippopotami are known for their relatively low food intake (Clauss et al. 2007), but they consume larger particle sizes because they chew less. Larger particle size still allows them to gain adequate nutrients from their food, presumably due to the longer digestion time in these species.

Common hippopotami mainly eat small shoots, grasses, and reeds, which are held by their lips and then cropped with "a lateral swinging movement" of the head (Field 1970: 82). Their foraging areas are easily recognizable as "hippo lawns," because they graze grasses down to ground level. As discussed previously, their consumption of meat is occasional and opportunistic. Although a hippopotamus only eats 1% to 1.5% of its body weight during a five- to six-hour grazing bout (compared to 2.5% for most other hoofed mammals), they can survive many weeks without food (Nowak 1999).

All three species of wetland-dependent deer consume grass. For the marsh deer of South America, grasses comprise the largest percentage of forage (22%), while legumes were the second-most-common prey species (11%). In Brazil's Pantanal wetland, however, they consumed more than forty species of plants from seventeen families (Tomas and Salis 2000). Wetland plants make up a large part of their diet, although they will also eat new growth on shrubs. In Asia, Père David's deer (*Elaphurus davidianus*) and the water deer also forage primarily on grasses and aquatic plants (Nowak 1999). Depending on the location, water deer forage mainly on herbaceous dicots, although consumption of woody plants appears to increase in island populations (Guo and Zhang 2005; Kim, Lee, and Lee 2011). The rumen pillars provide physical separation of the internal structure of the rumen of many ungulates, but they are poorly developed in water deer. Therefore, there is reduced digestibility of carbohydrates from plant material (Jung et al. 2016). To compensate, water deer preferentially chose foods lower in fiber (e.g., herbs, forbs, and young grasses); consumption of woody stems is more likely associated with availability versus preference. The southern and Nile lechwe also target succulent grasses and water plants. The southern lechwe (*Kobus leche*), in particular, forages in water up to its shoulders to access aquatic plants (Nowak 1999).

Sitatungas spend most of their time in the tropical wetlands of sub-Saharan Africa, which is reflected in their dietary preferences of riparian and wetland plants. They mainly forage on herbs, grasses, and shrubs. A recent study in western Uganda identified thirty-four species of plants eaten by the sitatunga, with the aquatic flowering plant *Cyperus papyrus* comprising the bulk (22%) of its diet (Ndawula et al. 2011). Along with wild forage, they also consume a large amount of domestic crops during the wet season. For all herbivores, seasons dramatically influence forage availability and thus, diet selection.

HERBIVORE/FRUGIVORE

There are four species with diet preferences fitting the herbivore/frugivore dietary niche: the southern bog lemming, the Tanzanian vlei rat, Bunn's short-tailed bandicoot rat (*Nesokia bunnii*), and the Australian swamp rat (*Rattus lutreolus*). These species primarily eat leaves but also consume reproductive parts of plants (fruits, storage roots) and fungi.

The southern bog lemming, despite eating grasses and sedges like the closely related northern bog lemming, also consumes a good quantity of mosses, fruit, fungi, roots, and even bark. Comparable to many semi-aquatic rodents, they opportunistically eat invertebrates (Stegeman 1930). Their powerful jaws are ideal for gnawing (Nowak 1999).

Otomys rodents, including the Tanzanian vlei rats, have a very similar diet of roots, shoots, bark, seeds, and berries, which is sometimes supplemented with insects (Nowak 1999). There is very little else known about this small freshwater herbivore of southern Tanzania and northern Malawi. Similarly, very few studies have been conducted on Iraq's endangered Bunn's short-tailed bandicoot rat (*Nesokia bunnii*); however, its diet of grasses, grains, roots, and cultivated fruits and vegetables makes it an occasional crop pest where its habitat still exists (Al-Robaae 1977). Somewhat better known is the Australian swamp rat, for which basal stems and young rhizomes of monocots comprise the bulk of its diet, although it also eats herbs, fungi, and seeds (Watts and Braithwaite 1978; Cheal 1987).

FRUGIVORE/GRANIVORE

Delany's swamp mouse (*Delanymys brooksi*) is the only species that fits the frugivore/granivore dietary niche. This category includes species with a diet comprised primarily of seeds and nuts. Delany's swamp mouse feeds exclusively on seeds (Nowak 1999).

The frugivore/omnivore dietary niche includes mammals with a diet primarily consisting of reproductive parts of plants (e.g., fruit, roots), while also including a broad selection of seeds, small invertebrates, and small vertebrates. This category includes seven semi-aquatic mammals from four genera.

The European and southern water voles consume succulent vegetation but also eat roots, bulbs, tubers, fruit, seeds, and animal matter (Nowak 1999). European water voles have an especially broad diet that also includes water snails, freshwater mussels, and mollusks, while the southern water vole hunts insects, mollusks, and small fish (references in Maté and Barrull 2012). The southern water vole has even been photographed attacking a viperine snake (*Natrix maura*) on the Iberian Peninsula in Spain (Maté and Barrull 2012). Although these vipers often prey on southern water voles, in this case, a female southern water vole staged a prolonged attack (about forty-five minutes) on the snake, which had been feeding on a fish. The two researchers, Isabel Maté and Joan Barrull, suspected that the female water vole was protecting her young or, alternatively, the water vole fancied the snake as a meal to add a good source of protein to its diet.

Crab-eating raccoons tend to be a little less exotic in their dietary preferences. They commonly eat palm fruit (*Allagoptera arenaria*), and in a study in the Espírito Santo state of Brazil, palm fruit was found in 80% of their scat (Gatti et al. 2006). They eat a wide variety of fruit, arthropods, and small vertebrates. Amphibians and fishes are infrequently eaten. The reference to crabs in its name is likely associated with its diet in mangrove ecosystems, where crabs are abundant. The hairy babirusa (*Babyrousa babyrussa*) is a more specialized feeder and mainly eats leaves, fallen fruit, and fungi (Nowak 1999), although much of its dietary preferences have been identified from studies of zoo animals. Although once thought to be a single species, there are now three recognized species, each with its own food preferences, likely because of geographic isolation.

The Sulawesi babirusa (*B. celebensis*) is omnivorous with a diet of fruit, foliage, roots, worms, and carrion (Pudjihastuti, Pangemanan, and Kaunang 2009). They are able to eat the poisonous fruit and seeds of the keluak tree (*Pangium edule*) and the nonpoisonous sugar palm (*Arenga pinnata*). Both the Sulawesi babirusa and Togian Islands babirusa (*B. togeanensis*) visit volcanic salt licks and ingest the soil (see references in Leus et al. 2016 and Macdonald et al. 2016). Along with wild fruit, Togian babirusas eat a wide variety of cultivated fruit (e.g., cassava, mango, coconut, bananas), as well as foliage, roots, and animal matter (Macdonald et al. 2016). Similarly, the water chevrotain eats a variety of leaves, fruit, and buds of woody plants, but it also preys on insects, crustaceans, and small mammals (Dubost 1984). Much of what we know about the water chevrotain's diet comes from the research of Gérard Dubost's early work in Gabon.

Summary

While sufficient information is known about the diet and adaptations of predacious semi-aquatic mammals, information gaps remain, which has broader implications for conservation of these species. There are several species of semi-aquatic mammals for which no dietary information is readily available. The Ucayali water rat (*Amphinectomys savamis*) is a cricetid rodent only known from two localities. Given the diet breadth of other cricetids (e.g., muskrats and web-footed marsh rats), it would be difficult to assign it to a specific niche. Similarly, very little is known of three other South American cricetid rodents: the Chaco marsh rat (*Holochilus chacarius*), Lagiglia's marsh rat (*H. lagigliai*), and the marsh rat (*H. sciureus*). One could assume they have similar diets as the web-footed marsh rat, but even that species is poorly understood. Also, as seen with some semi-aquatic genera, closely related species can have different foraging strategies. Similarly, the western water rat (*Hydromys hussoni*) is known only from a single location in the Wissel Lakes region of Papua Province, Indonesia. Its close relative the rakali, however, has a relatively wide distribution that reflects its diverse diet. Foraging studies provide a window into the most important aspect of a mammal's life—what it eats so it can survive to mate and propagate its species. Lack of information about prey selection makes effective conservation strategies difficult, if not impossible. To adequately manage a species, it is necessary to adequately manage its prey, while being simultaneously aware of its predators.

9

The Prey: Predator-Prey Interactions

Food is the burning question in animal society, and the whole structure and activities of the community are dependent upon questions of food-supply.

Charles Elton, 1927, *Animal Ecology*

As winter was slipping into spring, I looked out across the frozen pond next to my house and saw a muskrat on the ice adjacent to an abandoned beaver lodge it was using as a den. It had just chewed its way through the ice and was eating a handful of semi-aquatic vegetation it had brought to the surface. For a mere second I turned my back, and when I looked again the shape and extent of the muskrat had changed. Through the spotting scope, I could see the muskrat was still there, but a great-horned owl was now standing on top of it among a pool of blood forming on the ice. The owl soon flew off with its catch. A few minutes later, I scanned the scene again, and to my surprise, a second muskrat was sitting in almost the same spot as the first. Sometimes, a long winter under the ice overrides vigilance, especially when nutritional gains outweigh the risk of predation. Just as predation of other organisms by semi-aquatic mammals reflects a balance of ecological costs and benefits, so do prey have to balance their energetic needs with costs to mitigate predation risk. This chapter examines the adaptations of semi-aquatic mammals as prey, from macro to micro scales.

Nature of Predation

Predation is not a random event. There is often a vulnerability about in-dividuals that succumb to predation, thus making them easier to detect, catch, and kill. Animals can be vulnerable at any age. Although young, sick, and older individuals are more often at risk due to physical or be-havioral limitations, juveniles and adults displaced through intraspecific social interactions are also exposed to predation risk. The dynamic in-terplay between habitat quality and availability, prey survival, and ulti-mately prey population adds further complexity to predation risk and the success of predation avoidance strategies.

Individual Level Effects

Individual condition of prey plays a key role in predation risk. In a classic study by Paul Errington in 1946, muskrats (*Ondatra zibethicus*) in well-established breeding territories tended to avoid predation by American mink (*Neovison vison*), while individuals without a territory, weakened by drought, or otherwise injured, were preyed upon regularly. Even if not killed by predators, those individuals were unlikely to survive to reproduce. Although healthy individuals are killed, more often it is the sick, young, homeless, and older individuals that are more susceptible to predation. The larger the animal, the lower the predation risk, but even the largest of the semi-aquatic mammals experience predation risk at times of vulnerability.

The early part of an animal's life is when it is most vulnerable. Young common hippopotami are predated on by lions (*Panthera leo*), hyenas (Hyaenidae), and crocodiles (Crocodylidae); however, adults have al-most no natural predators, other than humans (Hayward and Kerley 2005). A 1936 study of marsh rabbits (*Sylvilagus palustris*) in Florida reported that young marsh rabbits were consumed much more often by water moccasins (*Agkistrodon piscivorus*) and diamondback rattlesnakes (*Crotalus adamanteus*) than adults were (Blair 1936).

Size is not the only vulnerability for young mammals; age-related behavior, such as dispersal, is also a factor. Life expectancy of both North American and Eurasian beavers can exceed fifteen years in the wild, but dispersing 1.5-to-2.5-year-old North American beavers (*Castor canaden-*

sis) have higher mortality rates than do 4.5-to-10.5-year-old adults (Payne 1984). Eurasian beavers (*C. fiber*) that delay dispersal beyond 3.5 years of age benefit more than those that dispersed sooner (Mayer, Zedrosser, and Rosell 2017a, 2017b). Being homeless equates to a dangerous lack of familiarity with possible escape routes, and for beavers, the lack of lodge or bank den prevents a dispersing juvenile from physically removing itself from potential attack. Given that dispersal can be one of the most dangerous times in a young animal's life (Lucas, Waser, and Creel 1994), waiting to increase one's competitive abilities and life experience might serve to increase survival (Mayer, Zedrosser, and Rosell 2017a). Social learning for long-lived colonial species plays a key role in development of these defensive skills, particularly for species that rely on aquatic habitat to provide escape terrain and cover.

However, for older adult animals, there are other vulnerabilities created due to intraspecific competition or changing environmental conditions. As an example, displacement caused by competition for mates can force an adult to find new territory, an overland journey involving increased predation risk. Dr. Frank Rosell and his students at the University of Southwest Norway in Bø have been tracking the same population of Eurasian beavers for over twenty years. A recent study determined that, despite their **social monogamy**, beavers sometimes switched mates; this most often occurred when a younger male entered an occupied territory and, through overpowering the resident male, caused a "forced divorce," after which the defeated adult left his territory for a more vulnerable, and likely lethal, fate (Mayer et al. 2017). For semi-aquatic mammals, droughts or unexpected drainage of their habitats can also result in overland dispersal, which then dramatically increases predation risk, regardless of age.

Population Level Effects

In addition to risks created by a weakening in social status and security, an animal's lapse in social vigilance can also result in increased vulnerability to predation. In some situations, age and social factors may combine with environmental conditions to influence predation risk of prey species at the population level. As an example, a study on Brazilian capybara highlights the importance of these interacting factors, which can combine with devastating effect on the prey population.

In 1977, world renowned wildlife biologist George Schaller and Brazilian biologist J. M. C. Vasconcelos investigated how jaguars killed capybara (*Hydrochoerus hydrochaeris*) in the state of Mato Grosso, Brazil (Schaller and Vasconcelos 1978). They determined that capybara in the Pantanal wetlands were a top prey item for jaguars, and they described a typical attack as follows.

> A capybara sat about two meters from the water's edge while a female jaguar approached at a walk, screened by a bush. When 15 m from her prey, she broke into a trot. Suddenly aware of its danger the capybara bolted for deep water. However, the jaguar grabbed it within 4 m, in 15 cm of water, and killed it with deep bites in the throat and the back of the head. After moving the carcass ashore, she apparently left her kill to walk back and forth along the beach. She then straddled the capybara's body with her forelegs, and, picking it up in her jaws, dragged it in typical cat fashion across the beach and into the forest for a total distance of 110 m. (Schaller and Vasconcelos 1978: 297)

Although the capybara is the world's largest rodent at over 25 kg (Fig. 9.1), it cannot overcome a killing bite through its braincase by a 90 kg jaguar. Adults as well as young can be at risk of predation, particularly if vigilance lapses. Unlike typical trends in age-related predation, in this study adult capybaras comprised the majority of kills; only five of the seventy-seven animals killed by jaguars were young. Some of this difference can be explained by biology, as there were simply fewer young within the population. Of the thirty to thirty-five surviving capybara observed during the study, only one female had any offspring with her (just one baby), which does not reflect a typical capybara group. In this case, other factors affected both population structure and the vulnerability of the adults to jaguar predation in this area.

In 1974, a major flood reduced the grassy margins along the wetlands where capybaras foraged. Overcrowding on the remaining patches of grass led to a major outbreak of equine trypanosomiasis (*Trypanosoma* spp.), which killed a large percentage of the population. Even three years later, when Schaller and Vasconcelos did their study, evidence of disease still lingered, with over 11% of the remaining animals still showing signs

Figure 9.1. Capybara in the San Diego Zoo. G. A. HOOD

of illness (e.g., sores, emaciation). A combination of disease and direct mortality by jaguars reduced this population by a third in two months. Consequences of disease were cumulative. There was a likely loss of vigilance and vigor for escape, and there were likely lower reproduction rates because of illness. For this capybara population, it became the prefect storm. A third confounding factor was ongoing habitat loss following the flood, which resulted in less nutritionally rich food sources. Reduced group cohesion could have influenced social behaviors as well. Where predation might be otherwise absorbed by **recruitment**, in a struggling population, reduced group cohesion may well push the population beyond recovery.

Compensatory Reactions

In healthy populations, impacts of predation are often limited by compensatory reactions. In species with high reproductive rates, such as the European water vole (*Arvicola amphibius*), a lower proportion of young survive to breeding age than in species that might only have one young per year (e.g., water chevrotain, *Hyemoschus aquaticus*) or just one calf every other year (e.g., common hippopotamus, *Hippopotamus amphibius*). In the case of the water vole, net recruitment is reduced naturally in crowded populations, and any predation after the population has reached **carrying capacity** is absorbed by the water vole's high reproductive rate, thus often speeding up population recovery (Fig. 9.2). It is when prey densities are low that the inability to compensate for high

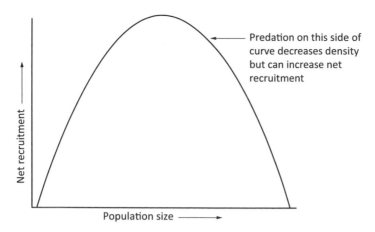

Figure 9.2. Net recruitment curve for a hypothetical wildlife population. G. A. HOOD

levels of predation might bring a population or species to the verge of extinction, as seen with the Brazilian example of capybaras and jaguars.

Predictable patterns in predator and prey populations are commonly reported. As shown previously (see Fig. 8.1), there appears to be a cyclic relationship between American mink and muskrats in the Peace Athabasca Delta (PAD) of northern Canada. It would be easy to assume that the mink are driving muskrat population declines; however, prey populations are more often determined by food availability and other habitat factors. As muskrat populations grow, shoreline vegetation is consumed faster, and eventually intense foraging increases interspecific competition and lowers food availability. Water levels are also critical to muskrats in northern latitudes. The decline of muskrats in the PAD has been associated with dramatic decreases in spring flooding since the construction of the Bennett Dam in neighboring British Columbia (Straka et al. 2018). There has been a subsequent decrease in winter water depths and the quality and quantity of herbaceous vegetation. Rather than mink causing reductions in muskrat populations, mink population dynamics appear to follow that of their prey. Interactions between predators and prey are often multifaceted, especially in the context of habitat loss and fluctuations in other competing wildlife populations.

The southern water vole (*Arvicola sapidus*) of Portugal, Spain, and France has as many as thirty-five predator species, including both native and introduced species (Mate et al. 2015). As with all semi-aquatic mammals, water voles are susceptible to aerial, terrestrial, and aquatic predators; some of these predators target juvenile water voles (mouse sized), and others target adult voles (~20 cm body length). Habitat loss, fragmentation, and modification are central to the southern water vole's population declines. Subpopulations of southern water voles naturally live in patchy distributions linked by habitat corridors that allow for gene flow among populations. However, if too many subpopulations are lost to predation, the remaining populations will become completely isolated, particularly if these habitat corridors are destroyed. Even if a predator does not specialize on the voles, if the predator's preferred prey declines, the southern water vole could then become a dominant prey species through a process called **prey-switching**.

Such is the case for southern water voles when the European wildcat's (*Felis silvestris*) preferred prey of European rabbits (*Oryctolagus cuniculus*) are lacking (see Mate et al. 2015). The same situation exists when river levels drop and fish, the preferred prey of Eurasian otters (*Lutra lutra*), become scarce in areas encompassing southern water vole habitat. Adding to the complexity is the introduction of non-native American mink into these same areas (Melero et al. 2012). When otters hunt fish instead of voles, there is a density-dependent compensatory increase in predation on voles by American mink to avoid competition. The presence of a novel predator is even more concerning because antipredator behavior of southern water voles has not evolved for avoiding American mink (Macdonald and Harrington 2003; Mate et al. 2015).

Similar concerns hold true for marsh rabbits and round-tailed muskrats (*Neofiber alleni*) in areas with red imported fire ants (*Solenopsis invicta*), which were introduced into the United States from South America in the early 1930s (Allen, Demarais, and Lutz 1994). Red imported fire ant infestations covered over 121 million hectares by 1999, including marsh rabbit and round-tailed muskrat habitats in the southern United States (Wojcik et al. 2001). Despite their size, red imported fire ants are a fierce predator capable of killing newborn rabbits (Ashdown 1969) and round-tailed muskrats (https://georgiawildlife.com/species). Round-tailed muskrats construct floating feeding platforms and dome-shaped houses of grasses and other vegetation, which can attract invading fire ants. Young mammals are especially susceptible to their stings until they acquire enough fur to provide some level of protection (Allen, Dema-rais, and Lutz 1994). Increased development of roads and other human structures in the Florida Keys, an area with several endangered endemic species, increases the probability of the presence of fire ants, thus high-lighting the cumulative effects of human activities on the susceptibility of prey to introduced predators (Forys, Allen, and Wojcik 2002). De-spite direct effects of predation (i.e., death) by red imported fire ants, indirect effects (e.g., reduced survival, reduced weight gain, behavioral changes, changes in foraging patterns, reduced food availability) can also have dramatic impacts on wildlife populations (Forys, Allen, and Wojcik 2002). Many semi-aquatic mammal species are killed by introduced predators, as indicated in Table 9.1. Other introduced predators exist (including Burmese pythons, *Python bivittatus*, released into wetland ecosystems), and there are surely other semi-aquatic mammals preyed upon by predators listed in Table 9.1, but they are not readily recorded as such in the literature. Of note, however, is the global effect introduced species have on vulnerable prey populations.

The outcomes of predation are varied for prey populations. In some cases, there is a somewhat predictable cycle. When prey abundance is high, predation pressure increases, thus reducing prey densities to a point where predation pressure relaxes. Generally, predation rate is low at low prey densities. However, predator-prey dynamics are complex, and there are often synergies that are hard to predict, especially when introduced species are in the mix. Even seemingly positive programs can impact endangered prey species, for example, recovery programs that introduce endangered predator species into the same area as endangered prey species. However, there is more than one type of predation that can affect semi-aquatic mammals and, as seen with the capybaras of the Pantanal, one form of predation can influence another.

Other Forms of Predation

Instantaneous killing of prey is just one form of predation. Parasitism and parasitoidism are special forms of predation that pose a threat to their prey. Unlike other predators, **parasites** and **parasitoids** do not kill their prey immediately, and sometimes they do not kill their prey at all. Instead this is a more intimate form of predation, with attacks involving

Table 9.1. Non-native predators of semi-aquatic mammals

Species	Cats	Dogs	Rats	Red fox	Mink	Invertebrates
platypus, *Ornithorhynchus anatinus*	X	X		X		
little water opossum, *Lutreolina crassicaudata*	X	X				
swamp rabbit, *Sylvilagus aquaticus*		X				
marsh rabbit, *Sylvilagus palustris*	X					fire ants
Cabrera's hutia, *Mesocapromys angelcabrerai*			BR			
European water vole, *Arvicola amphibius*			NR		X	
southern water vole, *Arvicola sapidus*			NR		X	
marsh rat, *Holochilus sciureus*		X				
round-tailed muskrat, *Neofiber alleni*						fire ants
Musso's fish-eating rat, *Neusticomys mussoi*				X		
southern bog lemming, *Synaptomys cooperi*	X	X		X		
Australian swamp rat, *Rattus lutreolus*	X			X		
false water rat, *Xeromys myoides*	X			X		
star-nosed mole, *Condylura cristata*	X	X				
Russian desman, *Desmana moschata*					X	
European mink, *Mustela lutreola*					X	
marine otter, sea cat, *Lontra felina*	X	X	X			
Neotropical otter, *Lontra longicaudis*	X	X	X			
southern river otter, *Lontra provocax*		X			?	
giant otter, *Pteronura brasiliensis*		X				

Source: Data from the IUCN Red List (2019).

Note: Other non-native predators exist but are not well documented (e.g., Burmese python in Florida). Mink = American mink, BR = black rat (*Rattus rattus*), NR = Norway rat (*R. norvegicus*).

one or very few prey individuals during the course of the predator's life cycle. The effect on prey can be death, but it is more often disease or overall weakness, affecting some or a large proportion of the prey population in an affected area.

Parasites cannot survive without their hosts and often live symbiotically with their hosts for longer periods of time. Generally, parasites only consume parts of their hosts, and they can be either internal or external parasites. Parasitoids, on the other hand, are insects that lay eggs on a host; these eggs then develop into parasitizing larvae that begin to consume parts of the host as they develop.

It would be rare to find any naturally existing organism lacking parasites, given the variety and adaptability of parasitic organisms. To cover all parasites and parasitoids affecting the semi-aquatic mammals covered in this book would be an impossible task; however, a survey of some of the common groupings provides insight into the ecological realities of these parasitic organisms and their effects on the semi-aquatic species that form their prey.

There are two general groups of parasites: (1) **microparasites** (e.g., viruses, bacteria, fungi, and protozoans), and (2) **macroparasites** (e.g., flatworms, roundworms, ticks, fleas, lice, flies, and mites) (Anderson and May 1979). Microparasites are small and have high rates of reproduction directly within the host. Infection by some microparasites (e.g., viruses) can induce immune responses within the host to guard against reinfection, if the host does not die from the initial infection. Macroparasites take longer to reproduce, and reproduction within the host is rare. Their growth within the host often results in the release of eggs, which then infect an intermediate host, only to start the infection cycle anew. Macroparasites include flatworms/tapeworms (platyhelminths), roundworms (nematodes), and various arthropods (fleas, flies, lice, mites, and ticks). Generation times of macroparasites are longer than most microparasites, and their infection of their hosts can be persistent in nature (Anderson and May 1979). Hosts are commonly reinfected (Anderson and May 1979).

Parasites are transmitted either through direct contact with the parasite or through a **vector** (intermediate host species). A classic vector-borne disease of emerging concern for human health is Lyme disease, which is caused by the bacteria *Borrelia burgdorferi* and is mainly transferred via a bite of ticks of genus *Ixodes*. In 2012, a Romanian study was the first to detect Lyme disease in two European mink (*Mustela lutreola*) (Gherman et al. 2012). It has also been detected in marsh rice rats (*Oryzomys palustris*) in North Carolina (Ryan et al. 2000) and in water deer (*Hydropotes inermis*) in Korea, where it is transmitted by the *Haemaphysalis* tick (VanBik et al. 2017). Some diseases, often transmitted through direct contact (e.g., consumption), are caused by prions (proteinaceous infectious particles). Prions lack a nucleic acid genome and instead are reconfigured proteins that then influence other proteins to change abnormally as well (Dalsgaard 2002). As such, they are not considered to be a microparasite, but they will be included with the microparasites here because of their size and emerging concern of infections across multiple taxa.

Another means of categorizing parasites is their location on their prey. Ectoparasites are found on the outer body of hosts, as seen with ticks, fleas, mites, and lice. They generally stay on the host for only part of their lifecycle, for example to take a blood meal, and then to drop off to complete the rest of their lifecycle elsewhere. Their bites can inadvertently transmit diseases to the host in the process. **Endoparasites** live within the body of the host, sometimes at the cellular level. Regardless of size and location, microparasites and macroparasites often play a key role in the transfer of **zoonotic disease** (transmission from other mammals to humans). Given their connection to both terrestrial and aquatic habitats, semi-aquatic mammals can carry a broad suite of parasites, many of which are only just being identified.

While conducting research for this book, I regularly used an academic search engine called Web of Science (https://clarivate.com /webofsciencegroup/solutions/web-of-science); it helped me identify all research papers on semi-aquatic mammals over the past twelve decades or more. This search engine identified over 580 papers published in the past twenty years alone on a variety of diseases affecting almost sixty species of freshwater-dependent mammals. Given the nature of scientific literature, this list is not exhaustive, but it does provide important insights into disease ecology and the microparasites and macroparasites that target mammalian hosts. Almost a third of the studies were equally focused on Eurasian otters, American mink, and capybaras (~30% combined); American mink and capybaras are often raised in captivity for fur farming, so outbreaks of disease can be costly. The research within these published studies provides some excellent examples of many common diseases and how they influence the overall health of semi-aquatic mammals.

Viruses contain genetic material but lack the cellular structure found in bacteria, fungi, and protists. From a conservation perspective, one of the most notable viruses affecting European mink and wild American mink is the Aleutian mink disease virus (AMDV). It is also of concern on mink farms around the world, and recently it has resulted in a reoccurrence of outbreaks on mink farms in Denmark, despite implementation of eradication measures in 1976 (Christensen et al. 2011). It has a high prevalence in many places in Europe and North America, with ~56% of free-living mink testing positive for AMDV DNA in Sweden (Persson et al. 2015). In Ontario, Canada, wild mink living in close proximity to mink farms have an increased risk of infection because of spillover events (Nituch et al. 2011). AMDV is caused by the *Carnivore amdoparvovirus 1* virus and can result in spontaneous abortions, respiratory distress in young mink, and potential illness and death at all ages (Bowman et al. 2014).

Mink, along with all mustelids, are also susceptible to canine distemper virus (CDV), an often deadly RNA virus that attacks several body systems, including the respiratory and central nervous system. The first documentation of CDV in captive Asian small-clawed otters (*Aonyx cinereus*) was in 2005 (Bosschere et al. 2005). There is increasing concern in areas where farmed mink escape and interact with otters (e.g., the North American river otter and the southern river otter, *Lontra provocax*). Invasive mink can act as a bridge host by transferring the virus from domestic dogs to wild otters (Sepúlveda et al. 2014), which could have disastrous consequences for threatened otter populations.

Another deadly RNA virus of concern in mammals is the rabies virus (*Rabies lyssavirus*). Although not well documented in semi-aquatic mammals, it has been found in North American river otters (Serfass et al. 1995) and North American beavers (Morgan et al. 2015). It can present differently in different mammals, but respiratory arrest is usually the final stage before death.

Other viruses, such as hantavirus in rodents, are of more concern for humans than for the rodents themselves, which likely makes hantavirus of great interest in the study of disease in some semi-aquatic rodents and shrews (e.g., *Arvicola amphibious, Holochilus chacarius, H. sciureus, Myocastor coypus, Neomys fodiens, Ondatra zibethicus, Oryzomys palustris, O. couesi*). The idea of parasites as predators, however, is best represented by viruses such as the deadly cowpox (*Orthopoxvirus*) in Eurasian beavers and European water voles (Hentschke et al. 1999; Essbauer et al. 2009), for which the animals have not developed relative immunity. Microparasitic infections can often present as multiple infections within the same individual, likely because of an already depressed immune system.

In a study of disease in wild water vole populations in the United Kingdom, 32% of forty-nine voles tested for a complete suite of pathogens were co-infected with more than one parasite, while many of the remaining seventy-nine voles that were not tested for all diseases also hosted multiple parasites (Gelling et al. 2012). The most common parasites ranged from bacteria (e.g., *Escherichia coli, Bartonella* spp., and *Leptospira* spp.) and protozoans (*Giardia* spp. and *Cryptosporidium* spp.), with a number of internal macroparasites (e.g., *Strongyle*, nematode species, and *Trichuris*) and external macroparasites (e.g., fleas) represented as well.

Other microparasites of concern in semi-aquatic mammals include the fungus *Mucor amphibiorum*, which is deadly in platypuses, given the disease's relatively new arrival in Tasmania and lack of resistance to it in native populations (Connolly 2009). There is also increasing awareness of prion diseases, including transmissible mink encephalopathy (TME) in farmed mink (Hadlow and Karstad 1968) and the potentially susceptibility of chronic wasting disease (CWD) in South American marsh deer (Falcão et al. 2017). For prion diseases, confirmation can only be achieved once an animal is dead, thus making surveillance programs somewhat more difficult, especially since there are no known vectors for these diseases that might provide visible clues to the disease's presence.

MACROPARASITES

Many of the arthropod macroparasites (i.e., ticks, fleas, lice, and mites) are closely associated with microparasites found in semi-aquatic mammals. Although ticks are strictly ectoparasites, the diseases they carry quickly enter the blood stream where they carry out all or parts of their lifecycle inside their hosts. Tick-borne diseases are especially prevalent and include the bacterial disease rickettsiosis in the Argentine swamp rat (*Scapteromys tumidus*; Venzal et al. 2008), Eurasian otter (Santoro et al. 2017), and capybara (Fortes et al. 2011). In South America, *Amblyomma* ticks are the main vector of this and similar diseases. The most deadly of the spotted fevers (*Rickettsia rickettsia*) is Brazilian spotted fever (Polo et al. 2017). The tick *Amblyomma cajennense* is the main vector and capybaras are its primary host. Increased proximity of humans to capybaras and their habitats is creating heightened concern of zoonotic infection. In North America, the marsh rice rat is host to the lone star tick (*Amblyomma americanum*), which carries numerous serious diseases, including ehrlichiosis, tularemia, and southern tick-associated rash illness. As the tick bites, it injects alpha-gal into the blood stream, which can then result in the human host developing a sometimes-fatal allergy to red meat.

Unlike ticks, some mites feed directly on the surface of an animal and cause damage to skin and fur, but they rarely carry secondary diseases. Instead, they can cause skin conditions such as scabies. However, blood-sucking mites, as with ticks, can also transmit a variety of diseases carried by viruses, bacteria, protozoans, and fungi. Lice are another ectoparasite found on semi-aquatic mammals and, as with ticks and mites, can be vectors of disease. In some cases, one animal can harbor numerous species of mites. In Poland, six Eurasian beavers had twenty species of mites from the genus *Schizocarpus*, eight of which were described as new species (Bochkov et al. 2012). Another study assessed reproductive health and physiology of muskrats relative to ectoparasitic mites and determined that muskrats appeared unaffected by mite infestations (Prendergast and Jensen 2012).

Studies on other parasitoids in many semi-aquatic mammals are rare, yet certainly studies do exist. For example, in marsh rats (*Holochilus sciureus*) of northern South America, warble flies (*Cuterebra apicalis*) live mainly on the ventral surface of older individuals (Twigg 1965). This particular fly is common to rodents, on which the fly creates a **furuncle** just under the skin of the rodent. Fly larvae develop in the furuncle before migrating from their host as adult flies. Ungulates are also regularly associated with parasitoids, often flies.

Flatworms (platyhelminths) and roundworms (nematodes) are the most common macroparasites living inside the body, although some mites can as well. There are two classes of entirely parasitic worms in Phylum Platyhelminthes: Class Trematoda and Class Cestoidea (Feldhamer et al. 2015). For many of these species, adult mammals are the **primary host**, and the adult parasites produce eggs that are usually released through the feces. The larvae then enter at least one **intermediate host** prior to being consumed again by the primary host. This life cycle allows for repeated reinfections over the host's lifetime. Trematodes include liver flukes (*Fasciola hepatica* and *F. gigantica*). Cestodes include tapeworms, known to reach substantial lengths (e.g., 25 m) (Feldhamer et al. 2015). There are numerous species of flatworms that use semi-aquatic mammals as their definitive (primary) hosts.

The diversity of parasite species associated with semi-aquatic mammals, and the complex interactions involved in their life cycle, can be seen in relation to the common hippopotamus. The trematode *Oculotrema hippopotami* is found on the nictitating membrane and under the eyelid of the hippopotamus and causes irritation (Stunkard 1924). Liver flukes (*Fasciola nyansae*) live in the livers of young hippopotami, although hippopotami become increasingly immune to the parasite with age (Eltringham 2002). A female hippopotamus in Kruger National Park was found with atrophied ovaries that might have been caused by blood flukes (Smuts and Whyte 1981), suggesting potential exceptions in immunity. Blood flukes (schistosomes) also parasitize hippopotami while living in the veins that direct blood from the gastrointestinal tract to the liver and while living in pelvic veins. Unlike liver flukes and *Buxifrons buxifrons*, free-swimming larvae of blood flukes re-enter the hippopotamus by penetrating its skin (Eltringham 2002). The parasitic trematode flatworm (*Buxifrons buxifrons*) is common in the fore-stomach and the first couple of meters of the small intestine of hippopotami (Guilbride et al. 1962). These parasites use the hippopotamus as a definitive host for the adult parasite, which then releases its eggs in the host's feces. Once back in the water, eggs hatch and the free-swimming larvae enter an intermediate host (e.g., snail), where they then mature and develop into another larval stage, which ultimately is released and attaches to vegetation. Vegetation is then ingested by the hippopotamus for the cycle to begin anew. For many of these macroparasites, repeated infection relies on the predictable behavior of the host and re-exposure to the larvae.

Coypus can also carry diverse parasite loads. In a 2012 study, Pablo Martino and his colleagues examined wild coypus from South America. They found their sample population hosted nineteen species of parasites: 82% of coypus were infected with nematodes, 46.1% with protozoans, ~33.3% with trematodes, and 12.8% with cestodes. Age and sex did not influence the presence of parasites except for the preference of young coypus (< one year old) by nematodes. They also detected the protozoans *Cryptosporidium* spp. and *Giardia* spp. (Martino et al. 2012).

It is difficult for hosts of parasites to avoid parasitism, yet some behaviors (e.g., grooming) do help reduce parasite loads. In the case of Père David's deer (*Elaphurus davidianus*), the loss of adaptive parasite-defense behaviors resulted in dramatic population declines due to parasites, thus contributing to the deer becoming extinct in the wild. Remaining individuals were raised in captivity. A 2014 study by Zhongqiu Li, Guy Beauchamp, and Michael Mooring determined that the loss of adaptive parasite-defense behaviors was a result of the genetic bottleneck experienced because of its meager population numbers and many decades in captivity. Fortunately, the researchers determined that some antiparasitic behaviors had been retained in the deer; this bodes well for their ability to withstand ticks following their recent release into the wild in a park south of the Yangtze River in China.

Although avoiding parasites and other predators can be difficult, semi-aquatic mammals have developed various behavioral strategies for avoiding and surviving predation if caught.

Predator Detection and Avoidance

Antipredator strategies and behaviors are essential for species to survive. Semi-aquatic mammals have the extra challenge of maintaining vigilance on land and in water, which can make avoiding predators all the more difficult. As discussed in Chapter 4, nocturnal activity might offer some advantages in avoiding predators, as long as prey species have efficient means of detecting predators in the dark. Avoiding habitats occupied by predators might also help an individual evade predation. In addition to these types of behavioral adaptations, there are morphological adaptations for avoiding detection. If an animal is caught, other strategies can help it escape a predator and live to see another day.

Detection

Several species of semi-aquatic mammals use some form of camouflage to blend into the environment. The fur of the web-footed tenrec (*Limnogale mergulus*) of Madagascar is brownish mixed with reddish and blackish hairs on the upper parts of the body, while fur on the ventral surface is a pale yellowish-gray (Nowak 1999). This combination of colors and tones provides the web-footed tenrec with camouflage from predators, while reducing detection by its prey. Similarly, the blackish-brown fur on the back and dull brownish-gray fur on the ventral side of Kemp's grass mouse (*Deltamys kempi*) of South America provides camouflage from its primary predator, the barn owl (*Tyto alba*), which hunts for these mice from above (González and Pardiñas 2002). Cryptic coloration is common in many species regardless of size. It is evident in the 40 g montane fish-eating rat (*Neusticomys monticolus*) from the Andean streams of South America and the 2 kg flat-headed cat (*Prionailurus planiceps*) of Southeast Asia. One is insectivorous, the other carnivorous, but both have adapted to blend into their environment in their roles as both predators and prey.

In the southeastern United States, the swamp rabbit (*Sylvilagus aquaticus*), with its rusty-brown to black back and white throat and belly, also uses its cryptic coloration and "freezing" behavior to avoid being detected by predators (Chapman and Feldhamer 1981). This is also a common behavior of the marsh rabbit (Blair 1936). As with many mammals, swamp rabbits use temporal avoidance to evade predators; they are most active in the evening and delay their activity until twilight as days grow longer in the spring (Holler and Marsden 1970). However, as spring and summer progress, they will risk predation by being more active prior to sunset. This behavior reflects the trade-off between energy intake from foraging (forage availability increases in late spring and summer) and predation risk. Another actively deceptive behavior used by swamp rabbits to confuse predators "is to climb onto a log or tree top, walk its length, then backtrack and leap to the side, moving away at right angles to the previous course" (Terrel 1972: 292). When in imminent risk, a swamp rabbit will enter the water and swim to a safe location, where it will then lie underwater with only its eyes and nose above the surface (Hunt 1959).

Predator avoidance tactics can include the use of aquatic habitat as cover, and therefore taking advantage of physiological adaptations for a semi-aquatic lifestyle. Similar to the swamp rabbit, the capybara also hides in the water with just its nostrils and eyes exposed (Nowak 1999). It can remain completely submerged for up to five minutes. When threatened, the Nimba otter shrew (*Micropotamogale lamottei*) can stay submerged for even longer periods of time (up to fifteen minutes, with an average ten minutes), which they accomplish by lowering their metabolic rate (Vogel 1983). Larger semi-aquatic mammals will also use aquatic habitats for cover. In central Africa, the sitatunga (*Tragelaphus spekii*) retreats to water when threatened and hides underwater with only its nostrils exposed (Nowak 1999). Such deceptive behaviors allow these mammals to take advantage of a semi-aquatic lifestyle. Despite being exposed to terrestrial, aerial, and aquatic predators, semi-aquatic mammals have water to provide protection from many of these threats. The use of burrows and lodges by several of these species provides additional escape from most forms of predation.

Chemical

A more active means of avoiding predation is to detect predators before being detected. In most semi-aquatic mammals, olfaction is highly developed. It is common to see beavers sniffing the air between foraging bouts, a form of chemical vigilance to locate and identify potential predators. To study the behavioral response of Eurasian beavers to predator odors, Frank Rosell and Andrzej Czech (2000) obtained samples of odors from river otters, red foxes (*Vulpes vulpes*), lynx (*Lynx lynx*), wolves (*Canis lupus*), and brown bears (*Ursus arctos*) (documented predators of beavers) from three Polish zoos. They also obtained domestic dog and human scents to use in the study. They then applied each of the odors to aspen stems and placed the sticks at ten beaver lodges for a series of summer and autumn trials. Sticks treated with European river otter scent were consistently avoided in summer and were avoided marginally less in autumn. Interestingly, all sticks treated with predator odors were more often avoided than those with human and dog odors. Odors from river otters, red foxes, lynx, wolves, and brown bears had the strongest effects. River otter odor remained the most avoided in the fall, although lynx, human, and red fox odors also were avoided more often at that time of year. In North America, Engelhart and Müller-Schwarze (1995) found that beavers most often avoided coyote (*Canis latrans*), Canada lynx (*Lynx canadensis*), and North American river otter (*Lontra canadensis*) odors.

Olfactory recognition has also been noted in the Australian swamp rat (*Rattus lutreolus*). Interestingly, swamp rats recognize odors derived from the integumentary system of the spotted-tail quoll (*Dasyurus maculatus*), a carnivorous marsupial native to Tasmania, but not to equally dangerous introduced predators. In a controlled experiment, the swamp rat did not respond to either cat or fox scent (both introduced predators), yet it showed an avoidance response to quoll scent (McEvoy, Sinn, and Wapstra 2008). A combination of species-specific ecology, life experience, and evolutionary history with a particular predator species influences the ability to detect this risk. In the case of the swamp rat, it appears that it might be better at recognizing coevolved predators (e.g., quoll) than novel ones, which raises concerns about the impact of introduced predators on native species (Hayes, Nahrung, and Wilson 2006).

Despite the majority of semi-aquatic mammals having reduced pinnae, some have excellent hearing, including the otters (Chanin 1985). During her PhD research on giant otters (*Pteronura brasiliensis*) in Suriname, Nicole Duplaix (1980) observed a pair of otters sleeping on the shore of a river. The otters heard a canoe that was being paddled upstream long before it came into view and could be detected by Duplaix. In response, the otters dove into the water. Their reduced pinnae, however, might impact their ability to locate the direction of the sound, even if the sound itself is easily detected (Duplaix 1980). In contrast, the marsh rabbit, with its much longer pinnae, has a keen sense of hearing and responds immediately to sound and location (Blair 1936). Despite variable abilities to hear underwater, as discussed in Chapter 5, most species rely on hearing for predator detection.

Another cue to the presence of predators comes from warning sounds produced by the prey themselves. The most iconic example of alarm sounds within semi-aquatic mammals is the warning slap of a beaver's tail on the water. Numerous times, I have experienced the loud response of a beaver's alarm as my canoe apparently drifted too close for comfort. In response, other members of the family group usually make a quiet dive and appear farther away when they resurface. Not only does the slap make me focus on the beaver at hand, it also causes an immediate distraction from my previous thoughts. Warning and distraction are both effective antipredator strategies. The equally gregarious capybara uses a distinct alarm bark, which it makes repeatedly when a predator is spotted (Yáber and Herrera 1994). Juveniles do not emit alarm barks, instead subadult and adult capybaras use them when there is a potentially dangerous situation (Barros et al. 2011). The individual making the alarm bark assumes an alert posture, elevates its head and ears, and raises its fur. Other capybaras tend to freeze in place as an alarm response. This bark not only serves to alert conspecifics—it also informs the predator of its detection.

Behavior

Social vigilance is an important behavioral defense adaptation of species living in groups. Although a solitary lifestyle might make an animal less conspicuous, it takes a lot of energy to be the only individual watching for predators. With group living, there is a higher chance that at least one individual will detect a predator and then alert the group. A more cynical view is that another member of the group might become prey instead. With social vigilance, not all group members need to be fully alert at all times. Instead they can eat, sleep, or tend to their young while another group member scans for intruders. As mentioned above, capybaras use alarm barks to alert the rest of the group when a predator is near. By living in stable groups of up to forty individuals, capybaras find greater protection from predation (Herrera and Macdonald 1989).

Group living is less common in semi-aquatic mustelids (other than a female with cubs), although smooth-coated otters (*Lutrogale perspicillata*), from the Middle East to Vietnam, are known to live in groups of up to nine or more individuals, with an average of 4.5 (Kruuk 2006). They can show vigilant behavior within groups, but because of the rarity of predators for smooth-coated otters in Malaysia and Indonesia, group living is more likely related to exploiting food (van Helvoort et al. 1996). Giant otters also live in large family groups of up to twenty individuals, although groups of two to seven otters are more common (Duplaix 1990). Rather than using vigilance to avoid detection from predators, these noisy, visibly active otters are more inclined to have a young male initiate an attack on an approaching predator (e.g., a caiman or jaguar).

Surviving Predation

Despite all the evolutionary adaptations to avoid predation, predation does occur. There are various means to survive predation once detected or caught. An unusual approach is that of the little water opossum (*Lutreolina crassicaudata*) of South America. It is immune to the venom of several snakes, including the pit viper jararaca (*Bothrops jararaca*), a trait that is not uncommon in didelphids (Jansa and Voss 2011). It is so immune to viper venom that it actually consumes the snakes, rather than the opposite, but more expected, result. Other species, rather than killing and consuming their potential predators, use various strategies to survive.

Intimidation

For some species, intimidation is effective at warding off predators. When threatened inside their nest boxes during a captive experiment, round-tailed muskrats rushed at intruders with front feet clenched to their chests while chattering their teeth (Birkenholz 1972). When I approached a muskrat on the front lawn of my home, it rose up on its hind feet, brought its forefeet to its chest, and starting making intimidating noises. It stood its ground and, once alone again, resumed eating. Intimidation works. Other species also use intimidation to deter predation. American mink are fearless when approached by a predator, even those larger than themselves.

Flight

Most semi-aquatic mammals flee to water once they detect a predator. Several examples were discussed earlier, including the swamp rabbit and capybara. Avoidance tactics are sometimes combined with retreat to water for more complete hiding cover. The water chevrotain, as with the swamp rabbit, immediately freezes, but if attack is imminent, it quickly retreats to the water where it dives and keeps most of its body below the

Figure 9.3. Pygmy hippopotamus in the San Diego Zoo. G. A. HOOD

surface in order to hide (Nowak 1999). Coypus, despite being clumsy on land, also run to the nearest water body, dive in, and stay submerged for several minutes (Feldhamer, Thompson, and Chapman 2003). Although it is reported to run toward vegetation rather than the water, the pygmy hippopotamus (Fig. 9.3) does indeed seek water when threatened (Nowak 1999). Smaller species, including the American water shrew, also dive and swim to avoid predators (Wilson and Ruff 1999).

Species highly adapted for aquatic locomotion are subject to aquatic predation at least as often as terrestrial predation. Many species have adapted additional behaviors to avoid predation in water. Otters are so well adapted to swimming that they use their agility to escape predation. The African clawless otter (*Aonyx capensis*) is so effortless in its aquatic movements, it almost looks serpentine while swimming. It is also very effective at avoiding predators in water. Its biggest predation risks are on land, where it is occasionally consumed by fish eagles (*Haliaeetus vocifer*) and Nile crocodiles (*Crocodylus niloticus*) (Larivière 2001).

Fighting Back

Along with social vigilance, several species exhibit aggression against a common predator. If captured, fighting back is another option. Common hippopotami have few predators and are not at all intimidated by other species. They readily kill humans, large vertebrates, and other hippopotami on land and in water (Dudley et al. 2016). The smooth-coated otter, as with most mustelids, can also be surprisingly aggressive. In 1992, Vangla Nagulu (1992: 41), of the IUCN Otter Specialist Group, reported that a fisherman died after an encounter with smooth-coated otters. A cub was caught in a fishing net and was brought to shore, all the while emitting continuous shrill cries. A group of otters successfully rescued the cub by biting and scratching the fisherman. The man later succumbed to injuries.

Less intimidating, but still impressive, is the swamp rabbit's behavior when under restraint. Charles Lowe (1958) described swamp rabbits he had caught in Georgia that were "striking out viciously with their hind feet even when the eyes [were] covered" (Lowe 1958: 120). He then described a method of handling that results in the rabbit being stretched between two hands, where it then remained quiet. Although these descriptions involve human-wildlife responses, these responses evolved from interactions with native predators, and it is not surprising that similar defensive behaviors are used against other predators.

Regardless of the source of predation, animals must either avoid detection or defend themselves so they can survive and produce reproductively viable offspring (**fitness**) to sustain future populations. A combination of physical and behavioral traits increases their chances. The most appropriate response should also result in the least energetically costly and most effective outcome. The diversity of predation defense mechanisms seen in semi-aquatic mammals is the result of many generations of evolutionary adaptation to both predators and the environmental context in which the semi-aquatic mammal lives. Yet, for many of these species today, humans remain the most dangerous predator, in part because more specific defense mechanisms have yet to be adequately refined to counter direct and indirect human impacts, as described in Part V, "Conservation Challenges and Management Approaches."

PART IV
REPRODUCTION

Mating and Offspring

Mating is not key to fitness, but the genetic contribution to a reproductive effort is.

Virginia Hayssen and Teri Orr, 2017, *Reproduction in Mammals: The Female Perspective*

Reproductive Biology

In 1927, a former vaudeville comedian and one of Australia's leading naturalists and experts on the platypus, Harry Burrell, published *The Platypus: Its Discovery, Zoological Position, Form and Characteristics, Habits, Life History, etc.* It remains a treasure-trove of detailed illustrations, narratives, research, and insights into the odd and wonderful history of this unique mammal. Burrell's dry wit complements his ability to document the chronology of endorsements and refutations of some of the world's leading biologists at the time as to whether the platypus was even a mammal at the broadest level, and more specifically **oviparous** or **ovoviviparous**. There were even some highly respected zoologists and anatomists, including Henri de Blainville and Georges Cuvier, who initially thought the platypus was **viviparous**. It is worth the read to see open scientific debate at its finest. Burrell is quick to note that indigenous Australians had known for millennia that the platypus laid eggs (oviparity), although that distraction did not quell the debate from the experts in Europe for several decades. The debate began with the first published description of the platypus in 1799, by Dr. George Shaw of the British Museum, and it quelled when there was broad agreement in 1884 that the platypus was an egg-laying mammal, after Cambridge zoologist W. H. Caldwell went

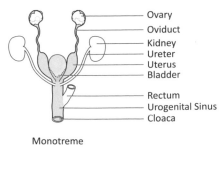

Monotreme

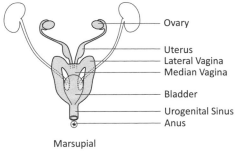

Marsupial

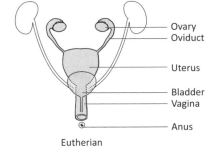

Figure 10.1. Three types of female reproductive organs: monotreme, marsupial, eutherian. M. BRIERLEY

Eutherian

to Australia and shot a platypus that had one egg in the nest and another egg still descending from its oviduct (Caldwell 1887).

It is now accepted that mammals fall into one of three groups relative to how offspring are produced following internal conception: monotremes, marsupials, and eutherians (Fig. 10.1). As discussed above, monotremes are oviparous and lay eggs from which tiny, altricial offspring are born. Marsupials, such as the water opossum (*Chironectes minimus*; Fig. 10.2), bear live but prematurely developed young that then continue development within the mother's pouch. The largest group of mammals, the eutherians, have longer gestation periods in which the

Figure 10.2. Water opossum.
M. BRIERLEY

embryo develops fully within the mother and is born at a much more advanced stage than either monotremes or marsupials.

To call eutherians "placentals" is a bit disingenuous to marsupial mammals (metatherians). Both groups have placentas, with the marsupial placenta being a simple structure within the uterus that provides contact between the yolk sac of the fetus and the mother's blood capillaries. Gestation times in marsupials are generally short (≤ two weeks), and once the young exit the birth canal, they follow a scent trail to the teats inside their mother's pouch (Kemp 2017). Inside the pouch, they latch on to a teat and the mother begins to lactate. The eutherian placenta allows the embryo to quickly embed in the wall of the uterus via numerous thin processes that develop between the embryo and uterine wall (Kemp 2017). These connections increase surface area for diffusion of oxygen and nutrients, and they increase the growth and development of the fetus. Since gestation requires less energy than lactation, the eutherian fetus can develop over a longer period of time within the uterus at a somewhat lower energetic cost for the mother. Therefore, some eutherian offspring are born almost fully developed and become relatively independent within a short time after birth. However, given the evolutionary success of the platypus, water opossum, and the many semi-aquatic eutherians, each form of reproduction has its advantages and disadvantages.

For the platypus, egg laying is an ancestral trait (Wourms and Callard 1992). The earliest mammals laid eggs, but now the only remaining egg-laying mammals are the platypus and the echidnas. The platypus has two separate uteri that, combined with the bladder and ureter, ultimately empty into a common urogenital sinus that then connects to a cloaca

(Pasitschniak-Arts and Marinelli 1998). Despite having two ovaries, the platypus only has a functional left ovary for egg development. Following conception, the zygote moves through the oviduct where secretions create a solid but expandable shell around the zygote (Hayssen and Orr 2017). Platypuses usually produce two eggs (Burrell 1927), which remain inside the mother for two to three weeks (Grützner et al. 2006). Eggs are laid in a nesting burrow and are assumed to be incubated for approximately ten days (Holland and Jackson 2002). The mother is believed to incubate the eggs close to her belly, although the eggs can also rest on nesting material (Burrell 1927; Pasitschniak-Arts and Marinelli 1998). Plate 18(2) in Burrell's book shows a brooding female platypus with a slight indentation in her abdomen that Burrell suspected replaced the pouch used by echidnas for incubating eggs. Given the difficulty in breeding and keeping platypus in captivity, little is known about this aspect of the reproductive cycle of the platypus.

Once the eggs hatch, the tiny embryos (9 mm to 17 mm) are able to grasp with their forelimbs, but their hind limbs resemble paddles. They use their forelimbs to pull themselves up their mother's fur to nurse on milk that exudes onto her skin from her mammary glands. Young platypuses are weaned at about four months of age (Holland and Jackson 2002). Remarkably, the mother does not leave the burrow during incubation and the first part of lactation and fasts for approximately forty days (Hayssen and Orr 2017).

Reproduction in marsupials is noted for its short gestation period and neonates that are distinguished by the "olfactory organ, the mouth, and the clawed forelimbs" (Kemp 2017: 27). Marsupials also have two-phased lactation within the pouch: a teat-attachment phase and a phase where the young suckle intermittently (Hayssen and Orr 2017). With the exception of immature marsupials, the teats of marsupials are hairless and very similar to the teats of eutherians (Koyama et al. 2013). Initially, milk is more dilute and contains vitamins, minerals, and antibodies to aid disease-resistance. It is commonly reported that later the milk becomes richer and higher in protein to increase growth rate in the young, although recent research has adjusted the timing of this process. In their 2013 study, Sachiko Koyama and her colleagues noted that "what is remarkable about marsupial nipples and mammary glands is that they can simultaneously produce milk of different nutritional composition

Figure 10.3. Mother common hippopotamus with young. M. BRIERLEY

termed asynchronous concurrent lactation" (2013: 122). Unusual for marsupials, the pouch of the water opossum opens backward to prevent water from entering it when the female swims while the young are inside. The *m. pars pudenda* sphincter muscle helps keep the pouch tightly closed, while fatty secretions help to keep the seal watertight and the neonates safe from drowning. The young can tolerate low oxygen levels and stop breathing for several minutes without issue while the mother is underwater (Marshall 1978a).

Semi-aquatic eutherians have diverse reproductive patterns over and above their longer gestation periods, more complex placenta, and generally shorter lactation periods (Hayssen and Orr 2017). The common hippopotamus (*Hippopotamus amphibius*; Fig. 10.3) would be an exception, with a lactation period of approximately one year (Nowak 1999). Conversely, young marsh rice rats (*Oryzomys palustris*) of the eastern and southern United States are weaned after approximately fourteen days. Gestation periods range from an estimated seventeen days for the Bunn's short-tailed bandicoot rat (*Nesokia bunnii*) in the marshes of southeastern Iraq to 324 days for the common hippopotamus. Size can be one determinant of parental care, but there remains a broad range in levels of parental care and dependence in eutherians. Level of care can relate to litter size, sociality, and even mating systems.

Mating Systems

There are five main mating systems reported for mammals: **monogamy** (exclusive pair), **polygyny** (male with several exclusive females), **polyandry** (female with several exclusive males), **polygynandry** (both males and females with several mates with social ties), and **promiscuity** (both males and females with several mates without social ties). At its most basic, monogamy represents the mating of a single female and single male, sometimes for just the mating season, sometimes for life. The pair live together and both care for the offspring. Within monogamy, there is **social monogamy**, where the pair cohabits for a specified period of time, copulates, and cooperates to raise offspring, yet extra-pair copulation is known to occur occasionally (Lukas and Clutton-Brock 2013). Although 90% of birds display social monogamy (Cockburn 2006), a recent review of 2,545 mammals by zoologists Dieter Lukas and Tim Clutton-Brock (2013) at the University of Cambridge confirmed that social monogamy is relatively rare in mammals (~9%). When social monogamy does occur in mammals, mate-guarding is the primary strategy to maintain the pair bond. This strategy is thought to have evolved when females lived a solitary existence in discrete home ranges and males had to guard against other male competitors. As such, the males were unable to access other females (Lambert, Sabol, and Solomon 2018). A previous study from 1977 determined that only 3% of mammals were socially monogamous (Kleiman 1977). Despite the small difference in percentages in the more recent study, monogamy remains uncommon in mammals. Another form of monogamy is **genetic monogamy**, in which offspring are genetically linked to only the exclusive female and male pair. In reality, complete genetic monogamy is extremely rare, especially in mammals (Lambert, Sabol, and Solomon 2018). Even within the same species, individuals might be socially monogamous in some situations and genetically monogamous in others (Klug 2018). As American biologist Hope Klug states, "monogamy is an evolutionary conundrum" (Klug 2018: 1) because increased genetic contributions through multiple matings are associated with increased fitness. However, Klug's research and other recent genetic research reveals that monogamy in mammals can be complex, as seen in both species of beavers.

Eurasian beavers (*Castor fiber*) are considered socially monogamous, with some evidence of low levels of extra-pair paternity (5.4% of 166 young and 7% of 100 litters), with the extra-pair father generally being the dominant male from the closest neighboring territory (Nimje 2018). Prior to Priynak Nimje's study in southwestern Norway, which obtained genetic samples from the same population of beavers for over twenty-one years, Eurasian beavers were considered to be genetically monogamous (Syrůčková et al. 2015).

One of the only studies conducted on North American beavers (*C. canadensis*) revealed extra-pair paternity in 56% of the litters tested, although samples size and study duration (two years) were significantly lower than the Norwegian study (Crawford et al. 2008). It would be interesting to determine whether latitude influences genetic monogamy. Southern populations are more mobile during warmer and shorter winters than beavers farther north. Beavers generally mate from January to February in most areas, which is when they would be unable to leave their ponds and mix with other beavers after freeze-up in the late fall and early winter. The beaver pond next to my property in Alberta, Canada, is usually frozen by mid to late October and does not open up until late April or early May. During that time, beavers have no access to other beaver colonies.

Another semi-aquatic mammal reported to exhibit social and genetic monogamy is the muskrat (*Ondatra zibethicus*) (Lambert, Sabol, and Solomon 2018). In a study from the University of Saskatchewan in western Canada, DNA fingerprinting of muskrats over two breeding seasons revealed no evidence of extra-pair paternity (Marinelli, Messier, and Plante 1997). With both species of beavers and the muskrat, fathers also provide care to the young (paternal care). Beaver pairs mate for life, or until the male is push out by a younger male (Mayer, Zedrosser, and Rosell 2017a). Female muskrats provide much of the care until the young are weaned (Errington 1963). Once the young leave the nest, the father is more active in their care, for example, carrying food, accompanying young to emergent vegetation, and providing additional vigilance near the burrow (Marinelli and Messier 1995). Nevertheless, there is plasticity in mating strategies, with other populations of muskrats following a polygynous mating system rather than monogamy (Caley 1987). Also, at the population or species level, a common mating system might be the norm, yet individual-level behaviors might differ from what is generally seen as standard behavior (Klug 2018).

Social monogamy has also been reported for several species of otters. Giant otters (*Pteronura brasiliensis*) often live in highly cooperative groups composed of a dominant reproductive pair and offspring of multiple litters (Duplaix 1980; Ribas et al. 2016). In reality, many family structures occur that differ from the mated alpha pair and their offspring, so often considered the norm. In research conducted in the Pantanal wetlands of Brazil by Carolina Ribas and her colleagues (2016), there were groups where the older dominant female (approximately eight to nine years of age) was replaced by a younger female subordinate of the same group. In another situation, a fight between two groups over territory resulted in the inclusion of a female and male insubordinate in the winning group (Ribas et al. 2012). Relatedness within giant otter groups is still very high but, as with muskrats, other ecological drivers can influence variability in pair-bonding and genetic monogamy.

Other monogamous otters include the smooth-coated otter (*Lutrogale perspicillata*) and Asian small-clawed otter (*Aonyx cinereus*) (Timmins 1971; Hussain 1996). As with most otters, they mate in the water but give birth on land. Smooth-coated otters form small family groups that include a mated pair and up to four offspring from previous seasons (Hussain 1996). In his classic study of smooth-coated otters in central India, Syed Hussain found smooth-coated otters to be more social than closely related species (*Lutra*). During his research, group size varied from one to nine animals, with the largest group sizes in the monsoon season (Hussain 1996). The much smaller Asian small-clawed otter forms groups of up to twelve to thirteen individuals (Timmins 1971). The Asian small-clawed otter is thought to be socially monogamous (see references in Larivière 2003), but there is still much to learn about its behavior in the wild (Kruuk 2006). Two additional species are reported to have short-lived monogamy for the duration of the mating season, the African clawless otter (*A. capensis*) (Dalerum 2007) and the Congo clawless otter (*A. congicus*) (Larivière 2001). The marine otter (*Lontra felina*), often thought to be a solitary species, has been reported as both monogamous and polygynous (Ostfeld et al. 1989).

Cooperative breeding, where young receive care from nonparental group members, can be seen in giant otters and beavers. Cooperative breeding is commonly associated with species that are socially monogamous over a sustained period, and extra-pair paternity is fairly low (Lukas and Clutton-Brock 2012). This behavior can help increase fitness benefits for the young as well as subordinates. Although Dieter Lukas and Tim Clutton-Brock did not regard giant otters as cooperative breeders in their 2012 study, they did acknowledge that both Eurasian and

North American beavers were cooperative breeders, given the multigenerational nature of the family group and shared roles in support of rearing of kits and contributions to the group as a whole. The 2016 study of giant otters by Carolina Ribas and her colleagues, however, documented shared territorial patrols, group care of otter cubs, paternal care, coordinated defensive behavior, and even instruction of young on how to fish. They determined that "kinship ties are associated with social cohesion and cooperation among giant otters" (Ribas et al. 2016: 68). Some discrepancies among studies are likely due to the metrics involved in fitting definitions to behaviors, while other differences might be because of the context of available studies and behavioral plasticity within species.

Monogamy has also been reported in smaller mammals, especially some species of water voles (Nowak 1999; Lukas and Clutton-Brock 2013); however, this behavior appears to be situational. Southern water voles (*Arvicola sapidus*) are **facultatively monogamous** in small habitat patches (Pita, Mira, and Beja 2010), but they are otherwise promiscuous (Román 2007). Flexibility in mating systems has its advantages. A monogamous system is advantageous when bi-parental care aids in reproductive success and mate availability is low or highly dispersed (Klug 2018); polygynous systems often have higher male reproductive success because of the male's direct association with a number of females within his territory (Mate et al. 2013). Fossorial forms of European water voles (*A. amphibious*) occur as a family group composed of an adult pair and two generations of young (Meylan 1977), although a Norwegian study concluded that the social structure of semi-aquatic European voles was promiscuous (Frafjord 2016), similar to southern water voles. Although reported as monogamous in some reviews (Lukas and Clutton-Brock 2012: SI Table 2), the North American water vole (*Microtus richardsoni*) does not appear to be monogamous, as evidenced by longer reproductive periods in males than females and the limitation of females to two litters per year (Ludwig 1984, 1988). Biologist Daniel Ludwig, who conducted extensive research on North American water voles in the Canadian Rockies, described the North American water vole as a polygamous species (Ludwig 1984).

The star-nosed mole (*Condylura cristata*), sometimes reported as solitary (Lukas and Clutton-Brock 2013), is more often noted as living in colonies consisting of several moles (Hamilton 1931; Banfield 1974).

William John Hamilton Jr. (1931) described star-nosed moles as "gregarious" and suggested that the male and female overwinter together, which Hamilton considered an "extended courtship." The male leaves the female at the start of the spring mating season (Hamilton 1931; Banfield 1974); however, given that mating season extends from January to June, one can assume that the pair mate prior to the male leaving the female. One of the first signs of breeding season is the dramatic enlargement of the male's testes and prostate gland to a mass approximately 10% of the male's body weight (Eadie and Hamilton 1956). The tails of both sexes enlarge during the winter and especially during mating season. Mixed groups of nonreproductive star-nosed moles are often reported (Petersen and Yates 1980), yet they do show some variability in their social structure.

Sociality

Group Living and Associated Breeding Strategies

Several semi-aquatic or freshwater-dependent species live in groups, as seen with the giant otter; however, group living outside of family groups is not common in many species of semi-aquatic mammals. Spotted-necked otters (*Hydrictis maculicollis*) form into mixed groups and then disperse into separate male and female groups of eight to twenty individuals after mating season (Nowak 1999; Kruuk 2006). More often, group living involves a group structure of several breeding females that share a common area where they carry out their daily activities (Lukas and Clutton-Brock 2013). These social structures typically result in polygynous or polygynandrous mating systems. During their extensive review, Dieter Lukas and Tim Clutton-Brock identified several freshwater mammals that live in nonmonogamous groups (Table 10.1).

Group living can aid reproductive success, as seen in capybaras (Herrera and Macdonald 1989). Capybaras live in stable home ranges (~5 to 16 ha) over a number of years, and the group defends these home ranges against nongroup members. These areas include a variety of habitat types, ranging from shrubs to ponds, which in turn provide a variety of resources as seasons change (Herrera and Macdonald 1989). Unexpectedly, this semi-aquatic mammal requires bushy banco (higher lying

Table 10.1. Mammals living in nonmonogamous groups

Species	Group size	Mating system	Reference
greater cane rat, *Thryonomys swinderianus*	1 male, 2 to 3 females + offspring	polygynous	Cox, Marinier, and Alexander 1988
capybara, *Hydrochoerus hydrochaeris*	5 to 100	polygynous to promiscuous	Herrera 2013
lesser capybara, *Hydrochoerus isthmius*	solitary or in groups	polygynous to promiscuous	Mones and Ojasti 1986
coypu, *Myocastor coypus*	~11, mainly females + subadults, some males	polygynous	Guichón et al. 2003
hairy babirusa, *Babyrousa babyrussa*	1 to >15, mainly maternal groups	polygynous	Clayton and Macdonald 1999
Sulawesi babirusa, *Babyrousa celebensis*	1 to >15, mainly maternal groups	polygynous	Clayton and Macdonald 1999
Togian Islands babirusa, *Babyrousa togeanensis*	1 to >15, mainly maternal groups	polygynous	Clayton and Macdonald 1999
common hippopotamus, *Hippopotamus amphibius*	~10 to 15 (up to 150)	polygynous to polygynandrous	Laws 1984; Pluháček and Steck 2015
southern lechwe, *Kobus leche*	5 to 70 females	polygynandrous	Nowak 1999

Note: As indicated by Lukas and Clutton-Brock (2013), excluding the sitatunga. Updated references for mating system and group size included.

land) for foraging and access to dry ground during the wet season. Bushy banco appears to be critical for the survival of newborns, and groups that try to rear young without this habitat component are generally unsuccessful (Herrera and Macdonald 1989).

As mentioned in previous chapters, group living also provides capybaras with defense against predators, especially for the young. Once a predator is detected, the adults gather the young into a group and surround them, facing out toward the oncoming threat (Macdonald 1981). Synchronous reproduction over a two-week period allows for communal nursing and allows the young to form cohesive groups (Herrera 2013). Swamp rabbits are also synchronous breeders, with **estrus** lasting approximately one hour (Sorensen, Rogers, and Baskett 1968).

Although not synchronized, several species of mammals have a seasonality to their breeding that corresponds to resource availability (e.g., forage). The southern lechwe (*Kobus leche*) exhibits lek-breeding, a very rare breeding behavior for ungulates in particular, and mammals in general (Nefdt and Thirgood 1997). It is better known in various species of birds, especially grouse. Lekking is a territorial behavior exhibited by males during the breeding season. The lek itself is a defined area, often used over multiple years, that a few breeding male southern lechwes use solely for mating (Schuster 1976). Obtaining the best lekking territory is critical for males because it influences females' choice of mates. Although the southern lechwe can mate throughout the year, there is a distinct increase in mating frequency from November to February, with a peak during the winter floods of December and January when herds are migrating from the low-lying areas (Schuster 1976).

In central Zambia's Lochinvar and Blue Lagoon National Parks, Richard Shuster (1976) observed from fifty to one hundred adult male southern lechwes in a 500 m diameter area over a number of days. The males spaced themselves approximately 15 m apart. Shuster described the lekking behavior as follows.

> Attachment was shown by the presence of the same males on the same patches of ground over successive hours or days and by spatial limitations when interacting with other lechwe. Chases of one male by another often ended when the pursuer drew up sharply and displayed, with head held high, feet prancing, tail wagging, and head shaking. The same was also seen when a female entered a male's territory. He typically ran up to and around the female, trying to prevent her escape without physical contact. If she left the territory, the male's chase ended abruptly in a display, thus allowing the next male to take over, and so on in relay until the female either chose to remain within a male's territory or left the lek. (Schuster 1976: 1241)

The presence of the other males in leks, and the associated defensive behavior among the males, concentrates much of the breeding into a relatively narrow timeframe, despite additional breeding throughout the year. As with all behaviors in mammals, there are energetic trade-offs in males defending leks and females visiting the leks. Lekking male southern lechwes had higher mating rates than males without leks. Rory Nefdt and Simon Thirgood (1997) determined that when mating was attempted within a herd environment, rather than on leks, males were disrupted by other males eight times more often than in lekking situations. Leks provide males with higher mating rates, and females in estrus can choose the "best" males on display, sometimes described as the "good-genes hypothesis" (Hamilton and Zuk 1982). Ultimately, all mating behavior, regardless of mating system, should favor successful production of viable offspring. The majority of mammals are neither monogamous nor highly social, yet they breed successfully through strategies of their own.

Although not included as a group-living mammal because it has only lived in captive herds until very recently, Père David's deer provides another breeding strategy that is common in various cervids. Males generally keep to themselves for much of the year, except just prior to and shortly after mating season. During the rut, they join groups of females and fight with other males to have exclusive mating rights to what is referred to as a harem (Nowak 1999). As the breeding season continues, rival males can drive out existing males. During this time, males rarely eat and only forage once they leave the harem. It is an energetically expensive time for the males.

Solitary Behavior

Relative to mammalian reproduction, approximately 68% of species exhibit solitary behavior in breeding females (Lukas and Clutton-Brock 2013), although this "solitary" existence does not account for the young that often accompany them for what can be long periods of care. As a result, Virginia Hayssen and Teri Orr note that "reproductive females are seldom solitary" (2017: 232). For example, the pygmy hippopotamus (*Choeropsis liberiensis*) has a gestation period from 196 to 201 days, and the young hippopotamus often stays with its mother until it matures at three to five years of age (Eltringham 2002). As Hayssen and Orr point out, "solitary" is more often a reality for nonbreeding males, although females without an adult male are regularly called "solitary" whether with or without young. This term, as used below, will address females of breeding age without young.

Solitary females tend to hold individual home ranges in which they forage and later raise young, only interacting with males to breed. A phylogenetic reconstruction by Dieter Lukas and Tim Clutton-Brock (2013) determined that, for the common mammalian ancestor, females were solitary and male territories overlapped those of several females. Lukas and Clutton-Brock also identified social monogamy as a derivative from a solitary lifestyle, which is generally an ancestral condition. In conjunction with their research on social monogamy, Lukas and Clutton-Brock documented forty-seven species that exhibit solitary behavior (Table 10.2). As seen previously, there are benefits to group or paired living, yet a relatively solitary existence can afford lower detection rates by predators, reduced food competition, decreased intraspecific aggression, and less exposure to disease and parasites. A polygynous mating system provides a higher probability of a male passing his genes to the next generation, thereby increasing his reproductive fitness.

There are four mating systems that apply to generally solitary species that do not employ short-term monogamy: polygyny, polyandry, polygynandry, and promiscuity. As with monogamous mating systems, genetic analyses are one of the few ways to fully confirm single or multiple paternity of offspring. Field studies are required to determine whether multiple male and female pairings are based on loose social bonds (polygynandry) or breeding without any social ties (promiscuity). Polyandry is the rarest mating system in mammals, although recent research indicates that it is more common than previously believed (Taylor, Price, and Wedell 2014).

The platypus has a single mating season (common to polygynous mating systems), and the females are the sole care providers (Grant and Temple-Smith 1998). Two main glands in males increase in size during mating season: the crural gland attached to the hollow spurs on each rear ankle and the testes (Temple-Smith 1973). Burrell (1927) indicated a dramatic increase in the male scent-glands. The spurs, which can inject an excruciatingly painful poison into an adversary, are primarily used to fend off rival males, who are equally aggressive during mating season. Once a male has gained the attention of a receptive female, the two undergo courtship behaviors, including circling each other in the water and the male holding the female's tail with his bill (Holland and Jackson 2002). As with many semi-aquatic mammals, the female uses a combination of burrow and nest during gestation and early care of her young. Unlike many other mammals, platypuses were notoriously difficult to

Table 10.2. Solitary semi-aquatic mammals

Species	Mating system	Reference
platypus, *Ornithorhynchus anatinus*	polygynous	Grant and Temple-Smith 1998
water opossum, *Chironectes minimus*	polygynandrous	Oliver 1976
swamp rabbit, *Sylvilagus aquaticus*	polygynous, polygynandrous	Chapman and Feldhamer 1981
marsh rabbit, *Sylvilagus palustris*	promiscuous	Chapman and Willner 1981
European water vole, *Arvicola amphibius*	polygynous	Nowak 1999
North American water vole, *Microtus richardsoni*	polygynous	Ludwig 1984
South American water rat, *Nectomys squamipes*	polygynous	Lima, Pinho, and Fernandez 2016
round-tailed muskrat, *Neofiber alleni*	polygynous	Nowak 1999
Coues's rice rat, *Oryzomys couesi*	unconfirmed	
marsh rice rat, *Oryzomys palustris*	likely promiscuous	Nowak 1999
Congo forest mouse, *Deomys ferrugineus*	unconfirmed	
false water rat, *Xeromys myoides*	unconfirmed	
Northern bog lemming, *Synaptomys borealis*	unconfirmed	
southern bog lemming, *Synaptomys cooperi*	unconfirmed	
swamp musk shrew, *Crocidura mariquensis*	unconfirmed	
Kahuzi swamp shrew, *Crocidura stenocephala*	unconfirmed	
Malayan water shrew, *Chimarrogale hantu*	unconfirmed	
Himalayan water shrew, *Chimarrogale himalayica*	unconfirmed	
Bornean water shrew, *Chimarrogale phaeura*	unconfirmed	
Japanese water shrew, *Chimarrogale platycephalus*	unconfirmed	
Chinese water shrew, *Chimarrogale styani*	unconfirmed	
Sumatran water shrew, *Chimarrogale sumatrana*	unconfirmed	
elegant water shrew, *Nectogale elegans*	unconfirmed	
southern water shrew, *Neomys anomalus*	unconfirmed	

Table 10.2. *continued*

Species	Mating system	Reference
Eurasian water shrew, *Neomys fodiens*	promiscuous	Cantoni 1993
Transcaucasian water shrew, *Neomys teres*	unconfirmed	
Glacier Bay water shrew, *Sorex alaskanus*	unconfirmed	
Pacific water shrew, *Sorex bendirii*	unconfirmed	
American water shrew, *Sorex palustris*	unconfirmed	
Pyrenean desman, *Galemys pyrenaicus*	polygynous inferred	Stone and Gorman 1985
fishing cat, *Prionailurus viverrinus*	unconfirmed	
marsh mongoose, *Atilax paludinosus*	polygynous inferred	Baker 1992b
aquatic genet, *Genetta piscivora*	unconfirmed	
European mink, *Mustela lutreola*	unconfirmed	
American mink, *Neovison vison*	polyandrous	Yamaguchi et al. 2004
Eurasian otter, *Lutra lutra*	polygynous	Chanin 1985
North American otter, *Lontra canadensis*	polygynous	Melquist and Hornocker 1983
marine otter, *Lontra felina*	monogamous, polygynous	Ostfeld et al. 1989
neotropical otter, *Lontra longicaudis*	polygynous, promiscuous	Oliveira et al. 2011
southern river otter, *Lontra provocax*	unconfirmed	
African clawless otter, *Aonyx capensis**	unconfirmed	
Congo clawless otter, *Aonyx congicus**	unconfirmed	
pygmy hippopotamus, *Choeropsis liberiensis*	polygynous, promiscuous	Pluháček and Steck 2015
water chevrotain, *Hyemoschus aquaticus*	polygynous inferred	Nowak 1999
marsh deer, *Blastocerus dichotomus*	polygynous	Oliveira et al. 2005

Note: Identified by Lukas and Clutton-Brock (2013), except for the star-nosed mole and spotted-necked otter, which are excluded (some group-living and monogamous behavior reported; Nowak 1999).

*Chanin (1985) reports short-lived monogamy for *Aonyx capensis* and *A. congicus*, followed by solitary behavior.

keep and breed in captivity until two successes in 1998, which then led to additional litters in other facilities (Hawkins and Battaglia 2009). Knowledge of basic reproductive biology and behaviors was essential for understanding the breeding success of platypuses, yet difficult to gain from field observations because of their nocturnal and secretive habits.

The water opossum is thought to be polygynandrous, with females coming into estrus more than once per year (**polyestrous**). Unlike other Neotropical marsupials, water opossums do not have a specific reproductive season, as evidenced by the presence of juveniles throughout the year (Galliez et al. 2009). The presence of males near, or even using, the same den as a female shows greater investment to possible mate-guarding and creation of social bonds than with a strictly promiscuous mating system. Water opossums also demonstrate precopulatory behaviors, including the male circling or following the female prior to initiating contact (Oliver 1976).

The pygmy hippopotamus, unlike the water opossum, is reported to be a promiscuous species (Eltringham 2002; Pluháček and Steck 2015). It is sexually **monomorphic** (both sexes are similar in size and appearance) and, outside of brief encounters during mating season, individuals seem to keep to themselves and even purposefully avoid other conspecifics. No social bonds are formed between the females and males, and the females are solely responsible for the care of their young (Nowak 1999).

Delayed Implantation

The only currently identified polyandrous semi-aquatic mammal is the American mink (*Neovison vison*) (Yamaguchi et al. 2004). Nobuyuki Yamaguchi, along with a research team from the Wildlife Conservation Research Unit at Oxford, conducted a study along 24 km of the River Thames west of Oxford City. They obtained DNA from twenty-seven male and twenty-four female mink. During the study, there were six litters from one dam (female mink) that revealed multiple paternity. The ability to mate with several males partially lies in the ability of female mink to support different embryos from different estrous cycles (**superfetation**); superfetation allows the females to continue mating with different males. American mink are also induced ovulators, and can begin ovulation when triggered by external stimuli (e.g., pheromones, contact). Because estrus can exceed a month, males do not remain with a female to ensure she does not mate with other males (Yamaguchi et al. 2004). As with some other mustelids, American mink can extend their reproductive period through **delayed implantation**, whereby low levels of progesterone delay the implantation of the ova by 27.5 days on average (Orr and Zuk 2014). The ova are implanted only after the mink ceases ovulation (Sundqvist, Ellis, and Bartke 1988).

Two studies that involve a closer examination of the reproductive physiology of the European mink (*Mustela lutreola*) reveal that this species does not exhibit delayed implantation (Amstislavsky et al. 2009; Nagl et al. 2015), despite earlier reports that it does (Nowak 1999; Thom, Johnson, and Macdonald 2004; Orr and Zuk 2014). Efforts to restore populations of this critically endangered mustelid have led to research into the reproductive physiology of captive European mink in the hope that they might one day breed more successfully in the wild. Nowak (1999) noted delayed implantation as "possible" in smooth-coated otters (*Lutra perspicillata*), but in two ancestral character state reconstructions by Michael Thom and his colleagues (2004), delayed implantation was not found.

Delayed implantation is found in North American river otters, and possibly Neotropical river otters (*L. longicaudis*) (Hamilton and Eadie 1964; Larivière 1999). North American river otters mate fairly soon following **parturition** (giving birth to young); however, with a delayed implantation of at least eight months, their total gestation period extends to twelve months (Hamilton and Eadie 1964). Less is known about the duration of the delay of implantation in the closely related Neotropical otter, which Larivière (1999) considers "facultative." The Neotropical otter's gestation of fifty-six days is considerably shorter than its northern cousin, and seems relatively short for other otters known to use delayed implantation. A key benefit of delayed implantation is to allow for increased mate choice for females and, thus, higher overall reproductive fitness (Sandell 1990).

As previously indicated, a species' dominant mating system is just one aspect of the reproductive biology of semi-aquatic mammals, and a variable one at that. Different population densities, habitats, and dominance structures influence how mates are selected. Yet, all the biological puzzle pieces combine to present a more complete picture of how wildlife populations remain viable over time and space. Once offspring are born, parental care becomes the next step in species persistence.

Parental Care

The duration and quality of parental care can be relatively short-lived, or it can extend over a few years. The terms altricial and precocial are common dichotomous descriptors of a neonate's developmental stage at birth, but it is important to acknowledge that developmental state represents a continuum that is also dependent on environmental context (Hayssen and Orr 2017). Some species of mammals have newborns that are relatively independent and capable of walking or running almost immediately after birth. Generally, these "precocial mammals produce heavier neonates and heavier litters, possess longer gestation periods, attain adult mass more slowly, and have smaller litter sizes" (Derrickson 1992: 57).

In David Alderton's book *Rodents of the World*, he describes young capybaras as "miniatures of their parents, able to follow their mother and graze grass as soon as they are born" (Alderton 1996: 74). Unlike altricial mammals, they are born fully furred with a full set of permanent teeth. Their early physical agility, however, does not make them completely independent. Young capybaras are not weaned for up to four months, although grass is one of the most important components of their diet during this time (Mones and Ojasti 1986). Despite their relative independence, newborn capybaras only weigh about 1.5 kg, while adults generally weigh from 35 to 65.5 kg (Mones and Ojasti 1986). As expected, capybaras do not build nests, but rather they hide their young in nearby vegetation when leaving them unattended.

The coypu (*Myocastor coypus*) also produces highly precocial young that can leave their mother after only five days of nursing, although six to ten days is the norm (Nowak 1999). Female coypus have a postpartum estrus of one to two days and, as with the American mink, are induced ovulators. These aspects of their reproductive biology complement their high **fecundity** (one to thirteen young per litter and two to three litters per year). Their reproductive output is somewhat balanced by an 80% mortality for coypus younger than one year of age (Willner et al. 1980). Such is the balance between high reproductive output and high juvenile mortality. High levels of recruitment are what ultimately drive population growth.

Rather than producing numerous young over short time periods, some semi-aquatic mammal species produce few young and care for them for longer periods of time (Table 10.3). Neonates from these species are often considered less developed than precocial neonates because these altricial young are highly dependent on one or both parents. However, Hayssen and Orr (2017) note that many neonates that require parental care for longer periods of time can have many of the same features as relatively independent neonates, further emphasizing the caution of applying altricial and precocial classifications as dichotomous extremes. For example, beavers are born completely covered in fur, as are coypu; however, newborn beavers are almost completely dependent on their parents for months after birth and tend not to leave the lodge until they are three or four weeks old. Adult beavers bring vegetation back to the lodge for the young beavers for several weeks after weaning. Although not always the case (e.g., capybaras sometimes only have one offspring), litter size and time between litters is also used to identify altricial species.

Every other year, the common hippopotamus gives birth to a single offspring; unlike many semi-aquatic mammals, common hippopotami sometimes give birth in shallow water. Young are not weaned until approximately one year of age, and they do not reach sexual maturity until seven to fifteen years of age for females and six to thirteen years of age for males (Nowak 1999). Also unlike most semi-aquatic mammals, young hippopotami nurse while underwater. Intense parental care and protection can pay off in that, compared to herds subject to habitat loss and poaching, calf-to-adult ratios can be twice as high in protected areas, such as the Mara River in Kenya, where habitat quality and poaching are less of a concern (Kanga et al. 2011).

Various reproductive strategies have evolved by responding to the advantages and disadvantages of having several offspring with limited care, having fewer offspring with longer periods of parental care, or some intermediate approach. Ultimately, population numbers for the varying levels of parental care might not be much different, it is the path taken to ensure ongoing replacement of reproductive adults that differs.

Reproduction and Conservation

Understanding reproductive biology of mammals is critical to their conservation, yet the mating system of many semi-aquatic mammals remains a mystery. Of the 140 species of mammals described in this book, the social nature of seventy-six of them (54%) remains unknown. For some species, captive breeding programs are the only means to gain insight into their reproductive biology, but environmental stimuli and social cues are often unknown and out of context. Molecular and genetic analyses are increasing our understanding of parentage and genetic health of populations; however, ensuring that a species' reproductive success continues to achieve ongoing recruitment of offspring into a healthy reproducing population is critical. These efforts, as seen in the next section, require a full understanding of the very basics of a species' biology, ecological interactions, and natural and external threats. Ensuring species persistence is a multifront effort that includes molecular biologists, ecologists, land managers, communicators, and a suite of others with specialized skills that combine to ensure that individual animals can provide effective genetic contributions that carry their species forward.

Table 10.3. Published reproductive features of semi-aquatic mammals

Species	Mating season	Gestation (days)	Offspring per litter
Ornithorhynchus anatinus	Varies	27*	1 to 3 (2)
Chironectes minimus	Varies		2 to 5 (3.5)
Lutreolina crassicaudata	Aug.–Oct.	61 to 65	2 to 5 (3)
Limnogale mergulus	Dec.–Jan.		3
Micropotamogale lamottei		~50	1 to 5 (2.6)
Micropotamogale ruwenzorii			1 to 2
Potamogale velox	Nonseasonal		2
Sylvilagus aquaticus	Varies	35 to 40	1 to 6 (3)
Sylvilagus palustris	Feb.–Sep.	28 to 27	2 to 5 (2.3)
Castor canadensis	Jan.–Feb.	128	1 to 4 (3)
Castor fiber	Jan.–Feb.	60 to 128	1 to 6 (3)
Thryonomys swinderianus	Wet season	137 to 172	1 to 12 (4)
Hydrochoerus hydrochaeris	Nonseasonal	~150	1 to 7 (3.5)
Hydrochoerus isthmius	Nonseasonal	104 to 111	1 to.8 (3.5)
Myocastor coypus	Nonseasonal	128 to 130	1 to 13 (5)
Arvicola amphibius	Mar.–Oct.	20 to 22	2 to 8 (4)
Arvicola sapidus	Mar.–Oct.	21	2 to 8 (3.5)
Holochilus brasiliensis	Spring/summer	26 to 30	3 to 6
Holochilus chacarius	Aug.–Nov.		4 to 7
Holochilus sciureus	Nonseasonal	29	1 to 8 (3.7)
Microtus richardsoni	May–Sep.		2 to 9
Nectomys squamipes	Aug.–Sep.		2 to 7
Neofiber alleni	Nonseasonal	26 to 29	1 to 4 (2.2)
Neotomys ebriosus			1 to 2
Ondatra zibethicus	Mar.–Sep.†	25 to 30	1 to 11 (4.7)
Oryzomys couesi	Nonseasonal	25	2 to 7 (~4)
Oryzomys palustris	Feb.–Nov.	25	1 to 7 (4.5)

Species	Litters per year	Age of weaning (months)	Age of sexual maturity	Life expectancy (years)
Ornithorhynchus anatinus	1	3 to 4	4 years	17
Chironectes minimus		1.5 to 2	10 months	2
Lutreolina crassicaudata	2	3	6 months	3
Limnogale mergulus				
Micropotamogale lamottei		40 days		
Micropotamogale ruwenzorii				
Potamogale velox				
Sylvilagus aquaticus	2 to 5	12 to 15 days	20 to 30 weeks	10 (1.8)
Sylvilagus palustris	≤4	12 to 15 days	31 weeks	4
Castor canadensis	1	2 weeks	1.5 to 2 years	10 to 20
Castor fiber	1	6 weeks	1.5 to 2 years	7 to 8
Thryonomys swinderianus	2		1 year	
Hydrochoerus hydrochaeris	1 to 2 (1)	3 to 4	1.5 years	10 (6)
Hydrochoerus isthmius	1			
Myocastor coypus	2 to 3	6 to 10 weeks	3 to 4 months	2 to 3
Arvicola amphibius	2 to 4	21 days	5 weeks	5.4 months
Arvicola sapidus	3 to 4	3 weeks	5 weeks	2 to 4
Holochilus brasiliensis			2 to 4.5 months	
Holochilus chacarius			2 months	
Holochilus sciureus		10 to 15 days	4 months	
Microtus richardsoni	2	< 6		
Nectomys squamipes				
Neofiber alleni	4 to 6	21 days	90 to 100 days	2
Neotomys ebriosus				
Ondatra zibethicus	5 to 6	21 to 28 days	1.5 to 2 years†	3 to 4 years
Oryzomys couesi				
Oryzomys palustris	7 (5)	11 to 13 days	7 weeks	<1

Table 10.3. *continued*

Species	Mating season	Gestation (days)	Offspring per litter
Rattus lutreolus	Oct.	23 to 25	1 to 11 (5)
Scapteromys aquaticus	Nonseasonal		2 to 5 (4)
Scapteromys tumidus	Nonseasonal		2 to 5 (4)
Synaptomys borealis	May–Aug.	23	1 to 8 (3)
Synaptomys cooperi	Apr.–Sep.	21 to 23	
Colomys goslingi	Mar.–Jul.		1 to 3
Dasymys incomtus	Nonseasonal		1 to 9 (3)
Hydromys chrysogaster	Sep.–Mar.	34	1 to 7 (3.5)
Nesokia bunnii	Nonseasonal	17	1 to 8
Pelomys fallax	Nonseasonal		1 to 9 (2.5)
Xeromys myoides	Nonseasonal		1 to 4
Delanymys brooksi	Spring		3 to 4
Chimarrogale hantu	May		5 to 7
Neomys fodiens	Apr.–Sep.	19 to 21	3 to 8 (5.7)
Neomys teres			5 to 9
Sorex bendirii	Jan.–Aug.	21 days	2 to 4
Sorex palustris	Feb.–Aug.	21 days	3 to 10 (6)
Condylura cristata	Jan.–Jun.	45	3 to 7 (5)
Desmana moschata	Jun. & Nov.	40 to 50	2 to 5
Galemys pyrenaicus	Jan.–May	30 days	1 to 5 (3.6)
Prionailurus viverrinus	Jan.–Feb.	63 days	1 to 4 (2.6)
Atilax paludinosus	Varies	69 to 80	1 to 3
Cynogale bennettii			2 to 3
Mustela lutreola	Feb.–Mar.	35 to 75	2 to 7 (4)
Neovison vison	Feb.–Apr.	39 to 78	2 to 10 (5)
Lutra lutra	Nonseasonal	60 to 63	1 to 5 (2.5)

Species	Litters per year	Age of weaning (months)	Age of sexual maturity	Life expectancy (years)
Rattus lutreolus	several		9 to 10 months	
Scapteromys aquaticus				
Scapteromys tumidus				
Synaptomys borealis	2 to 6 (2)	16 to 21 days	5 weeks (m)	up to 2.5
Synaptomys cooperi		21 days	5 weeks (m)	
Colomys goslingi				
Dasymys incomtus		30 days		
Hydromys chrysogaster	2 to 3	29 days	4 months	6.2
Nesokia bunnii				
Pelomys fallax				
Xeromys myoides				2 to 3
Delanymys brooksi				
Chimarrogale hantu				
Neomys fodiens	2 to 3	38 to 40 days	3 to 4 months	≤3
Neomys teres	up to 3		<1 year	2
Sorex bendirii	2		in 2nd year (m)	<2
Sorex palustris	2 to 3		in 2nd year	18 months
Condylura cristata	1	30 days?	10 months	3 to 4
Desmana moschata	2	1		
Galemys pyrenaicus	1 to 3		2 years	3.5
Prionailurus viverrinus			10 months	12
Atilax paludinosus	2	1 to 2	255 days	<19
Cynogale bennettii				<5
Mustela lutreola		10 weeks	11 months	10
Neovison vison	1	5 to 6 weeks	12 months	10
Lutra lutra	1	3 to 4	2 to 3 years	17

Table 10.3. *continued*

Species	Mating season	Gestation (days)	Offspring per litter
Lutra sumatrana	Nov.–Dec.	60	
Hydrictis maculicollis	Jul.	60	1 to 3
Lutrogale perspicillata	Varies	60 to 63	1 to 5 (2.5)
Lontra canadensis	Dec.–Apr.	290 to 380‡	1 to 5 (2)
Lontra felina	Dec. or Jan.	60 to 65	2 to 4 (2)
Lontra longicaudis	Varies	57 to 87‡	1 to 5 (2.5)
Lontra provocax	Jul.–Aug.		1 to 4 (1.5)
Pteronura brasiliensis	Jul.–Aug.	65 to 70	1 to 5 (2)
Aonyx capensis	Nonseasonal	63	1 to 3
Aonyx cinereus	Nonseasonal	60 to 64	1 to 7 (4.4)
Aonyx congicus	Nonseasonal	~60	2 to 3
Procyon cancrivorus	Jul.–Sep.	60 to 73	2 to 7 (3.5)
Babyrousa babyrussa	Nonseasonal	155 to 158	1 to 2
Choeropsis liberiensis	Nonseasonal	184 to 204	1
Hippopotamus amphibius	Varies	227 to 240	1 to 2 (1)
Hyemoschus aquaticus	Nonseasonal?	186 to 279	1
Blastocerus dichotomus	Varies	271	1
Elaphurus davidianus		288	1 to 2
Hydropotes inermis	Nov.–Jan.	170 to 210	1 to 8 (2.5)
Kobus leche	Nonseasonal	230 days	1
Tragelaphus spekii	Nonseasonal	247	1

Note: Averages are in parentheses. Data compiled from Nowak (1999), IUCN species notes, and *Mammalian Species* journal. "Nonseasonal" indicates there is no set breeding season, but there might be a peak season. "Varies" indicates variation by location. Symbols: * includes incubation of eggs, † in northern habitats, and ‡ includes delayed implantation. Only scientific names are provided.

Species	Litters per year	Age of weaning (months)	Age of sexual maturity	Life expectancy (years)
Lutra sumatrana	1			
Hydrictis maculicollis	1		2 years (f)	22.9 (captivity)
Lutrogale perspicillata	1	3 to 5	2 to 3 years	20 (captivity)
Lontra canadensis	1	4	2 years	13
Lontra felina	1	leave at 10		
Lontra longicaudis	1			
Lontra provocax	1			
Pteronura brasiliensis	1	~ 4	2 years	14.5
Aonyx capensis	1	1.5 to 2	1 year	14 (captivity)
Aonyx cinereus	2		13 months	16 (captivity)
Aonyx congicus			1 year	
Procyon cancrivorus	1	2.5 to 4	1 year	
Babyrousa babyrussa	2		5 to 10 months	24 (captivity)
Choeropsis liberiensis	1	6 to 8	4 to 5 years	>43.8 (captivity)
Hippopotamus amphibius	0.5	6 to 8	6 to 13 (m), 7 to 15 (f)	61 (captivity)
Hyemoschus aquaticus	1	90 to 186	9 to 26 months	8 to 13
Blastocerus dichotomus	1			
Elaphurus davidianus	1	10 to 11	2.25 years	23.25 (captivity)
Hydropotes inermis	1		5 to 6 months (m), 7 to 8 months (f)	13.9 (captivity)
Kobus leche		4	1.5 (f), 2.5 years (m)	15 years
Tragelaphus spekii	~1	4 to 5	1 (f) to 1.5 years (m)	21.5 (captivity)

PART V
CONSERVATION CHALLENGES & MANAGEMENT APPROACHES

Status and Threats

> The only hope for the species still living is a human effort commensurate with the magnitude of the problem.
>
> Edward O. Wilson, 2016, *Half-Earth, Our Planet's Fight for Life*

Conservation: A Background

At one point the hairy-nosed otter (*Lutra sumatrana*) was thought to be extinct (Aadrean et al. 2015). There were no captive populations for future reintroductions and no known sightings throughout its range in Southeast Asia for several years. In February 1999, Thai biologist Dr. Budsabong Kanchanasaka learned of three otter cubs rescued by a man in a remote Thai village. As she examined the otters in the village, their fur-covered noses provided immediate confirmation that they were hairy-nosed otters (Wright, Olsson, and Kanchanasaka 2008). Shortly thereafter, other biologists intensified their monitoring efforts and began to find hairy-nosed otters in other countries, including Vietnam (Nguyen, Pham, and Le 2001), Sumatra, Indonesia (Lubis 2005), northern Myanmar (Duckworth and Hills 2008), and Cambodia (Heng et al. 2016).

The population of hairy-nosed otters is currently known from seven disjoint locations in what had once been a larger, relatively continuous, distribution (Wright, Olsson, and Kanchanasaka 2008; Duplaix and Savage 2018). Historic records (i.e., 1830 to 1910) identify the hairy-nosed otter as fairly widespread, with a distribution also including Singapore and Brunei

Darussalam (Sasaki, Nor, and Kanchanasaka 2009). The IUCN currently lists the hairy-nosed otter as extinct in India and Myanmar and possibly extinct in Brunei Darussalam (Aadrean et al. 2015). There was a single male individual held in the Phnom Tamao Wildlife Rescue Centre in Phnom Penh, Cambodia that was obtained from an illegal zoo that had been closed (Wright, Olsson, and Kanchanasaka 2008). Unfortunately, the otter died in February 2010, but another hairy-nosed otter was rescued in July 2010 and continues to live at the Centre (https://www.thainationalparks.com/species/hairy-nosed-otter). Fishers are known to keep cubs as pets, but "they often die after short time in captivity (probably due to inappropriate care), or they are eventually killed and skinned" (Wright, Olsson, and Kanchanasaka 2008: 51; see also Poole 2003).

In some ways, the hairy-nosed otter presents a story of hope, it seems to have survived against all odds and shows some capacity for **ex situ conservation**. However, more progress is required to establish a captive breeding program. Also, the stressors and threats that brought it to such a precarious state by the late 1990s still exist. Fortunately, its distribution includes two small protected areas in Vietnam, but considerably more protection is required as human populations and resource extraction continue to encroach into remaining habitats. These otters not only represent a slight glimmer of hope for species persistence against all odds, they also highlight the challenges confronting so many semi-aquatic mammals.

The short list of issues threatening the persistence of hairy-nosed otters includes a restricted range, habitat loss, pollution, depletion of prey due to overfishing, bycatch in nets, poaching, the pet trade, trade in traditional medicine, and climate change (Aadrean et al. 2015; Gomez et al. 2016). As pressing as many of these issues are in today's world, the origin has a much longer history. More than a century ago, the conservation movement became part of the public psyche at a time when many North American species that were once commonplace, such as beavers and river otters, experienced stark declines. Protecting wild areas in many parts of the world became more widespread after Yellowstone National Park became the world's first national park in 1872. To address human impacts on the natural environment, the International Union for the Protection of Nature (now the IUCN, International Union for Conservation of Nature) was established during an international meeting of government representatives and conservation organizations in Fontainebleau, France, in 1948 (Christoffersen 1997).

Even before Rachel Carson wrote *Silent Spring* in 1962, the IUCN noted the negative impacts of pesticides on the environment (Christoffersen 1997), although it was Rachel Carson's book that really turned the tide of public awareness. The IUCN's greatest asset was the integration of governmental, private, and nongovernmental organizations to meet a common goal. Through the involvement of scientific experts, in 1964 the IUCN created the *Red Data Book* on the conservation status of species (now known as the "IUCN *Red List of Threatened Species*"). Following the 1972 United Nations Conference on the Human Environment, which met in Stockholm, the IUCN helped draft and implement the *Convention on International Trade in Endangered Species of Wild Fauna and Flora* (1974), which is now known as CITES. CITES is an international agreement among 183 parties (as of 2019) that aims to ensure that international trade of wild animals and plants does not impact species survival in the wild. Enforcement is achieved mainly through species protection laws developed by individual countries. The work of the IUCN and the scientists who produce the species reports and help categorize species based on conservation status has been a critical resource for this book. For example, in 2018 Nicole Duplaix and Melissa Savage published the IUCN report *The Global Otter Conservation Strategy* to provide the most up-to-date information and conservation strategies for the global suite of otters. Without the expertise and hard work associated with the IUCN, the biology and population status of many of the mammals described here would be virtually unknown.

Categories of Threatened Species and Population Trends

The IUCN *Red List of Threatened Species* identifies a species' conservation status as belonging to one of nine categories: extinct (EX), extinct in the wild (EW), critically endangered (CR), endangered (EN), vulnerable (VU), near threatened (NT), least concern (LC), data deficient (DD), and not evaluated (IUCN 2017). The categories of extinct and extinct in the wild are self-explanatory and are the most dire classifications. The IUCN also describes species population trends as declining, stable, increasing, or unknown.

In addition to four extinct species and one that is still considered extinct in the wild (Père David's deer), there are 139 other semi-aquatic mammal species—that is, all of the remaining freshwater mammals discussed in this book—that fall into the other IUCN categories, as discussed in greater detail later in this chapter. The majority of them, sixty-three species comprising 45% of the extant semi-aquatic mammals, are of least concern; however, fewer than 40% of them have stable population trends, and 24% are declining, while 35% are unknown. Only one species of the 140 described in this book (the Eurasian beaver, *Castor fiber*) has an increasing population trend (IUCN 2019). Semi-aquatic mammals are found in every one of the IUCN categories (Fig. 11.1).

Population trends are determined from either a census or extrapolation from some other data on abundance obtained from field sampling (IUCN 2017). The categories are self-explanatory: increasing, stable, decreasing, and unknown (Fig. 11.2). Population trends are very helpful in predicting population viability and perhaps even future changes in the conservation status of a species. With more than 80% of semi-aquatic mammals in either declining populations or in populations where the population status is unknown, ongoing conservation and research are critical. The Eurasian beaver, which as recently as 2002 was considered near threatened in Eurasia (IUCN 2019), has seen increased growth over the past two decades. It is only through intense regulation and reintroduction campaigns that it has established wild populations from Mongolia to the United Kingdom (Sjöberg and Ball 2011).

Extinct and Extinct in Wild

There are four known semi-aquatic mammals that are noted as extinct on the IUCN Red List: the Malagasy hippo (*Hippopotamus lemerlei*), the Madagascan dwarf hippo (*Hippopotamus madagascariensis*), Nelson's rice rat (*Oryzomys nelsoni*), and Schomburgk's deer (*Rucervus schomburgki*). Both species of hippopotami were endemic to the island of Madagascar, but they went extinct sometime after human colonization. More recently, Nelson's rice rat, endemic to María Madre Island in Mexico, likely went in extinct by the 1950s, following colonization of introduced black rats (*Rattus rattus*). In Thailand, Schomburgk's deer was originally found in swampy plains with long grass, but it was extinct by the 1930s, follow-

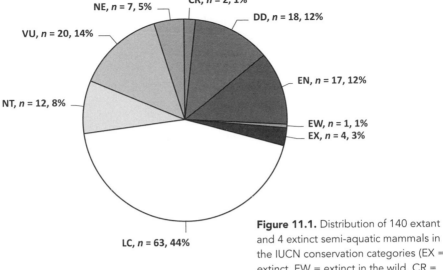

Figure 11.1. Distribution of 140 extant and 4 extinct semi-aquatic mammals in the IUCN conservation categories (EX = extinct, EW = extinct in the wild, CR = critically endangered, EN = endangered, VU = vulnerable, NT = near threatened, LC = least concern, DD = data deficient, and NE = not evaluated). DATA FROM THE IUCN 2019

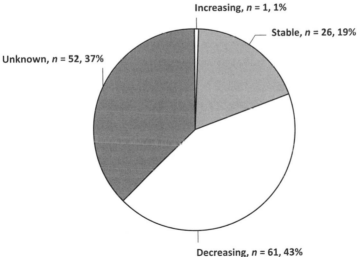

Figure 11.2. Population trends of 140 extant semi-aquatic mammals categorized by the IUCN (increasing, stable, decreasing, and unknown). DATA FROM THE IUCN 2019

ing conversion of its habitat for rice production, paired with overhunting (IUCN 2019). Although still considered to be extinct in the wild by the IUCN, Père David's deer (*Elaphurus davidianus*) has recently been reintroduced in China in the hope of establishing viable wild populations.

Conservation of Père David's deer presents a fascinating account of how past actions inadvertently aid current conservation initiatives. This distinctive deer originally lived in boggy wetland habitats of northeastern and east-central China until approximately 1,200 years ago (Yuan et al. 2017). It is also known locally as the milu, which means "four unalikes," because it has the "tail of a donkey, hooves of a cow, neck of a camel, and antlers of a deer" (Maddison, Zhigang, and Boyd 2012: 29). Père David's deer ultimately disappeared from the wild. The only remaining herd had been housed in the Nanyuang Royal Hunting Garden, south of Beijing, from 1205 to 1900.

It was 1895 when Père Armand David first saw them at the hunting park and, shortly thereafter, some live deer were brought to Jardin d'Acclimatation in Paris. By 1898 there were eighteen animals in various European zoos (Maddison, Zhigang, and Boyd 2012). The British Duke of Bedford ultimately obtained all remaining Père David's deer in Europe and housed them at his Woburn Abbey estate. It was a lucky bit of chance because in 1895 all remaining Père David's deer in China roamed from their enclosure when the wall of the Imperial Hunting Park collapsed. Most deer were poached and the rest otherwise perished. Today's population represents descendants from those remaining eighteen deer protected by the Duke of Bedford and his son. Of those eighteen deer, only seven males and seven females were able to breed, thus resulting in a breeding population of only fourteen animals.

Low population numbers resulted in an extreme **population bottleneck** that should have reduced their genetic variation. However, a recent molecular study (Zhu et al. 2018) revealed that the largest current population of Père David's deer has extensive genetic diversity, and that extended inbreeding might have actually aided the loss of harmful recessive genes. In the mid-1980s, the first herd of Père David's deer returned to captive facilities in China (Jiang and Harris 2016). Subsequent reintroductions have occurred, and now there are four wild populations composed of approximately six hundred individuals in the Hubei and Hunan provinces (Zhu et al. 2018). Whether they can survive the same threats they faced over a thousand years ago—hunting, urbanization, land-use change, and conflicts—remains to be seen. However, this time, there are teams of biologists and conservationists who are closely monitoring their well-being and who are continuing captive management of this remarkable species, which the IUCN still cautiously classifies as extinct in the wild.

Critically Endangered

Species within the critically endangered category have a very high risk of extinction and are in a particularly precarious state of vulnerability. The European mink (*Mustela lutreola*) and Ethiopian amphibious rat (*Nilopegamys plumbeus*) fit this category (Fig. 11.3). The Ethiopian amphibious rat is only known from a specimen that was trapped near the source of the Little Abbai River in 1927 and could very well be extinct (Lavrenchenko and Bekele 2017). There have been two attempts to find it without success. Even in the 1920s its highland riparian habitat was degraded through human disturbance and overgrazing by livestock (Kerbis Peterhans and Lavrenchenko 2008).

The European mink has experienced a 50% population decline at a minimum. It was considered to be vulnerable in 1998, endangered in 2008, and has been considered critically endangered since 2011 (IUCN 2019). Despite both **in situ** and ex situ conservation efforts, the IUCN

Figure 11.3.
Ethiopian amphibious rat (*Nilopegamys plumbeus*), photograph of a painting by Leon Pray. B. D. PATTERSON, FIELD MUSEUM OF NATURAL HISTORY, USED WITH PERMISSION

Small Carnivore Specialist Group predicts its population to decline an additional 80% given ongoing habitat degradation and loss and due to negative impacts from introduced American mink (*Neovison vison*) (Maran et al. 2016). Populations of the European mink are severely fragmented into at least three populations in northeast, southeast, and western Europe (Michaux et al. 2005) and, combined with several other threats, fragmented populations make its decline a multifaceted problem. Overexploitation and habitat loss have countered any attempts to successfully reintroduce the European mink into the wild, except perhaps on an island in Estonia. As described in Chapter 12, there are ongoing attempts to reintroduce the European mink to its native habitat while concurrently trying to address conservation challenges. Despite both the status of the Ethiopian amphibious rat and the European mink, neither species is protected under international legislation (e.g., CITES), outside of the European Union, or international management and trade controls (IUCN 2019).

Endangered

Endangered species have a high probability of extinction in the near future if their situation does not improve. Seventeen semi-aquatic mammals are classified as endangered by the IUCN, of which five are rodents, two are shrews, one is a desman, six are carnivores, and three are ungulates (Table 11.1). As expected, most populations are in decline, and two population have unknown trends. The status of these species represents a global decline in mammal populations, with distributions covering five of the eleven zoogeographic realms. Although efficacy of protection varies, all but four species—the Ecuador fish-eating rat (*Anotomys leander*), Bunn's short-tailed bandicoot rat (*Nesokia bunnii*), Cabrera's hutia (*Mesocapromys angelcabrerai*), and the Mexican water mouse (*Rheomys mexicanus*)—are confirmed residents of at least one protected area within their distribution (IUCN 2019). Only the pygmy hippopotamus (*Choeropsis liberiensis*) and Nile lechwe (*Kobus megaceros*) benefit from ex situ conservation, despite other species, such as the giant otter (*Pteronura brasiliensis*), being held in zoos where education strategies help raise awareness of their ecological and conservation needs. Nine species are protected by international legislative initiatives, mainly through listings on either CITES Appendix I or CITES Appendix II, although some have additional protection under the Bonn Convention (*Convention on Migratory Species of Wild Animals*). Other legislative protections for species at risk include the *Bern Convention on the Conservation of European Wildlife and Natural Habitats* and the EU Habitats Directive. Appendix I species under CITES are considered species threatened with extinction, while Appendix II species are not considered endangered, but international trade restrictions should be in place. Similar designation hierarchies exist for the Bonn Convention.

Five semi-aquatic mammal species classified as endangered by the IUCN are on Appendix I of CITES: the flat-headed cat (*Prionailurus planiceps*), the marine otter (*Lontra felina*), the southern river otter (*Lontra provocax*), the giant otter, and the Togian Islands babirusa (*Babyrousa togeanensis*). The marine otter and southern otter are also listed on Appendix I of the Bonn Convention. The pygmy hippopotamus, otter civet (*Cynogale bennettii*), and hairy-nosed otter are all on Appendix II of CITES, despite their endangered status. The otter civet requires immediate protection from timber harvesting and hunting to slow its population decline (Veron et al. 2006). Cabrera's hutia represents a suite of insular Caribbean mammals threatened with extinction (Turvey et al. 2017), yet is not covered by any international legislation. The other remaining endangered semi-aquatic mammals also lack legislative protections at the international level.

Vulnerable

Vulnerable species are likely to become endangered if their reproductive output or ability to survive does not improve, and they usually require human intervention in order for this to happen. Twenty semi-aquatic mammals are classified as vulnerable by the IUCN, of which two are African shrews, nine are rodents, one is a desman, three are carnivores, and five are ungulates (Table 11.2). Populations of all species are declining, except for the common hippopotamus (*Hippopotamus amphibius*), which is considered stable. However, the common hippopotamus is considered vulnerable due to habitat loss and illegal hunting for meat and ivory, particularly in areas experiencing social unrest (IUCN 2019). They are targeted for their canine teeth to counter the 1989 trade ban on elephant

Table 11.1. Semi-aquatic mammals categorized as endangered and their population trends

Species	Trend	Species	Trend
Ecuador fish-eating rat, *Anotomys leander*	D	otter civet, *Cynogale bennettii*	D
Cabrera's hutia, *Mesocapromys angelcabrerai*	U	hairy-nosed otter, *Lutra sumatrana*	D
Mexican water mouse, *Rheomys mexicanus*	U	marine otter, *Lontra felina*	D
montane shaggy rat, *Dasymys montanus*	D	southern river otter, *Lontra provocax*	D
Bunn's short-tailed bandicoot rat, *Nesokia bunnii*	D	giant otter, *Pteronura brasiliensis*	D
Kahuzi swamp shrew, *Crocidura stenocephala*	D	Togian Islands babirusa, *Babyrousa togeanensis*	D
Bornean water shrew, *Chimarrogale phaeura*	D	pygmy hippopotamus, *Choeropsis liberiensis*	D
Russian desman, *Desmana moschata*	D	Nile lechwe, *Kobus megaceros*	D
flat-headed cat, *Prionailurus planiceps*	D		

Data are from the IUCN Red List (2019).

Note: D = decreasing, U = unknown.

ivory. In 2008, the IUCN's Hippo Specialist Group reported a 125% increase in the export of canine teeth from 1989–1990 to 1991–1992, after the ban was implemented, with illegal trafficking continuing today.

As with species in the endangered category, vulnerable species live in geographically diverse zoogeographic realms. They are found from the Australian realm, the false water rat (*Xeromys myoides*), to the Palearctic, the southern water vole (*Arvicola sapidus*), the Pyrenean desman (*Desmana moschata*), and the water deer (*Hydropotes inermis*). Two species of otter, the Asian small-clawed otter (*Aonyx cinereus*) and the smooth-coated otter (*Lutrogale perspicillata*), are found in Southeast Asia, within the Oriental zoogeographic realm, where they suffer from extensive habitat loss combined with poaching for the wildlife trade (Gomez et al. 2016). All but three species, the Venezuelan fish-eating rat (*Neusticomys venezuelae*), the Cape marsh rat (*Dasymys capensis*), and Robert's shaggy rat (*D. robertsii*), are found in at least one protected area. Up to 60% of

Table 11.2. Semi-aquatic mammals categorized as vulnerable and their population trends

Species	Trend	Species	Trend
web-footed tenrec, *Limnogale mergulus*	D	Delany's swamp mouse, *Delanymys brooksi*	D
Nimba otter shrew, *Micropotamogale lamottei*	D	Pyrenean desman, *Galemys pyrenaicus*	D
southern water vole, *Arvicola sapidus*	D	fishing cat, *Prionailurus viverrinus*	D
Pittier's crab-eating rat, *Ichthyomys pittieri*	D	smooth-coated otter, *Lutrogale perspicillata*	D
Musso's fish-eating rat, *Neusticomys mussoi*	D	Asian small-clawed otter, *Aonyx cinereus*	D
Venezuelan fish-eating rat, *Neusticomys venezuelae*	D	hairy babirusa, *Babyrousa babyrussa*	D
Cape marsh rat, *Dasymys capensis*	D	Sulawesi babirusa, *Babyrousa celebensis*	D
Robert's shaggy rat, *Dasymys robertsii*	D	common hippopotamus, *Hippopotamus amphibius*	S
Tanzanian vlei rat, *Otomys lacustris*	D	marsh deer, *Blastocerus dichotomus*	D
false water rat, *Xeromys myoides*	D	water deer, *Hydropotes inermis*	D

Source: Data from the IUCN Red List (2019).

Note: D = decreasing, S = stable.

the population of fishing cats live in protected areas (IUCN 2019). Only five species are subject to ex situ conservation actions: the southern water vole, the Sulawesi babirusa (*Babyrousa celebensis*), the common hippopotamus, the marsh deer (*Blastocerus dichotomus*), and the water deer.

All three species of babirusa are listed on Appendix I of CITES, with the Sulawesi babirusa under continued hunting pressure even in protected areas. There are currently 190 individuals in thirty-four institutions worldwide (Leus et al. 2016), which could aid in situ recovery efforts if habitat conditions improve. Additionally, the marsh deer of South America and false water rat are also Appendix I species. Both species of otters, the common hippopotamus, and the fishing cat are on Appendix II. Pyrenean desmans are protected under Appendix II of the *Bern Convention on the Conservation of European Wildlife and Natural*

Habitats and the EU Habitats Directive (Annexes II and IV). The remaining species, many of which are rodents, lack international protection.

Near Threatened

Species considered to be at a lower risk of extinction are classified as near threatened and least concern. Near threatened species do not currently fit the threatened category (e.g., vulnerable to extinction or currently endangered), but they could in the future if conditions do not improve and population trends continue to decline. Of the twelve semi-aquatic mammal species in this category, all but two are experiencing population declines (Table 11.3). Population trends for Thomas's water mouse (*Rheomys thomasi*) and Issel's groove-toothed swamp rat (*Pelomys isseli*) are unknown. Half the semi-aquatic mammals classified as near threatened are found in the Afrotropical realm, including three of the five otters on in Table 11.3. The two other otters are from the Palearctic realm (Eurasian otter) and the Neotropical realm (Neotropical otter, *Lontra longicaudis*). Other species extend from the Australian zoogeographic realm (platypus) to the Oriental realm (Malayan water shrew) to the Panamanian realm (Thomas's water mouse). All but two species have declining population trends, while population trends of two others (Thomas's water mouse and Issel's grooved tooth swamp rat) are unknown. The platypus (*Ornithorhynchus anatinus*), Eurasian otter, and southern lechwe (*Kobus leche*) benefit from ex situ conservation in various off-site facilities. Despite past difficulties in breeding platypuses in captivity, the Healesville Sanctuary (part of Zoos Victoria) was the first facility to successful produce offspring in 1999. The Taronga Zoo in Sydney Australia has also bred platypuses in captivity successfully since the early 2000s (Hawkins and Battaglia 2009). Both facilities also work to raise awareness of habitat conservation in the wild. Challenging these efforts, however, are the combined effects of severe multiyear droughts and associated wildfires in Australia.

Half of near threatened semi-aquatic mammals are covered by international legislative protections. African clawless otters (*Aonyx capensis*) in Cameroon and Nigeria are listed on Appendix I of CITES, while all other populations are on Appendix II, including the Congo clawless otter (*A. congicus*) (IUCN 2019). Similarly, the spotted-necked otter (*Hydrictis*

Table 11.3. Semi-aquatic mammals categorized as near threatened and their population trends

Species	Trend	Species	Trend
platypus, *Ornithorhynchus anatinus*	D	Eurasian otter, *Lutra lutra*	D
crab-eating rat, *Ichthyomys hydrobates*	D	spotted-necked otter, *Hydrictis maculicollis*	D
Thomas's water mouse, *Rheomys thomasi*	U	Neotropical otter, *Lontra longicaudis*	D
Issel's grooved-toothed swamp rat, *Pelomys isseli*	U	African clawless otter, *Aonyx capensis*	D
Malayan water shrew, *Chimarrogale hantu*	D	Congo clawless otter, *Aonyx congicus*	D
aquatic genet, *Genetta piscivora*	D	southern lechwe, *Kobus leche*	D

Source: Data from the IUCN Red List (2019).
Note: D = decreasing, U = unknown.

maculicollis) is listed on Appendix II of CITES. The Neotropical and Eurasian otters are listed on Appendix I. The European otter has been bred in captive populations through a European breeding program since 1985, and these captive populations contribute to the re-establishment of wild populations (Prigioni et al. 2009). The southern lechwe, listed on CITES Appendix II, is particularly at risk of losing critical wetland habitats in central Africa and requires immediate habitat protection and increased population monitoring (IUCN 2019). All other near threatened species are only protected from state or country-specific legislation.

Least Concern

Species classified as least concern have been assessed by the IUCN and evaluated as not requiring current conservation focus. They are, however, reported on the IUCN Red List and their population trends are monitored and status re-evaluated over time. At sixty-three species, this category, by far, has the largest number of semi-aquatic mammals described in this book (Table 11.4). Approximately 40% of these species have stable populations, ~24% are in decline, 35% are unknown, and

Table 11.4. Semi-aquatic mammals categorized as least concern and their population trends

Species	Trend	Species	Trend
water opossum, *Chironectes minimus*	D	Common water rat, *Nectomys rattus*	S
little water opossum, *Lutreolina crassicaudata*	U	South American water rat, *Nectomys squamipes*	U
Rwenzori otter shrew, *Micropotamogale ruwenzorii*	U	round-tailed muskrat, *Neofiber alleni*	D
giant otter shrew, *Potamogale velox*	D	Andean swamp rat, *Neotomys ebriosus*	S
swamp rabbit, *Sylvilagus aquaticus*	D	montane fish-eating rat, *Neusticomys monticolus*	D
marsh rabbit, *Sylvilagus palustris*	U	Peruvian fish-eating rat, *Neusticomys peruviensis*	S
North American beaver, *Castor canadensis*	S	muskrat, *Ondatra zibethicus*	S
Eurasian beaver, *Castor fiber*	I	Coues's rice rat, *Oryzomys couesi*	U
greater cane rat, *Thryonomys swinderianus*	U	marsh rice rat, *Oryzomys palustris*	S
capybara, *Hydrochoerus hydrochaeris*	S	Goldman's water mouse, *Rheomys raptor*	S
coypu, *Myocastor coypus*	D	Underwood's water mouse, *Rheomys underwoodi*	S
European water vole, *Arvicola amphibious*	S	Argentine swamp rat, *Scapteromys aquaticus*	S
Kemp's grass mouse, *Deltamys kempi*	U	Waterhouse's swamp rat, *Scapteromys tumidus*	S
web-footed marsh rat, *Holochilus brasiliensis*	S	Alfaro's rice water rat, *Sigmodontomys alfari*	S
Chaco marsh rat, *Holochilus chacarius*	D	northern bog lemming, *Synaptomys borealis*	U
marsh rat, *Holochilus sciureus*	S	southern bog lemming, *Synaptomys cooperi*	U
Lund's amphibious rat, *Lundomys molitor*	D	Shaw Mayer's water rat, *Baiyankamys shawmayeri*	S
North American water vole, *Microtus richardsoni*	S	African wading rat, *Colomys goslingi*	S
western Amazonian nectomys, *Nectomys apicalis*	S	earless water rat, *Crossomys moncktoni*	U
Trinidad water rat, *Nectomys palmipes*	U	African marsh rat, *Dasymys incomtus*	U

Table 11.4. *continued*

Species	Trend	Species	Trend
West African shaggy rat, *Dasymys rufulus*	U	Eurasian water shrew, *Neomys fodiens*	S
Congo forest mouse, *Deomys ferrugineus*	S	Transcaucasian water shrew, *Neomys teres*	U
rakali, *Hydromys chrysogaster*	U	Pacific water shrew, *Sorex bendirii*	U
waterside rat, *Parahydromys asper*	U	American water shrew, *Sorex palustris*	S
creek groove-toothed swamp rat, *Pelomys fallax*	U	star-nosed mole, *Condylura cristata*	S
Australian swamp rat, *Rattus lutreolus*	D	marsh mongoose, *Atilax paludinosus*	D
swamp musk shrew, *Crocidura mariquensis*	U	American mink, *Neovison vison*	S
Himalayan water shrew, *Chimarrogale himalayica*	U	North American river otter, *Lontra canadensis*	S
Japanese water shrew, *Chimarrogale platycephalus*	D	crab-eating raccoon, *Procyon cancrivorus*	D
Chinese water shrew, *Chimarrogale styani*	U	water chevrotain, *Hyemoschus aquaticus*	D
elegant water shrew, *Nectogale elegans*	U	sitatunga, *Tragelaphus spekii*	D
southern water shrew, *Neomys anomalus*	D		

Source: Data from the IUCN Red List (2019).

Note: D = decreasing, I = increasing, S = Stable, U = unknown.

only Eurasian beaver populations are increasing. Interestingly, despite being invasive and considered a pest species where it is an introduced species, populations of the coypu are declining in its native habitats in South America. As expected, semi-aquatic mammals classified as least concern are found in all of the eleven zoogeographic realms.

There are several species where it is uncertain whether their range includes protected areas—the European water vole (*Arvicola amphibious*), Shaw Mayer's water rat (*Baiyankamys shawmayeri*), the earless water rat (*Crossomys moncktoni*), the West African shaggy rat (*Dasymys rufulus*), the creek groove-toothed swamp rat (*Pelomys fallax*), and the swamp

Figure 11.4. Pacific water shrew (*left*) and American water shrew (*right*). M. BRIERLEY

musk shrew (*Crocidura mariquensis*)—and others where such protection is presumed, but not confirmed—the African marsh rat (*Dasymys incomtus*), the Himalayan water shrew (*Chimarrogale himalayica*), and the Chinese water shrew (*C. styani*). Even for closely related species that share some parts of their range, the confirmed level of protection varies. For the Pacific water shrew (*Sorex bendirii*), the IUCN assumes that its habitat falls within protected areas, while the American water shrew (*S. palustris*) has habitat confirmed within several protected areas (Fig. 11.4).

Three species, the American mink, North American otter, and sitatunga (*Tragelaphus spekii*), are subject to ex situ conservation through captive breeding programs within zoos and other facilities. For some species, for example, beavers, American river otters, and marsh rabbits (*Sylvilagus palustris*), there are ongoing education programs to increase public awareness and decrease conflicts. Beavers in particular are the focus of research and management assessments to help reduce human-wildlife conflicts with these important wetland species (Siemer et al. 2013). The only semi-aquatic mammal species in the least concern category that is covered by international legislation is the North American river otter (CITES Appendix II).

Data Deficient

The data deficient category describes species that have been evaluated, but there is insufficient information on their abundance or distribution to fulfill a proper assessment. There are eighteen species of semi-aquatic mammals categorized as data deficient by the IUCN, all of which have unknown population trends, except for Oyapock's fish-eating rat (*Neusticomys oyapocki*), which is in decline (Table 11.5). Seven are found in protected areas: the lesser capybara (*Hydrochoerus isthmius*), the Las Cajas water mouse (*Chibchanomys orcesi*), the Chibchan water mouse (*C. trichotis*), Ferreira's fish-eating rat (*Neusticomys ferreirai*), Oyapock's fish-eating rat (*N. oyapocki*), the western water rat (*Hydromys hussoni*), and the Glacier Bay water shrew (*Sorex alaskanus*). It is possible that the mountain water rat (*Baiyankamys habbema*) is found in at least one protected area. None of these species are subject to ex situ conservation efforts or education campaigns. Given how little is known about most of these semi-aquatic mammals, they are not protected by any international legislation. The data deficient category does not imply that these populations are not at risk, there is simply not enough information to assess their status until additional research is conducted to provide a more comprehensive assessment.

Not Evaluated

Species that have yet to be assessed by the IUCN fall under the not evaluated category. There are seven species that have not been evaluated by the IUCN: Lagiglia's marsh rat (*Holochilus lagigliai*), the Voss fish-eating rat (*Neusticomys vossi*), the plateau swamp rat (*Scapteromys meridionalis*), and four African marsh rats (*Dasymys griseifrons, D. longipilosus, D. medius,* and *D. shortridgei*). Several species of the African marsh rats (*Dasymys* spp.) have only recently been reclassified into their own species (Mullin, Pillay, and Taylor 2005), and the IUCN has noted in its species reports that the taxonomy of the *Dasymys* is still being reviewed. As new species are found, like the Voss fish-eating rat (Hanson et al. 2015), and others are reclassified, the list of species requiring evaluation regarding their population status and conservation needs will also grow.

Table 11.5. Semi-aquatic mammals categorized as data deficient
and their population trends

Species	Trend	Species	Trend
lesser capybara, *Hydrochoerus isthmius*	U	Fox's shaggy rat, *Dasymys foxi*	U
Ucayali water rat, *Amphinectomys savamis*	U	Angolan marsh rat, *Dasymys nudipes*	U
Las Cajas water mouse, *Chibchanomys orcesi*	U	western water rat, *Hydromys hussoni*	U
Chibchan water mouse, *Chibchanomys trichotis*	U	New Britain water rat, *Hydromys neobritannicus*	U
Stolzmann's crab-eating rat, *Ichthyomys stolzmanni*	U	Ziegler's water rat, *Hydromys ziegleri*	U
Tweedy's crab-eating rat, *Ichthyomys tweedii*	U	Hopkin's groove-toothed swamp rat, *Pelomys hopkinsi*	U
Ferreira's fish-eating rat, *Neusticomys ferreirai*	U	Sulawesi water rat, *Waiomys mamasae*	U
Oyapock's fish-eating rat, *Neusticomys oyapocki*	D	Sumatran water shrew, *Chimarrogale sumatrana*	U
mountain water rat, *Baiyankamys habbema*	U	Glacier Bay water shrew, *Sorex alaskanus*	U.

Source: Data are from the IUCN Red List 2019.
Note: D = decreasing, U = unknown.

Threats

Every two years since 1998, the World Wildlife Fund for Nature (WWF) publishes the *Living Planet Report*, which provides a science-based assessment of global ecological health and human impacts on the environment (WWF 2018). The latest report indicated an 83% decline in population size and/or abundance of freshwater species since 1970. The Freshwater Index representing this change includes all freshwater species, not just mammals. As seen previously, however, many freshwater mammals are experiencing dramatic declines. Not surprisingly, the WWF identified freshwater systems as the world's most threatened ecosystems due to habitat loss and modification, invasive species, overfishing, pollution, and climate change, among other factors (WWF 2018). Few impacts occur in isolation, and more often a suite of stressors produces synergistic effects that create complex conservation challenges. Some species (e.g., the western water rat), despite their restricted range, live in relatively high elevation or otherwise isolated habitats that are not immediately threatened (IUCN 2019). However, most semi-aquatic mammals are exposed to at least one anthropogenic threat. The IUCN species accounts associated with the IUCN Red List provide regularly updated conservation concerns for the species included in this book, where available.

It is rare that one species would only be impacted by just one environmental stressor, although in some cases the IUCN might only note one particular issue of concern for a specific species. For example, pollution resulting from gold mining in French Guiana and other parts of northern South America is noted as a key threat to the water opossum (Pérez-Hernandez et al. 2016). In other circumstances, the impacts on one species of semi-aquatic mammal are also experienced by another semi-aquatic mammal sharing the same range, while the very reduction of the first species also impacts the second. Congo clawless otter populations are negatively impacted by roads, deforestation, and pollution; however, the concurrent loss of common hippopotami because of similar stressors indirectly impacts the otters. Common hippopotami are ecosystem engineers and enhance wetland connectivity and channelization. As their populations decline, their "engineered" habitats are no longer available to otters and other freshwater species (Jacques et al. 2015). Similar interactions occur when beavers are locally extirpated and no longer maintain dams and dig channels. In this case water shrews, voles, muskrats, mink, and river otters are negatively impacted due to associated habitat loss. There are several ways to classify stressors on natural systems (see Feldhamer et al. 2015). General categories often include human population growth, habitat loss, habitat degradation, overexploitation, human perceptions of wildlife, and climate change, all of which can work synergistically to create a complex suite of problems that become increasingly difficult to solve with standard management approaches.

Human Population Growth

The human population has increased by two billion people since 1992 (Fig. 11.5), a 35% increase since the original *World Scientists' Warning to Humanity* was written in 1992 (Ripple et al. 2017). With a current human population exceeding 7.6 billion people, the demand for increased food

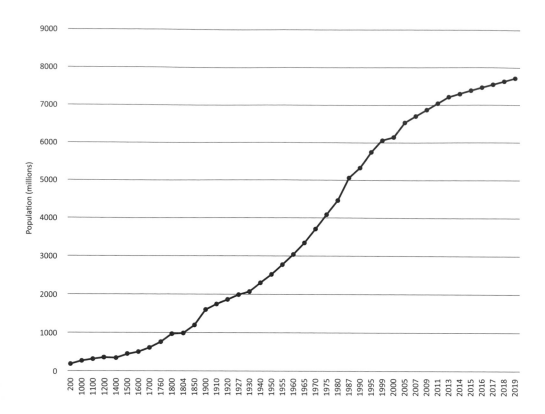

Figure 11.5. World population by year.
DATA FROM WWW.WORLDOMETERS.INFO

production, water, forest products, and space has increased in kind, as has the production of garbage and human waste. Sharing resources with other species has not been a priority for us, despite repeated warnings of rapid species declines due to direct and indirect impacts from human activities. In their recent article, William J. Ripple and his colleagues note that "we have unleashed a mass extinction event, the sixth in roughly 540 million years, wherein many current life forms could be annihilated or at least committed to extinction by the end of this century" (Ripple et al. 2017: 1026).

The myriad of human impacts on the environment are directly linked to overconsumption of resources. The terms "reduce," "reuse," and "recycle" are hierarchical; the most effective means to reduce our **ecological**

footprint is to reduce consumption of resources (Wackernagel and Rees 1996). Interestingly, some of the countries with the lowest population growth rates (e.g., Canada and United States) have some of the highest per capita rates of resource consumption in the world. In the conservation of semi-aquatic mammals, reducing global water consumption, poaching, and overfishing would have immediate, and positive, impacts on population declines. The WWF reports that "only a quarter of land on Earth is substantively free of the impacts of human activities," a number that is expected to decrease by 15% by 2050 (WWF 2018: 6). Reversal of these trends requires similar sociological catalysts that prompted global reductions in ozone-depleting chemicals and acid rain. Many of the other stressors discussed below are directly linked to not only human population growth but ecological awareness within society as a whole.

Habitat Loss

The fate of freshwater mammals lies in the state of their habitats. Wetland loss at a global scale exceeds 90% in areas such as California, USA, and southern Ontario, Canada, while agricultural areas in the western Prairie Provinces in Canada have lost as much as 70% of their wetlands since European settlement and agricultural expansion (Schindler and Donahue 2006). In Europe, approximately 60% of wetlands have disappeared over the past century, although that estimate increases to 90% in some regions (Mitsch and Gosselink 2007; Junk et al. 2013). In some areas such as China, where 22% of the country's natural wetlands have been destroyed in the past fifty years (Junk et al. 2013), several species of freshwater mammals are listed as data deficient and their status and much of their biology are unknown. Where data do exist, many of these studies are unpublished or out of date (Vié, Hilton-Taylor, and Stuart 2009). Of the freshwater animals, invertebrate, amphibian, and fish species tend to receive the greatest research focus. The total loss of wetlands globally is approximately 50% (Mitsch and Gosselink 2007). In many areas, wetland loss continues despite the ecological and socioeconomic importance of wetlands (Clare et al. 2011).

Habitat loss is the most common threat noted by the IUCN species specialist groups; much of this loss is associated with wetland drainage, agricultural land conversion, and deforestation. Wetland loss in particular has negatively impacted both species of water voles in Europe (European water vole and southern water vole). Recent precipitous declines of European water voles in the United Kingdom are directly linked to habitat loss, combined with predation by non-native American mink (MacPherson and Bright 2011). Southern water voles are particularly influenced by a decline in suitable freshwater habitats of adequate size and connectivity to other habitat patches (Pita, Mira, and Beja 2010). For both species, larger habitat patches dramatically improve patch occupancy and species persistence. Southern water vole populations are especially prone to isolation in the dry season, which is offset somewhat by irrigation ditches in agricultural areas. Agricultural areas without herbaceous riparian vegetation tend to be avoided.

Both species of voles conform to a **metapopulation** model, in which several spatially separated subpopulations live in relative isolation, yet have enough connectivity to allow some dispersal and gene transfer. In addition, connectivity among differing sizes of patches facilitates the local extinction and colonization of occupied and unoccupied patches (Levins 1969). In the case of the European vole, however, American mink can accelerate population extinctions, especially in smaller patches where cover and reduced patch size cannot buffer against mink incursions (MacPherson and Bright 2011). Larger patches can act as refuges against extreme weather events or other perturbations. For species that follow a metapopulation model, including voles and beavers, not only is patch size important but aquatic connectivity among the patches is also necessary (Hood and Larson 2015). In seasonally dry climates, southern water voles are limited in their dispersal ability in summer if the matrix between patches lacks water refuges to provide ease of movement, escape cover, and cooler environments.

As with other aquatic habitats, river and stream systems represent a delicate interplay between the aquatic environment and surrounding riparian vegetation. The meeting of these two habitat types supports a disproportionate amount of biodiversity relative to that found in fully terrestrial or open ocean ecosystems (Richardson et al. 2010). As with wetlands, rivers throughout the world have been dramatically altered through riparian deforestation, channelization, and dam construction. For many species of semi-aquatic mammals, riparian forests provide critical foraging areas and sites for bank dens. Both North American and

Eurasian beavers can cut over two hundred trees and saplings per year to sustain their family groups (Müller-Schwarze 2011). Access to riparian forests is essential for construction of lodges, dams, and food caches. For the European otter, riparian deforestation, channelization of rivers, and the construction of hydroelectric dams have contributed to its global status as near threatened (Kruuk 2006). Within Western Europe, otter populations are recovering, but they face ongoing population declines in Asia due to habitat loss and poaching (IUCN 2019).

The effects of deforestation are particularly significant on freshwater systems. Approximately 27% of disturbance to forested ecosystems from 2001 to 2015 was associated with commodity-based deforestation, with a global rate of deforestation of five million hectares per year (Curtis et al. 2018). Forestry, agricultural expansion, urbanization, and increasing numbers of wildfires all contribute to global deforestation. Consequently, freshwater systems can be impacted through increased insolation, sedimentation, and associated shifts in prey communities.

Reasons for deforestation vary by region. In temperate and boreal forests, large wildfires have become the dominant driver of forest loss, while agricultural expansion and commodity-driven deforestation dominate in tropical regions (Curtis et al. 2018). Agricultural expansion is the key driver of deforestation in Southeast Asia and sub-Saharan Africa, with palm oil plantations identified as an emerging issue by the IUCN for several species (e.g., the New Britain water rat, flat-headed cat, fishing cat, otter civet, hairy-nosed otter, and pygmy hippopotamus). Cattle grazing and creation of croplands is a major driver of forest loss in Central and South America (Curtis et al. 2018), where both marsh deer and southern river otters are two of several affected species identified by the IUCN. In Madagascar, the web-footed tenrec requires permanent, clean, fast flowing streams and rivers for its survival (Benstead, Barnes, and Pringle 2001). Deforestation has caused notable shifts in stream invertebrate communities, which in turn has certainly altered prey communities of the web-footed tenrec, a top predator in affected streams (Benstead, Douglas, and Pringle 2003).

Habitat Degradation

While habitat loss involves a direct reduction in the quantity of habitat, habitat degradation involves an impairment of habitat quality through the introduction of foreign chemicals, non-native species, and other inputs. Some biologists also consider habitat fragmentation as a form of degradation (Feldhamer et al. 2015), while others classify it as habitat loss. In reality, it is a bit of both. Habitat must be lost to reduce remaining habitat into smaller patches, while the lack of connectivity among the patches lowers their overall quality for species persistence. Regardless, species in anthropogenically fragmented landscapes have a higher risk of extinction than those in contiguous habitat.

Swamp rabbits provide a classic example of how habitat loss paired with habitat fragmentation can reduce population numbers. In Illinois, swamp rabbits historically were found in swamps and river bottoms throughout southern Illinois, but now they have a limited distribution in the state (Whitaker and Abrell 1986; Watland, Schauber, and Woolf 2007). Nationally, it has a broader distribution ranging from southern Illinois to the Gulf of Mexico. Logging and agricultural land conversion have dramatically restricted the distribution of swamp rabbits into increasingly isolated habitat patches in Illinois and other parts of its northern range. In a 2007 study, Angela Watland, Eric Schauber, and Alan Woolf noted an 80% decline in the forested acreage of the Mississippi alluvial floodplain because of agricultural expansion, logging, and human developments. As seen in Chapter 12, "Introductions and Reintroductions," translocations are one management approach used to counter population declines.

Pollution comes from numerous sources: mine waste, oil spills, acid rain, and human effluent. The Mexican water mouse, Malayan water shrew, Pyrenean desman, and web-footed tenrec all require pristine, clean water to survive. Industrial pollution is an immediate threat to vulnerable Pyrenean desman populations, as is siltation from forestry practices for the web-footed tenrec. As top carnivores, otters are especially susceptible to pollution. Giant otters are increasingly at risk from effluent from gold mines, while some populations of North American otters are exposed to acid mine drainage from coal mines (IUCN 2019). Not only does acid mine drainage cause immediate health issues for mammals, it also reduces the pH to a level that makes the water uninhabitable for fish, plants, and invertebrates. Along with mining, urban sources of pollution impact freshwater systems. Many top carnivores, including hairy-nosed otters, bioaccumulate organochlorines and other biocontaminants from urban sources.

The IUCN species reports note contaminants of various origins as a threat for all species of otters. In particular, oil spills have multiple impacts on otters and other semi-aquatic mammals, including interference with waterproofing qualities of the fur and the ingestion and absorption of toxic petrochemicals. The impacts of such spills can persist for years after remediation and cleanup, even if done correctly. In 2003, an oil spill from the Trans-Ecuadorian Oil Pipeline System released twenty-two thousand barrels of crude oil into Papallacta Lake and two nearby streams; "much of the biological community was extirpated" (Araújo et al. 2014: 90). This lake serves as a water reservoir for the city of Quito, and it is also one of only two known locations for the endangered Ecuadoran fish-eating rat. So little is known about this species that the impacts of the oil spill cannot be fully determined.

Biological pollutants, in the form of introduced pathogens and non-native species, also degrade habitat quality. Diseases, as discussed previously in this book, can lead to poor reproductive output, compromised health, or death. Non-native species, another form of biological pollutant, either successfully persist in the environment (e.g., introduced black or brown rats) or are continually released into the environment from fur-farms (e.g., American mink). In some cases, their owners no longer wish to keep them as pets (e.g., pythons, cats, and dogs). In Florida's Everglades National Park (ENP), Burmese pythons (*Python bivittatus*) are particularly problematic. During a recent study, Robert McCleery and his colleagues (2015) radio-tracked ninety-five adult marsh rabbits, some of which they released into the national park where there was a well-established python population. They determined that Burmese pythons accounted for 77% of mortalities of marsh rabbits in ENP and caused notable reductions in the park's marsh rabbit populations. Outside the park, where there were no pythons, native mammals were the main predators of marsh rabbits, and they had a lower impact on rabbit populations.

Overexploitation

Historical fur trades nearly decimated beaver, muskrat, mink, and river otter populations, in particular because of the thick, durable fur that these animals have. The history of hunting and trapping of various freshwater mammals is well documented, especially relative to beavers (e.g., Innis 1930). The North American fur trade saw the near extinction of the North American beaver by the end of the 1800s in many areas of Canada and the United States (Hood 2011). In Europe, the Eurasian beaver was extirpated from most of its former range between the sixteenth and nineteenth centuries, when it was hunted for its fur, meat, and castoreum. By the beginning of the twentieth century, the European beaver had a global population of only 1,200 individuals in the more northern parts of Eurasia and, until recently, was considered a threatened species (Halley and Rosell 2002). With the exception of subpopulations in Mongolia and China, Eurasian beaver populations have recovered thanks to intense conservation efforts, and they are now found in many parts of Eurasia. Similar efforts allowed North American beaver populations to experience a more rapid recovery beginning in the 1940s.

The demand for fur as a fashion item has waned to some degree; however, there is still both a legal and illicit trade in wildlife products and wildlife as pets (Gomez et al. 2016). The common hippopotamus, water deer, southern lechwe, flat-headed cat, fishing cat, and several species of otters are regularly targeted by poachers. Along with otter cubs (e.g., hairy-nosed otter and giant otter), the crab-eating raccoon is sold in the pet trade (IUCN 2019). Water deer are poached for meat and semidigested milk in the rumen of fawns, which is thought to cure indigestion in children (Won and Smith 1999). As with otters, the flat-headed cat is poached for its fur, while the fishing cat is used for fur and consumption.

In their 2018 report, *Illegal Otter Trade in Southeast Asia*, Lalita Gomez and Jamie Bouhuys stated that "overall, the pet trade emerged as the most pressing threat to the survival of otters, particularly in Indonesia and Thailand" (Gomez and Bouhuys 2018: vii). Their report only covers four species of otters (Eurasian otter, hairy-nosed otter, Asian small-clawed otter, and smooth-coated otter), but the numbers are concerning. From 2015 to 2017, there were thirteen seizures of fifty-nine otters in Indonesia, Malaysia, Thailand, and Vietnam. In approximately 560 online postings, there were up to 1,189 otters for sale, mostly from Indonesia (Gomez and Bouhuys 2018). Asian small-clawed otters were most frequently encountered, followed by smooth-coated otters. Although less common, the Eurasian otter and the extremely rare hairy-nosed otter were also advertised for sale as pets. A more extensive examination of the otter trade from 1980 to 2014 revealed at least 964 registered trade incidents involving 43,692 individual otters from these four Southeast Asian species of otter, some involving pelts and others live otters. The trade is global, with otter furs and live otters being trafficked to several countries, including the United States, Japan, South Korea, United Arab Emirates, Denmark, Hong Kong, Kuwait, and the Czech Republic (Gomez et al. 2016). Beyond the wildlife trade, several species of semi-aquatic mammals are hunted for personal consumption and medicinal uses (e.g., greater cane rat, capybara, fishing cat, aquatic genet, hairy-nosed otter, Neotropical river otter, Congo clawless otter, all three species of babirusa, both species of hippopotamus, water chevrotain, Chinese water deer, and lechwes).

Cultural Perceptions of Wildlife

One of the greatest challenges of conservation is not the physical securement of habitat, or even the reintroduction of species to previously impacted, but restored, environments. The greatest challenge is working within a broad context of coexisting cultures with different ecological, religious, and socioeconomic perceptions and realities. Effective ecosystem management requires that standard Euro-American science-based management and conservation ethics are flexible, integrative, and replaceable where appropriate. As previously noted, semi-aquatic mammals live in a global suite of environments that are subject to a myriad of cultural beliefs and economic realities. Traditionally, much of what average Europeans and colonial North Americans know of these species is based in childhood stories written from a Eurocentric perspective, often supplemented by visits to local zoos. As such, global management of these species runs the risk of being neocolonial in character (Garland 2008). The Fishing Cat Conservancy, an organization with collaborators ranging from Southeast Asia to North America, has developed a community-based model for research and conservation (www.fishingcatconservancy.org). Not only do members of the organization train local people (including youth) to conduct fieldwork, they also rely on traditional hunters to help them find and monitor fishing cats in the wild. Their education programs actively engage local schools and villages, and they also have an online presence to increase awareness at a global scale.

The Fishing Cat Conservancy serves as a model for working with local communities to overcome perceptions some people have of competition for resources with local wildlife (e.g., fishing cats competing for fish). In some areas, people believe that fishing cats pull fish from farmer's ponds; farmers respond in kind with retaliatory killings. An ongoing education campaign is having some success in changing those perceptions. For spotted-necked otters, otter specialists have documented a belief among fishers in Uganda that "otters eat the eyes from fish caught in their nets" (Reed-Smith et al. 2010: 88). Similar perceptions of North American river otters eating prize game fish also contribute to conflicts. Such perceptions potentially could be countered by grassroots community-based research and education campaigns styled after ones developed for fishing cats. Culturally sensitive, multifaceted approaches, rather than top-down management, are essential to counter beliefs that intensify impacts on species and their habitats.

Other species of semi-aquatic mammals are used in traditional medicines. In their 2016 report on the illegal otter trade, Lalita Gomez and her colleagues noted various uses of species of Southeast Asian otters being used for medicinal purposes. In India, there are records of epilepsy being treated with otter blood (Kruuk 2006), while oil rendered from otter body fat is used as medicine for joint pains and pneumonia (Meena 2002). In northern India, otter oil sells for just under US$0.05 per 10 g, with an adult otter producing approximately 500 g of oil (~$25 worth of oil per otter). In Cambodia, the penis bone of otters is sometimes crushed and mixed with coconut milk to act as an aphrodisiac (Dong et al. 2010). These issues, along with many others (e.g., "pest" management) result in population declines throughout species' ranges. Multifaceted problems require multifaceted responses, and solutions work best if education is reciprocal and flexible.

Climate Change

Climate change, driven by anthropogenic greenhouse gas concentrations, involves more than just an increase in global temperatures. Increases in extreme weather events, including severe droughts and floods, have immediate and often enduring impacts on freshwater systems (Döll et al. 2015). Associated increases in sea level also have negative impacts on coastal wetlands and other freshwater systems. Consequently, changing environments can significantly influence species distributions. In North America, beavers are expanding their range into tundra habitats in the far north, and their pond-building activities are accelerating the melting of permafrost at local scales (Tape et al. 2018). Although generally seen as having a positive influence on aquatic biodiversity, the impact of beavers on northern ecosystems is unknown. For other species, such as the western water rat of the Wissel Lakes region of Papua Province in Indonesia, current geographic isolation could be exacerbated by additional loss of freshwater connectivity and aquatic habitats.

Just as with many of the other threats to wildlife species, rising temperatures invariably will affect community composition and food webs, along with habitat structure related to nesting sites and travel corridors (Feldhamer et al. 2015). Physiologically, the delicate thermoregulatory requirements of many semi-aquatic mammals will also be challenged. The Intergovernmental Panel on Climate Change (IPCC), consisting of more than 1,300 international scientists, notes that global mean surface air temperatures will continue to increase over the twenty-first century, and it predicts average temperature increases between +1.3°C and +1.8°C by mid-century, from 2046 to 2065 (Meehl et al. 2007). As with other threats, climate change can have synergistic interactions with existing environmental stressors.

Synergistic Effects

Two species that exemplify the interactive nature of conservation challenges are the European mink and the giant otter. The geographic isolation of European mink immediately increases their risk of extinction. Early on, overexploitation of mink for furs and ongoing habitat loss initiated the start of their decline and fragmentation into isolated populations (Maran et al. 2016). In the Spanish population, there are fewer than five hundred individuals. This population is increasingly exposed to ongoing habitat loss, water pollution, genetic isolation, human-caused mortality, competition with the introduced American mink, and exposure to the often deadly Aleutian mink disease (Palazón and Melero 2014). Despite a previously extensive distribution throughout Europe, the European mink currently occupies less than 20% of its former range (Palazón and Melero 2014), and it has experienced population declines of more than 90% since the early twentieth century (Maran et al. 2016). These population declines cannot be accounted for by just one factor. As populations decrease, their genetic resilience also declines, which in turn reduces their resistance to other stressors, including disease and predators.

As with the mink, subpopulations of the giant otter in South America are fragmented and small (Groenendijk et al. 2015). Giant otter social behaviors make them more conspicuous and, therefore, easier to hunt as either a pest species or for the wildlife trade. Their reproductive output is lower than most other semi-aquatic mammals and, therefore, their populations are less resilient to population declines (Duplaix 1980). Although their overall distribution in South America appears extensive, they tend to live in relatively isolated habitats, thus making them vulnerable to nearby land-use changes (e.g., roads and mining).

South America's human population growth rate continues to increase above the global average, which puts additional pressures on natural resources and intact ecosystems. Gold mining, water contamination, overfishing, fisher-otter conflicts, deforestation, canine distemper, large infrastructure projects, urbanization, transportation networks, and tourism all contribute to population declines of giant otters (Groenendijk et al. 2015). If the impact of these threats cannot be slowed, ecologist Jessica Groenendijk and her colleagues predict at least a 50% decline in current populations over the next twenty-five years. Synergistic effects create what ecologists often call "wicked problems," which require integrated social, ecological, and economic approaches to help solve them.

Future Directions

Despite the daunting task of conserving freshwater habitats and the species they support, there have been various successes. Awareness has been increased, and illegal trafficking of wildlife has been curbed, as a result of increased legislation and global enforcement of restrictions on wildlife trade through CITES protections and country-specific legislative measures. Some zoos provide funding, educational programming, and expertise to support in situ conservation projects. The giant otter project supported by the San Diego Zoo's Institute for Conservation Research is one example. This project involves a team of Peruvian and other international researchers who provide increased research and awareness of giant otters within their natural habitats of South America. The following two chapters, Chapter 12, "Introductions and Reintroductions," and Chapter 13, "Management Approaches," highlight some of ongoing challenges and various successes in the conservation of semi-aquatic mammals.

12

Introductions
and Reintroductions

Why, sometimes
I've believed
as many as six
impossible things
before breakfast.

Lewis Carroll, 1872,
*Through the Looking
Glass and What
Alice Found There*

Translocations

In 2013, the IUCN and Species Survival Commission (SSC) released their *Guidelines for Reintroductions and Other Conservation Translocations*, a document that provides direction to ensure that the movement of organisms from one location to another is thoroughly evaluated and done with risk assessments and well-founded decision making in mind. Their definition of term **translocation**, "the human-mediated movement of living organisms from one area, with release in another," broadly encompasses accidental and intentional releases of organisms from both wild and captive origins (IUCN/SSC 2013: 1). Within the context of **conservation translocations** (translocations for deliberate conservation benefits), the goal of restoring wildlife populations can be accomplished through **reinforcement translocation** (release into an existing population of the same species) or **reintroduction** (release into a former but unoccupied range).

A third form of translocation is **conservation introduction**, which involves the release of a species into an area outside its natural range. Within the realm of conservation introductions, **assisted colonization** endeavors to prevent extinction of a species of concern by placing it in a novel range, while **ecological replacement** aims to have a translocated species perform a specific ecological function that most often was previously accomplished by a now extinct species. Traditionally, the term

introduction involved releasing a species (captive or wild) into an area outside its natural range, either through accidental or deliberate means (Kleiman 1989). There have been notable translocations involving fifteen species of semi-aquatic mammals, some of them intentional and focused on conservation, others less deliberate, yet resulting in dramatic ecological consequences. The inadvertent release of the common hippopotamus (*Hippopotamus amphibius*) into Colombia from the estate of the late drug lord Pablo Escobar (as described in Chapter 3) represents a classic accidental introduction of a non-native species, but its introduction had less obvious consequences than the introductions of the species described in this chapter.

For conservation translocations to be successful, there must be adequate funding, long-term availability of suitable habitat, public support, and the ability to safely move and care for the target animals (Kleiman 1989). Most importantly, there needs to be a self-sustaining source population, either from captive breeding programs or viable wild populations of similar genetic stock as the original populations being augmented or replaced. The research team involved in a program called the Scottish Beaver Trial (in Knapdale Forest in western Scotland) took a precautionary approach and used a Norwegian population of Eurasian beavers (*Castor fiber*) as the source for beaver reintroductions in Scotland, given their close proximity to the United Kingdom (Gaywood 2018). The reintroduction program included extensive preplanning, execution, and follow-up. Follow-up monitoring and research are critical, yet are often the least supported aspects of many translocation initiatives. The Scottish Beaver Trial provides an excellent example of a comprehensive and well-monitored translocation program.

Conservation translocations of species into their native range have involved eleven species, including the North American beaver (*C. canadensis*) within North America. Some translocations aim to reinforce existing populations, while others are reintroductions with the hope of re-establishing wild populations in areas where the species was previously extirpated. The platypus (*Ornithorhynchus anatinus*) and the European mink (*Mustela lutreola*) have been introduced to islands immediately adjacent to their natural ranges to establish "back-up" populations to some degree.

Prior to modern taxonomic approaches, some North American species were introduced well outside their range under the assumption that they were the same species as their rapidly declining European counterparts (Morgan 1868). Such was the case for some North American beaver and American mink (*Neovison vison*) introductions of the 1920s and 1930s, both of which present ongoing conservation challenges for indigenous Eurasian beaver, European mink, and many other species. Other than the aforementioned localized introduction of the common hippopotamus, there are four species of semi-aquatic mammals that have been introduced outside their range onto more than one continent: the North American beaver, coypu (*Myocastor coypus*), muskrat (*Ondatra zibethicus*), and American mink. Each species has more than one reason that initiated its introduction, but all are of conservation concern. Regardless of the goals and outcome of translocations, whether through the introduction of non-native species or the reintroduction of native ones, there are lessons to be learned that can help guide our future management of freshwater habitats and the species they support.

Introductions

Surprisingly, most introductions go unnoticed and are unsuccessful (Dodds and Whiles 2010); however, those species that do succeed tend to share common characteristics. As seen in the reproductive outputs of coypus and muskrats, successful invasive species are able to produce large numbers of offspring that are relatively independent at an early age. Each of the invasive species of semi-aquatic mammals had human assistance in establishing themselves outside their range. Sometimes they benefited from several release attempts before becoming firmly established. Their novel habitats often lacked effective predator populations, and the species' own characteristics allowed them to become effective competitors. In the case of coypus, the vegetation they were expected to eat and control as "weed eaters" was less palatable than the native vegetation that managers hoped to restore, and the predators that it was assumed would keep coypu populations in check (alligators) were generally ineffective (Bertolino, Guichón, and Carter 2012).

Unlike many other invasive species that do succeed in establishing themselves but integrate into the environment without major ecosystem effects (Dodds and Whiles 2010), the four species of invasive semi-aquatic mammals have had ecosystem-wide impacts that will be difficult, if not impossible, to reverse. Top predators, including American mink, tend to have community-level effects, and in areas where existing native assemblages are disturbed, establishment of invasive species is easier. As seen with southern water voles (*Arvicola sapidus*) in Spain, larger habitat patches appear to be more resistant to the invasion of American mink than smaller patches (Pita, Mira, and Beja 2010). Translocations, regardless of intent, succeed for a myriad of reasons that must be carefully balanced by wildlife managers.

North American Beaver

Once widespread in Eurasia, Eurasian beaver populations were near collapse by the end of the nineteenth century, with only 1,200 individuals remaining within eight isolated populations (Nolet and Rosell 1998). Introductions of North American beavers into Eurasia began in the early twentieth century with unsuccessful introductions of ten beavers into Russia between 1927 and 1933, mainly to augment hunting and trapping for furs (Safonov 1975). These ten animals were combined with eight Eurasian beavers from the Voronezh region in the southern Steppes of Russia. At the time, there was no realization they were two distinct species of beavers. Established in 1923, the Voronezhsky Nature Reserve continues to serve as a research and breeding facility for beavers (Fig. 12.1); it is a United Nations Educational, Scientific, and Cultural Organization (UNESCO) Biosphere Reserve (Romashova 2016). Despite early setbacks, introductions of more than ten thousand beavers continued from the 1930s to the early 1960s with greater success. Both Eurasian and North American beavers were introduced to the Amur River in eastern Russia, in Kamchatka, and in other poorly documented locations (Halley and Rosell 2002).

Elsewhere in Europe there were other releases. A few North American beavers were released into Poland in the 1930s, with populations intact until at least the 1950s, after which these populations are believed to have gone extinct (Nolet and Rosell 1998; Halley and Rosell 2002). Seven North American beavers from the United States were introduced into Finland in 1935 and 1937, resulting in as many as 1,200 colonies being established in the South Savo Game Management District by 1997

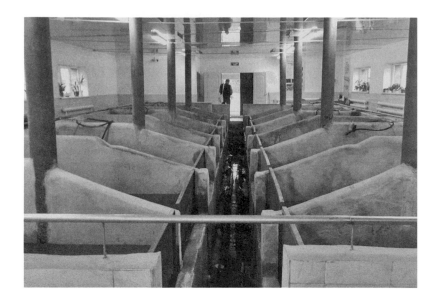

Figure 12.1. Breeding facility for North American and Eurasian beavers at the Voronezhsky Nature Reserve, Russia, in 2015. G. A. HOOD

(Härkönen 1999). Following the success in Finland, North American beavers were translocated from there to other locations in northwestern Europe (Fig 12.2).

Currently in Finland, North American beavers occupy 168 municipalities and Eurasian beavers occupy only twenty-nine; however, Eurasian beaver populations are increasing, while North American ones are declining (Brommer et al. 2017). In addition to being independently released into Russia, North American beavers have also dispersed from Finland into Russian Karelia (Halley and Rosell 2002). Additionally, three North American beavers were released into Paris in 1985, but were later removed from the wild (Moutou 1997), and fifteen beavers were released into the Danube in Austria (Nolet and Rosell 1998). The Austrian population is presumed extinct (Halley and Rosell 2002). A paper by Howard Parker and his colleagues (2012) provides an excellent historical and current assessment of invasive North American beavers in Eurasia.

It was once presumed that higher fecundity of North American beavers might give them some competitive advantages. As discussed later in

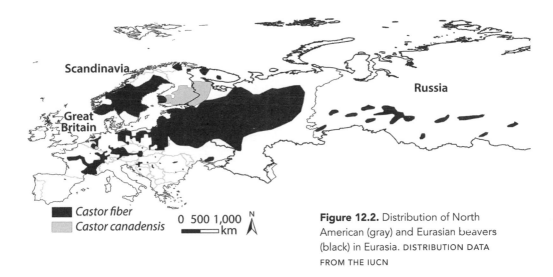

Figure 12.2. Distribution of North American (gray) and Eurasian beavers (black) in Eurasia. DISTRIBUTION DATA FROM THE IUCN

this chapter, reintroductions of Eurasian beavers have helped re-establish native populations in many parts of Eurasia, and these populations even appear to outcompete North America animals in some areas (Danilov, Kanshiev, and Fyodorov 2011). Fortunately, given their chromosomal differences, the two species cannot interbreed and produce viable off-spring. There remains, however, the problem of well-established populations of North American beavers in Europe, particularly in Finland where there were an estimated 10,500 North American beavers and only 1,500 Eurasian beavers detected during a 1998 survey (Ermala, Lahti, and Vikberg 1999). There are ongoing concerns that North American beavers could move into northern Sweden (G. Hartman, pers. comm.), but increased research and monitoring are required to fully assess this risk. Currently, northern Fennoscandia is still vulnerable to future colonization by North American beavers (Parker et al. 2012).

North American beavers have also been inadvertently introduced into Argentina and Chile in South America. Although distinct species in Europe, the ecological behaviors and roles of both North American and Eurasian beavers in Eurasia are almost identical. As such, ecosystem responses are well adapted to their presence. In South America, evolutionary adaptations of plants and other animals are completely lacking. The environmental impacts of beavers there have been devastating. Fifty North American beavers from Manitoba, Canada, were originally introduced by the Secretary of the Navy of Argentina in 1946 to Fagnano Lake, Tierra del Fuego (TDF) archipelago. The intent was to enrich native fauna and establish a local fur industry (Pietrek and Fasola 2014). Since then, beavers have expanded their range as far north as Puerto Natales, Chile, with potential to colonize continental Patagonia (Pietrek and Fasola 2014). Christopher Anderson and his colleagues call the beaver's invasion of southern South America the "largest alteration to TDF's forested biome in the Holocene" (Anderson et al. 2014: 214).

In South America, there are no endemic equivalents to the beaver's foraging and dam-building behaviors. Ecosystem engineering by beavers has completely transformed hydrological networks wherever beavers are found in South America, and in TDF beavers have transformed 20% to 43% of these networks (Anderson et al. 2014). In these subtropical ecosystems, which are normally composed of flowing streams mixed with forests and grasslands, beavers have since built dams and their associated pond networks. Southern beech trees (*Nothofagus pumilio*) are the beavers' primary food source, and their felling by beavers has completely changed the structure and composition of riparian plant communities (Graells, Corcoran, and Aravena 2015). Various eradication and control programs, mainly based on commercial incentives for fur, have been established, with varying results. Recently, new approaches have emerged in hopes of eradicating all beavers from their current South American range and recovering/restoring affected environments (Menvielle et al. 2010). These approaches require multijurisdictional efforts and agreements. Now that beavers have started to colonize mainland South America, the threat of further expansion poses increasingly significant environmental concerns.

Coypu

As with many other semi-aquatic mammals, the coypu is valued for its fur, with France being the first county to attempt farming coypus in the early 1880s (Carter and Leonard 2002). The extensive fur farming that was established in the coypu's native South America in the 1920s then expanded to North America, Europe, Asia, Africa, and the Middle East. Along with fur farming and related releases/escapes, coypus were intentionally released into non-native habitats to help control aquatic

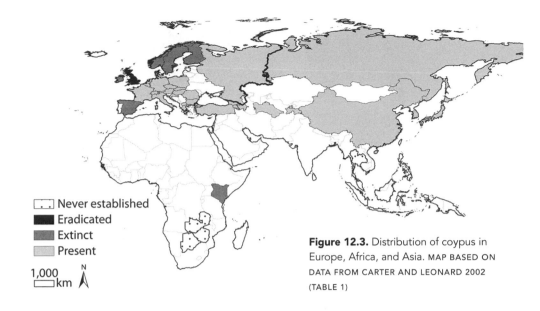

Never established
Eradicated
Extinct
Present

1,000 km

N

Figure 12.3. Distribution of coypus in Europe, Africa, and Asia. MAP BASED ON DATA FROM CARTER AND LEONARD 2002 (TABLE 1)

vegetation (Dozier 1952). They are now a significant pest species on all continents except Australia and Antarctica; they are also absent in New Zealand (Carter and Leonard 2002). In many places, coypus escaped or were released from fur farms when fur prices collapsed. Some non-native populations survived and flourished, while others either failed to establish, were extirpated, or went extinct (Fig. 12.3).

In North America, the first coypus were imported for fur farming at Elizabeth Lake, California, in 1899, and they were brought into Canadian zoos by 1900 (Bounds, Sherfy, and Mollett 2003). Captive populations were maintained at various fur farms in California, Washington, Oregon, Michigan, New Mexico, Louisiana, Ohio, and Utah in the 1930s. Given its propensity for heavily grazing riparian vegetation, the coypu was promoted for vegetation control in the 1930s and 1940s and was then transplanted throughout the southeastern United States (Dozier 1952). As such, coypus were released into wetland areas of Alabama, Arkansas, Georgia, Kentucky, Maryland, Oklahoma, inland Texas, and Louisiana, where they were also trapped for furs (Willner, Chapman, and Pursley 1979; Carter and Leonard 2002). Of those states, Kentucky is the only one where coypus never established despite introduction into the wild (Fig. 12.4). They are still hunted for furs in many states where they maintain viable populations.

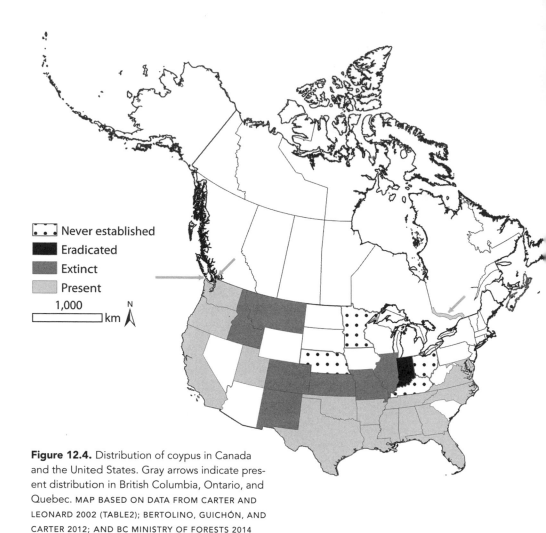

Figure 12.4. Distribution of coypus in Canada and the United States. Gray arrows indicate present distribution in British Columbia, Ontario, and Quebec. MAP BASED ON DATA FROM CARTER AND LEONARD 2002 (TABLE2); BERTOLINO, GUICHÓN, AND CARTER 2012; AND BC MINISTRY OF FORESTS 2014

Their introduction from their native habitat in South America into to wetland ecosystems outside their natural distribution has created significant ecological issues (Bounds, Sherfy, and Mollett 2003). They have experienced differential survival following introduction but remain widely distributed globally. Coypus are well known for significant overgrazing of riparian plant communities, destabilizing natural embankments along waterbodies, and crop predation (Bounds, Sherfy, and Mollett 2003). Alteration of wetland vegetation by coypus also displaces native populations of muskrats and promotes the growth of invasive plant species that further degrade these aquatic habitats (Carter and Leonard 2002).

In Europe, they outcompete southern water voles and Russian desmans (*Desmana moschata*) for breeding sites. Although there has been tremendous effort to remove introduced populations, it is no easy task.

After a dedicated program was established in Britain in 1971, wildlife managers were able to fully eradicate coypus by 1987 (Gosling and Baker 1989). In the United States they were eradicated from California in the 1970s, but several reproducing populations have re-established, as indicated by the discovery of a pregnant female in 2017 and subsequent sightings of other animals. The coypu has a limited distribution in three Canadian provinces (British Columbia, Ontario, and Quebec); it failed to establish viable populations in Nova Scotia (see Fig. 12.4). Recent reports note that they cannot yet successfully colonize in Ontario because of cold winter climates (Bertolino, Guichón, and Carter 2012). Although Quebec is equally cold, coypus continue to be seen in southern areas of the Ottawa River, likely through ongoing northern dispersal. With warming winter climates, non-native range expansion to the north is more likely.

Muskrat

Over a century ago, muskrats were released into several locations in Europe, often several times into the same areas. In 1905, Prince Colloredo Mannsfeld released five muskrats (two males and three females) onto his estate near Dobris (Dobříš) in what is now the Czech Republic (Becker 1972). These muskrats originated from southeastern Canada. As with other semi-aquatic mammals, fur farming was the central reason for their introduction outside of their native distribution. Subsequent escapes into the wild were inevitable. In his 1972 article, "Muskrats in Central Europe and Their Control," German zoologist, Kurt Becker states that "These animals must have been very prolific. During the first ten years they spread out from Dobris in concentric circles. Up to 1913, the radius of expansion increased by between four and thirty km annually" (Becker 1972: 18). By 1914, they had spread to southern Germany (Becker 1972), and by 1920 they had dispersed into Poland (Brzeziński et al. 2010).

Several other introductions and natural colonization events by feral muskrats occurred around the same time period. Five pairs were also in-

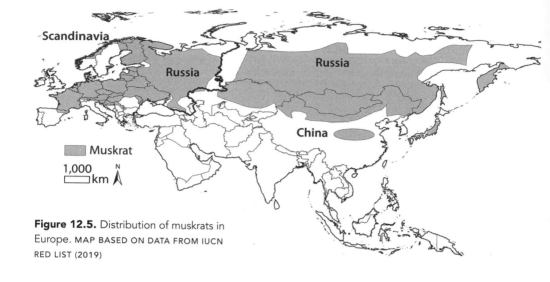

Figure 12.5. Distribution of muskrats in Europe. MAP BASED ON DATA FROM IUCN RED LIST (2019)

troduced to Austria and Hungry in 1905, with muskrat distribution covering approximately 40% of Austria by 1927 (Skyrienė and Paulauskas 2012). Following World War I, there appears to have been an increased effort to introduce muskrats to Finland (1920s to 1930s), the Vosges Mountains of France (1920s), Britain (1927), Belgium (1928), and Russia (1928 to 1932) (see references in Skyrienė and Paulauskas 2012). From these introductions, muskrats expanded their distributions into Switzerland and Alsace by 1930. Intentional releases started to involve very large numbers of muskrats, particularly in Russia, with eighty thousand animals released in many different areas (Danell 1996). There were an estimated 160,000 muskrats in the former Soviet Union by 1955. Finland also experienced repeated muskrat introductions, and it also supplied muskrats for introduction to Sweden and Norway (Danell 1996). Muskrats are now found throughout Europe (Fig. 12.5), including populations in Estonia (1947), Lithuania (1954), Latvia (1961), and Denmark (1989) (Skyrienė and Paulauskas 2012); however, early management actions allowed the British to eradicate muskrats by 1939.

Muskrats are also an invasive species on Navarino Island in the Cape Horn Biosphere Reserve in Chile (Schüttler, Cárcamo, and Rozzi 2008). As in their native habitat, muskrats are an important prey species for introduced American mink in Chile. A similar situation occurs in Poland where muskrats were found throughout the entire country by the end of the 1950s, followed by steady population increases until the mid-1980s, when American mink expanded their range into Poland (Brzeziński et al. 2010). Fur harvests in Poland went from 66,416 muskrats in the 1987–1988 hunting season to 4,567 muskrats in the 2007–2008 season. Marcin Brzeziński and his colleagues (2010: 341) determined that "mink predation is one of the most important factors in the decline of the muskrat population in Poland."

Elke Schüttler and her team (2008) suggest that the presence of muskrats in the Cape Horn area could have facilitated the arrival of mink to the area in reference to Simberloff and Von Holle's 1999 "Invasional Meltdown" hypothesis. This hypothesis states that non-native species aid each other's invasions, which in turn increases their impact on the native community. The Invasional Meltdown hypothesis is difficult to demonstrate in the wild, but it sets an interesting context for biological invasions. Even without the combined effects with American mink, muskrats (as with coypus) are known to compete for breeding sites with native semi-aquatic mammals such as European water voles (*Arvicola amphibius*) and Russian desmans (IUCN 2019). They also consume large amounts of aquatic vegetation and can change the structure and composition of vegetation communities.

A more difficult issue related to invasive species involves the other organisms they bring with them. When muskrats were introduced from North America into the former Soviet Union, four of a potential twenty-eight species of parasites survived the translocation—two trematodes (*Notocotylus quinqueserialis* and *Echinostoma armigerum*) and two nematodes (*Longistriata dalrymplei* and *Rodentocaulus ondatrae*) (Skyrienė and Paulauskas 2012). In Britain, muskrats arrived with two species of helminth trematodes (*N. quinqueserialis* and *Hymenolepis evaginata*). Similar situations exist elsewhere in Europe. Parasitic ticks also accompanied North American muskrats upon release into Lithuania (i.e., *Laelaps multispinosus*). Moving species and their natural parasite loads outside their native distributions is very much like "moving the zoo" (B. Samuel pers. comm.). Even when the primary invasive species has been eradicated, its hitchhikers very likely remain.

American Mink

The American mink traditionally has been highly valued for its fur and, as such, has been established in fur farms ranging from its native North America to Europe, South America, and Asia. The 2019 IUCN Red List notes thirty-four countries in which it is now considered an introduced species (extant and introduced: Belarus, Czech Republic, Denmark, Estonia, Finland, France, Germany, Iceland, Italy, Latvia, Lithuania, Montenegro, Norway, Poland, Portugal, Russian Federation, Serbia, Spain, Sweden, Ukraine, and United Kingdom; presence uncertain and introduced: Belgium, Hungary, Luxembourg, Netherlands, Slovakia, and Slovenia; extant and introduced (resident): Argentina, Austria, Chile, Greece, Ireland, Japan, and Romania). As with the other invasive semi-aquatic mammals, escapes and releases of American mink from fur farms have become routine. If fur prices falter, vigilance does as well. Although it is widespread within its introduced range, the American mink's presence and abundance differs depending on location. For example, they do not appear to breed in Belgium and the Netherlands, yet there are viable populations in nearby France and Germany (Bonesi and Palazon 2007).

American mink were first introduced to Europe in the 1920s (Dunstone 1993), with Sweden being one of the earlier countries to establish fur farming (Gerell 1967). Because of the production of higher fur quality in colder climates, most mink farms are in more northerly countries in Europe. With 2,200 fur farms in the year 2000, Denmark has the highest number of fur farms by far (Bonesi and Palazon 2007). Despite having an active mink farming industry in the United Kingdom since the 1950s, fur farming was banned there in 2003. There are now widespread feral populations of mink in the United Kingdom, but with notable declines as Eurasian otter (*Lutra lutra*) populations recover (Bonesi and Macdonald 2004). Biologists Laura Bonesi and Santiago Palazon (2007) provide extensive documentation of first records of mink in Europe and ecological impacts of mink in "The American Mink in Europe: Status, Impacts, and Control."

Outside of Europe, American mink are present in Japan on the major islands of Hokkaido, Honshu, and potentially Kyushu (Shimatani et al. 2010a, 2010b). In China, most mink farms, and possibly some feral populations, are found in the northeastern part of the county (Bonesi and Thom 2012). American mink in South American are currently confined to Argentina and Chile, but their range is expanding since their introduction from fur farms from the United States in the 1940s (Jaksic et al. 2002). As in other locations, feral mink have become problematic. A recent study along the Andean Patagonian forests of Argentina determined that mink now inhabit a contiguous range extending at least 800 km, along with colonizing Tierra del Fuego (Fasola et al. 2011). The researchers estimated an average straight-line rate of expansion from south to north at 6.9 km/year, which translates to 152 linear kilometers in twenty-two years. Such expansion rates have important ecological implications given the generalist diet of mink and its ability to live in coastal and inland habitats.

There are numerous ecological impacts resulting from the introduction of American mink onto other continents (see Bonesi and Palazon 2007: Table 4). Similarly, in its native distribution, interactions between feral mink from mink farms and wild mink contribute to outbreeding depression and the spread of disease (Bowman et al. 2007). Feral mink, regardless of location, predate on a variety of native prey, including mammals (e.g., European water voles and Pyrenean desmans, *Galemys pyrenaicus*), birds (e.g., waterfowl, ground-nesting birds, and gull and tern colonies), amphibians (e.g., common frog, *Rana temporaria*), and crustaceans (e.g., stone crayfish, *Austropotamobius torrentium*). They are also in direct competition with the European mink and Eurasian polecats (*Mustela putorius*), both of which are similar in size (Bonesi and Palazon 2007). For many species, including the European water vole and the critically endangered European mink, predation and competition by American mink can result in dramatic population declines (Bonesi and Palazon 2007; MacPherson and Bright 2011).

Reintroductions and Assisted Colonization

Conservation translocations, including reinforcement and reintroductions, have had varying success for some species of semi-aquatic mammals. Although technically a form of conservation introduction, assisted colonization has been used to augment platypus and European mink populations with some tentative success (Maran 2003; Furlan et al. 2012). One of the challenges with conservation translocations involving native species is that population declines likely result from several known and unknown factors, and many of these factors might still exist in their original environment. Wildlife professionals might be able to mitigate some of these issues, but often cost, logistics, and public/political support present additional challenges.

Conservation translocations work best when there are clearly defined objectives ("strategy") that guide management actions ("tactics") before, during, and after the translocation (Batson et al. 2015). Understanding species behavior is central to whether the animals will adapt to their new environment. Too often I have heard of **hard releases** of beavers and muskrats into the wild that are done without any understanding of their long-term foraging needs or social behaviors. Putting a beaver into an already occupied pond is an immediate trigger for territorial behavior in the resident beavers; this often results in injury or death for one or both parties, yet so often it is done for fear that the beaver being released will be alone. Preplanning also allows for an assessment of the carrying capacity of the site, which in turn can help determine how many individuals of each sex to release, how much food is available, and future capacity for dispersal. Setting up a conditioning enclosure to aid in a **soft release** can also increase the success of translocations of captive-born animals, but conversely it can induce stress in wild-born animals (IUCN/SSC 2013).

There are several considerations specifically related to the animals targeted for release. As seen with the Scottish Beaver Trial, selecting the best individuals for both genetic and physical suitability is important (Gaywood 2018). Preconditioning the animals for their new environment, determining the correct demographic characteristics of the new population, and postrelease management (e.g., intervention, dispersal, and reproduction) are important tactics within a well-planned strategy. Habitat suitability is critical, as is ensuring there are enough long-term resources (either naturally or augmented) and that various threats are controlled. Testing the animals for diseases that might transfer to the new environment is essential. Finally, how and when animals are released should be based on their behavioral biology relative to cover and timing (e.g., age of dispersal and time of day). For beavers, releasing individuals younger than a year old often results in their death, given their natural dependence on their parents until age two (McKinstry and

Anderson 2002). In many cases, especially depending on the era of the translocation, all or none of these considerations were applied to reintroductions of semi-aquatic mammals.

Platypus

The platypus is found within much of its pre-European range, except for reduced populations in the lower reaches of the Murrumbidgee and Murray River systems in South Australia (Grant and Temple-Smith 2003). In particular, agricultural practices, urban development, and forestry have impacted the ecological health of stream and riparian habitats required by the platypus. Major metropolitan areas have reduced water quality, degraded streambanks, developed weirs and other means of flow regulation, and introduced exotic species (Grant and Temple-Smith 2003). More recently, ongoing drought and unprecedented bushfires have become immediate threats.

For reasons not fully documented, one female and two male platypuses were introduced to Kangaroo Island off the south coast of Australia in 1928 (Furlan et al. 2012). They are generally assumed to have come from both Tasmanian and Victorian individuals. In 1941, five males and five females from the Healesville Sanctuary in Victoria were also released on the island, and another three males and three females were added in 1946 (Furlan et al. 2012). These platypuses developed into a self-sustaining population along the Rocky River at the western end of the island. In 2012, Elise Furlan and her colleagues estimated that there were 110 individuals on the island and that the current habitat was likely at carrying capacity.

Genetic results of this assisted colonization reveal that the small population size on Kangaroo Island and limited opportunity for founders have resulted in a loss of ~30% of the population's original genetic diversity. Interestingly, the genetics of the island population are shared with individuals from east-central Victoria's upper Yarra River but not with Tasmanian populations (Furlan et al. 2012). Increasing genetic diversity through additional reintroductions would be tricky given the presence of platypus mucormycosis disease in wild populations in Tasmania (Gust and Griffiths 2009). Conducting additional assisted colonization to other rivers on the island from island-born platypuses appears to be the better choice for increasing populations and reducing inbreeding (Furlan et al. 2012).

Elsewhere, the Australian Platypus Conservancy has conducted the only successful reintroduction of platypuses to Cardinia Creek in 2004, after large bush fires damaged the riverine habitat in 1983 (https://platy-pus.asn.au).

Swamp Rabbit

Swamp rabbits (*Sylvilagus aquaticus*) have a limited distribution in the bottomland forests of the southeastern United States, with a northern distribution as far as southern Illinois and Indiana. The species as a whole is not of conservation concern, despite declining population trends. However, there is some concern in southern Illinois, where habitat loss and fragmentation have resulted in population declines and patchy distributions (Watland, Schauber, and Woolf 2007). In an effort to maintain connectivity and recolonization potential among remaining habitat patches, Angela Watland and her colleagues (2007) conducted a feasibility study followed by a translocation of swamp rabbits in southern Illinois. After identifying suitable habitat and release sites for swamp rabbits, they released nine male and eight female swamp rabbits along the Little Wabash River in Wayne County.

Given the wild origin of the source animals, they released the translocated animals within twenty-four hours of their capture, after which they used telemetry to monitor postrelease survival. Fourteen of the seventeen translocated rabbits died following translocation, eight of them within seven days postrelease (Watland, Schauber, and Woolf 2007). In total, predators killed ten rabbits. The lack of success of this translocation might be related to the rabbits being in poor condition prior to release, which might have been detected if soft release had been used instead. Predation rates were also high, perhaps because of the rabbits' lack of familiarity with the new habitat (Watland, Schauber, and Woolf 2007). Preconditioning the site (e.g., predator reduction) might have helped overall survival, but it would be controversial and could result in unforeseen ecological impacts. At the core of the issue is reduced habitat availability and connectivity, which cannot be mitigated with translocations.

Marsh Rabbit

The translocation of marsh rabbits (*Sylvilagus palustris*) in Florida has had some success (Faulhaber et al. 2006). Currently these rabbits are threatened with habitat loss and fragmentation, as well as rising sea levels resulting from climate change. Reintroductions into areas of suitable habitat might increase resilience to changing environments. Translocation of twelve Lower Keys marsh rabbits (*S. p. hefneri*) into Little Pine Key in 2002 and seven into Water Key in 2004 resulted in an 81% and 100% success rate, respectively. As with the swamp rabbit translocation, wildlife biologist Craig Faulhaber and his research team used wild-caught marsh rabbits for translocation, which can increase translocation success (Fischer and Lindenmayer 2000). All animals (except one female and one male that were killed by predators) survived for more than five months. Since Hurricane Wilma in 2006, there has been patch abandonment in various areas within the Florida Keys (Schmidt et al. 2011), which presents additional considerations for translocation plans as extreme weather events increase. Additionally, any translocations must be preceded by removal of exotic predators, such as free-ranging cats and pythons (Cove et al. 2018).

North American Beaver

The North American fur trade began in earnest by the seventeenth century after early trading conducted by explorer Jacques Cartier from 1534 to 1542 (Müller-Schwarze 2011). While describing previous beaver numbers in his 1908 book *Artificial Preservation of Timber and History of the Adirondack Beaver*, Harry V. Radford stated that "they seem to have completely possessed the land, and to have been abundant almost beyond our present conception" (Radford 1908: 396). After almost three hundred years of intensive trapping and hunting, most beaver populations on the continent were at a breaking point. The records of the Hudson's Bay Company provide a stunning glimpse into the number of beaver pelts exported just to Britain. Over 4.7 million pelts were sold to the British auction houses between 1769 and 1868 (Hood 2011). There were, of course, similar auction houses elsewhere in Europe.

By the early twentieth century, the absence of beavers was reflected in drying ponds across the continent and dramatic decreases in the species they supported. There was also a desire to re-establish beavers as a game species. Efforts to restore extirpated beaver populations began by the early 1900s. Beaver populations that were nearly or completely extirpated from many areas of the east coast by the early 1700s were restored through reintroductions to several US states with a release of twenty Ontario beavers into the Adirondacks between 1901 and 1906 (Müller-Schwarze 2011). Seven of those beavers were purchased from the Canadian exhibit at the Louisiana Purchase Exposition in St. Louis (Radford 1908). By 1915, the Adirondack population was estimated to be at least fifteen thousand beavers. Similar reintroductions occurred in many parts of the United States and Canada.

In Canada, concerted beaver reintroductions began in the 1940s, including the translocation of seven beavers from Banff National Park to Elk Island National Park in 1941. Shortly thereafter, in 1948, there was another translocation of beavers from Prince Albert National Park to Wood Buffalo National Park in Canada's subarctic (Hood 2011). Similar translocations occurred elsewhere in the country, but most are poorly documented. There appears to have been a culture of translocating beavers in the 1930s and 1940s in North America, as reflected in reinforcement releases in Kansas between 1930 and 1950. Today there is a move to use beavers as an "ecological restoration tool," with legal and illegal assisted colonization occurring in many states and provinces, although such projects have mixed results and follow-up for many of them is rare (Petro, Taylor, and Sanchez 2015). Released animals do not always stay where intended and their ecological impacts require long-term monitoring.

Releases also occur following rehabilitation, such as the successful rehabilitation and release of three bitumen-contaminated beavers in east-central Alberta, Canada (Fig. 12.6; Hood 2015). Two of these beavers produced at least two litters in just over two years.

Eurasian Beaver

Only eight remnant populations of Eurasian beaver remained in Europe and Asia by the start of the twentieth century. These were small isolated populations in France (~30 individuals), Germany (~200 individuals),

Figure 12.6. Release of three rehabilitated beavers following bitumen contamination in Alberta, Canada.
G. A. HOOD

southern Norway (~100 individuals), Belarus and Ukraine (~290 individuals), southern Russia (~70 individuals), western Siberia (~30 to 40 individuals), and China (~100 to 150) (Nolet and Rosell 1998). By the late 1990s, there were approximately fifteen reintroductions throughout Europe. As in North America, some reintroductions occurred mainly in the interwar period of the 1930s to 1940s, for example, in Finland (1935–1937), Germany (1936–1940), Latvia (1927–1952), Norway (1927–1932), Poland (1943–1949), Russia (1927–1941), and Sweden (1922–1939). Other translocations range from the late 1940s to present day, for example, in Austria (1970–1990), Belgium (1998–1999), Bulgaria (2001–2002), Croatia (1996–1998), Czech Republic (1991 and 1996), Denmark (1999), Estonia (1957), Finland (1995), France (1958–1995), Germany (1966–1989; 1999–2000), Hungary (1980–2000), Lithuania (1947–1959), Luxembourg (2000), Mongolia and China (1959–1985), the Netherlands (1988–2000), Norway (1952–1965), Poland (1975–1986), Romania (1998–1999), Russia (1946–1964), Scotland (2009), Slovakia

(1995), Slovenia (1999), and Switzerland (1956–1977) (Nolet and Rosell 1998: Table 1; Halley and Rosell 2002: Table 1). There have also been unsanctioned and accidental releases of beavers throughout their current range.

The establishment of the Voronezhsky Nature Reserve in 1923 (Fig. 12.7), along with the Berezinsky Nature Reserve in Belarus (1925) and the Kondo-Sosvinsky Nature Reserve in Western Siberia (1927), created dedicated beaver reserves designed for the "protection and restoration of beaver populations in natural habitats of Russia" (Romashova 2016: 10). These reserves received approximately fifty animals found in the Usman, Ivnitsa, and Voronezh rivers in the southern steppes of Russia (Romashova 2016). The Voronezhsky Reserve was the primary source of breeding animals used in the reintroduction of beavers into the former Soviet Union and other countries. The first group of beavers, consisting of eight individuals, was relocated to Lapland Biosphere Reserve in 1934 (Romashova 2016). By the 1950s and 1960s, a large-scale translocation operation was in place, with beavers sent to Poland and Germany in the west and as far east as Mongolia. By 1977, approximately three thousand beavers had been relocated from the Voronezhsky Reserve to fifty-two regions of the former Soviet Union and many other countries in Europe (Romashova 2016). The reserve still maintains an active breeding facility for both Eurasian and North American beavers. The scientific contributions resulting from studies at the reserve are truly understated.

The outcomes of these translocation efforts are as broad as the geographic context of beaver translocations in Eurasia. Göran Sjöberg and John Ball's 2011 book, *Restoring the European Beaver: 50 Years of Experience*, highlights the positive influence the reintroduction of Eurasian beavers has had on aquatic macroinvertebrates and riparian and aquatic plant communities. The book also covers issues regarding human-beaver conflicts resulting from increasing beaver populations on a continent with high human population densities and associated land conversion.

From its listing as a near threatened species in 1996 to a species of least concern today, the Eurasian beaver has benefited from well-coordinated translocation programs, mostly supported by science-based management.

Figure 12.7. Entrance to the offices of the Voronezhsky Nature Reserve, Russia, in 2015. G. A. HOOD

Russian Desman

The Russian desman faces numerous threats, including wetland drainage, dams and diversions, removal of riparian vegetation, agricultural expansion, and bycatch in fishing nets, to name a few (IUCN 2019). Its population numbers showed a marked decline in the early twenty-first century, despite being considered a common game species in the nineteenth century (Bakka et al. 2018). In response, fifty-one desmans were reintroduced to nine lakes on the Kerzhenets floodplain in Kerzhenskiy State Nature Reserve in northwest Russia from 2001 to 2002 (Bakka et al. 2018). After the release of eleven animals into Krasnyy Yar Lake in 2001, there was some overwintering success. In 2002, ten of the forty desmans released into six lakes were radio-tagged and monitored daily. Five tagged animals died, and another four could not be detected within the first weeks of the study. Four other desmans that had dispersed from their release site were later found dead in fishing nets (Bakka et al. 2018).

Despite these early setbacks, fifteen years of monitoring (2005–2017) determined that a resident population was present along 50 km of the

Kerzhenets River. Populations remain low, with at least twenty-five animals in 2005 declining to between six and ten individuals in 2017. The low success rates are certainly associated with ongoing mortalities in illegal fishing nets, as well as competition for breeding habitats with invasive muskrats, declines in prey species, drought, and eutrophication of shallow lakes (Bakka et al. 2018). Desmans are also preyed upon by domestic cats (Khlyap, Bobrov, and Warshavsky 2010). Desmans were reintroduced to Ukraine in the 1950s and rediscovered in the Desna River (a tributary of the Dnepr). There have been recent reintroductions attempted in the area of Bryansk Forest Nature Reserve on the border of Ukraine and Russia (Kennerley and Turvey 2016), but follow-up assessments, if done, are not readily available.

European Mink

As one of the most endangered carnivores in Europe (Maran 2003), the European mink has been the focus of international conservation actions. Unlike the Russian desman, mink populations have been in decline since the 1800s. By the mid-nineteenth century, European mink had been extirpated from areas of Germany, all of Switzerland, and likely from Austria (Maran 2003). From the 1930s to 1950s, mink were extirpated from Hungary, Czech Republic, and Slovakia; they were extirpated from Finland, Estonia, Latvia, and Lithuania by the end of the twentieth century. Similar extirpations continued in more than 50% of the European mink's former range (Lodé, Cormier, and Jacques 2001). Wild populations still exist in southern France, Romania, Russia, Spain, and Ukraine. As mentioned previously, the European mink's decline is multifaceted, with habitat loss and degradation, overharvesting, introduced species (e.g., disease and American mink), and prey declines being key factors. Although introduced American mink currently have a significant impact on remaining populations of European mink, the European mink was in decline long before American mink arrived in Europe. The introduction of the American mink has just exacerbated an already dire situation. Ultimately, the combined effects of various human activities, including introduction of invasive species, are at the core of the species' decline.

The European Mink Conservation and Breeding Committee established an intensive, well-coordinated conservation breeding program for

European mink beginning in the early 1990s (Maran 2003). By 1998, a special breeding facility was constructed at the Tallinn Zoo in Estonia, which can house over two hundred European mink. There are seventeen associated institutions that also breed European mink to support ex situ conservation efforts. From these captive animals, a small assisted colonization project of seventeen animals was established in 2000 on Hiiumaa Island in Estonia (Maran 2003). Prior to the release of the seventeen European mink, all American mink were removed from the island (a total of fifty-three animals), and the European mink were preconditioned to hunt for wild prey, avoid humans and predators, and use aquatic habitat.

Following a well-design follow-up plan, researchers radio-tracked the translocated mink and noted a 33% survival rate (Maran 2003). Natural predation was the main source of mortality. All other mink were in good condition, and they are now breeding in the wild. The conservation team continues to conduct ongoing research and habitat enhancement projects. The Estonian reintroduction is a good example of pre- and postrelease planning, which likely accounts for its success.

Another conservation reintroduction program was established at a nature reserve in Saarland, Germany (Peters et al. 2009). Beginning in May 2006, forty-eight captive-bred mink were released into the nature reserve over a two-year period. As with the Estonian release, biologists used a soft release and acclimatized the mink to the release site for two weeks prior to release. Researchers radio-tracked the mink for two years. Although some mink left the original release site, others established distinct home ranges and persisted over the study period (Peters et al. 2009). Other than some unmonitored releases of European mink in Russia, the only other program is in France. A special breeding facility was constructed in France in 2014, but to date there appear to be no documented reintroductions (Maran et al. 2016).

Eurasian Otter

As with many other semi-aquatic mammals, Eurasian otters underwent dramatic population declines beginning in the mid-twentieth century (Roos et al. 2015). Along with habitat loss, hunting and pollution (e.g., polychlorinated biphenyl, or PCB) were important factors in their decline (Mason and Macdonald 1993; Hung and Law 2016). Eurasian otters continue to face a number of threats, including bioaccumulation of pollutants, water acidification and associated loss of prey species, habitat alterations (e.g., mining, dams, and canalization), and poaching (Roos et al. 2015). To counter population decline, there has been a European breeding program since 1985 that aims to develop self-sustaining populations of Eurasian otters that can then be released to the wild (Vogt 1995).

A combination of eleven wild Norwegian otters and twenty-five Swedish captive-bred otters were reintroduced into south-central Sweden between 1989 and 1992 (Sjöåsen 1996). There was a differential survival rate of reintroduced otters from wild populations (79% survival) versus captive-bred otters (42% survival). When released into unoccupied areas, only 28% of eighteen otters established home ranges (Sjöåsen 1997). The remainder either died or moved to another area to establish a home range. Key determinants of success include all-season access to fish and high habitat quality, especially for breeding females. Overall, otter reintroductions have helped restore viable otter populations in Sweden.

In Britain, the Otter Trust raised thirty otter cubs to adulthood from 1976 to 1981 (Chanin 1985), which allowed for contributions to other captive breeding programs. In anticipation of releasing otters to the wild, the Otter Trust built special prerelease pens to reduce human contact and acclimatize the otters to their environment. In 1983, a male and two females were released in East Anglia. The reintroduction was a success and established a wild breeding population (Chanin 1985). Following this success, at least seventeen otters were reintroduced to the wild between 1983 and 1989 (Sjöåsen 1997). There are now viable otter populations in the United Kingdom after steep population declines by the mid-twentieth century. Eurasian otters have also been successfully reintroduced to the Netherlands (2002) and Spain (1998) (Hung and Law 2016); however, a 1975 otter reintroduction in Switzerland failed due to high levels of PCBs in the water (Weber 1990). As of 2009, Eurasian otters are once again in Switzerland through habitat improvements and natural colonization. With these reintroductions, a top aquatic predator is once again an active part of European ecosystems.

North American River Otter

North American river otters (*Lontra canadensis*) suffered a similar fate as the Eurasian otter. Once distributed across most of North America, their distribution was less than 75% of its original range by 1977 (Melquist, Polechla, and Toweill 2003). By 1980, they were extirpated in eleven US states and declining in several others. In Canada, they are extirpated from the province of Prince Edward Island (Serfass, Evans, and Polechla 2015). Key drivers of population declines were habitat loss and degradation and overharvesting. However, North American river otters have since recovered to nearly 90% of their historical range, in part due to reintroduction programs (Melquist, Polechla, and Toweill 2003).

Concentrated reintroduction efforts began in the 1970s in twenty-one US states and the province of Alberta, Canada (Melquist and Dronkert 1987). From 1976 to 1998, 4,018 otters were released into twenty-one states and one province (Raesly 2001). All locations conducted postrelease evaluations, and the reintroductions are generally seen as a positive contribution to the restoration of otter populations in North America. Along with reintroductions, improved furbearer management techniques and the relative improvement in water quality also aided in population recovery. To date, twenty-nine states and all Canadian provinces except Prince Edward Island have viable otter populations.

Marsh Deer

The marsh deer (*Blastocerus dichotomus*) originally had a much larger distribution in South America than it does today. It is primarily threatened by wetland loss, agricultural and livestock production, and construction of hydroelectric dams (Figueira et al. 2005). To address these population declines, the Porto Primavera Marsh Deer Project was initiated, from which the Marsh Deer Reintroduction Project was developed for the Jataí Ecological Station (Figueira et al. 2005). Researchers radio-tracked five deer from their release in December 1998 until April 2000. Two of the three females died during the study, while both males and a female survived. Although not a complete success, this reintroduction provided new behavioral and ecological information on what is otherwise an elusive and poorly understood species.

Père David's Deer

As described in detail in the previous chapter, Père David's deer (*Elaphurus davidianus*) presents an example of how a species that was removed from the wild almost 1,200 years ago can still benefit from translocation opportunities today (Yuan et al. 2017). Its release into the Hubei and Hunan provinces of China has resulted in four wild populations (Zhu et al. 2018). Ongoing monitoring and protection will be essential to the ongoing conservation of this species in the wild.

Conservation Considerations

As seen with beavers and otters, some translocations can be a success, as long as they are within the species' native habitat. Once invasive species take hold, they can be nearly impossible to eliminate and even more difficult to manage in their new environments. As evidenced by the mixed results of Russian desman and marsh rabbit reintroductions, the presence of invasive species creates an almost impossible environment for reintroduction efforts. Invasive predators, in particular, present an immediate threat to any species reintroduction effort. As was done for European mink in Estonia, a great deal of effort is required to rid the target habitat of invasive species prior to reintroducing native species. Although European mink are surviving on Hiiumaa Island, there is ongoing vigilance to prevent American mink from recolonizing the island. Reintroductions can be a costly, time-consuming, and logistically challenging means to conserve species at risk. As seen in Chapter 13, there is a suite of conservation actions that can aid species recovery and habitat restoration.

13

Management Approaches

There is still so much in the world worth fighting for. So much that is beautiful, so many wonderful people working to reverse the harm, to help alleviate the suffering. And so many young people dedicated to making this a better world. All conspiring to inspire us and to give us hope that it is not too late to turn things around, if we all do our part.

Jane Goodall, *Dr. Jane Goodall's New Year message for 2018*

Population Recovery

Since 1937, a beaver has been on the Canadian nickel (five cent coin), and it has remained so throughout my life. As a very young girl, I remember asking my mother about the coin and what the animal was on the side opposite Queen Elizabeth II. She told me it was a beaver and that when she was a girl, and really for much of her life, beavers were very rare and had almost gone extinct because of the fur trade. It was an amazing statement that has stuck with me to this day. The North American beaver (*Castor canadensis*), a species that had almost gone extinct, is now present in practically all of its former range, and its populations have recovered and now flourish in wetlands and streams across much of North America. It represents one of the rare success stories, and it provides some hope given the daunting challenges facing so many other semi-aquatic mammals. The population recovery of beavers succeeded because of legislative measures, harvesting restrictions, advocacy and education, targeted and nontargeted reintroductions, its own biological characteristics, and, to some degree, sheer luck. Similar success stories are reflected in the population recovery of Eurasian beavers (*C. fiber*), Eurasian otters (*Lutra lutra*), and North American river otters (*Lontra canadensis*). So many other species, however, remain in a precarious position, mainly due to human actions and inaction.

At the heart of population recovery are two essentials: (1) access to suitable habitats and viable connections to the ecological communities they support, and (2) the ability of a species to live in those habitats without threat of overexploitation and unsustainable predation pressures. Although the rate of human population growth has slowed, the growth rate is highest in areas known as **biodiversity hotspots**, particularly in Upper Amazonia, the Guiana Shield, the Congo River Basin, New Guinea, and the Melanesian Islands (Cincotta, Wisnewski, and Engelman 2000). In 2000, human population growth rates in these regions were significantly higher than for the world as a whole. The same trends exist today.

Continued increase in population growth rates is just one aspect of human impact on the environment. Even in low growth-rate countries, such as Canada and the United States, per capita consumption rates, as reflected by a country's **ecological footprint**, are some of the highest in the world (WWF 2018). Humans simply must consume (and waste) fewer resources. Current trends toward renewable energies and waste reduction provide some hope, but an ever-increasing population is a more difficult issue to address. Relative to the conservation of semi-aquatic mammals, consumption is directly linked to our management and conservation of ecological resources, especially in aquatic and riparian systems. Despite the daunting nature of current environmental trends, several strategies are presently in use or being developed to mitigate human impacts on the environment.

Habitat Protection and Enhancement

Protected Area Management and Aquatic Connectivity

The consequences of habitat loss, habitat degradation, and loss or impairment of connectivity among aquatic habitats pose the greatest threat to the persistence of semi-aquatic mammals and the habitats on which they depend. From a broader perspective, many other species, including humans, rely on these ecosystems for their wellbeing, and the loss of freshwater ecosystems has global implications both ecologically and economically (United States Environmental Protection Agency 2015). To maintain healthy ecosystems, the "ecological character" of these waterbodies must be maintained through the integration and recognition of biological, chemical, and physical processes (Brooks et al. 2011: 49). Several management tools, including water-related laws and regulations, are increasingly applied as key components within a broader context of wildlife conservation. At the broadest scale, international treaties aim to tackle global issues by encouraging multijurisdictional support.

The Convention on Biological Diversity (CBD) is an international treaty that was presented at the United Nations Conference on Environment and Development ("Earth Summit") in 1992. It became effective on December 29, 1993, and to date there are 168 signatories, including all United Nations member states (and the European Union), except the United States (https://www.cbd.int). The goals of the Convention include the conservation and sustainable use of global biodiversity. Within the Convention, Aichi target 11 advocates that 17% of all terrestrial land and inland waters and 10% of coastal and marine areas be protected by 2020 (Di Minin and Toivonen 2015). In 2019, the world sat very close to that target, with terrestrial protected areas at 14.7% and marine protected areas exceeding the target at 17.6%.

Despite the increase in protected terrestrial areas, freshwater biodiversity is in decline, and inclusion of freshwater habitats in protected areas is low, especially for large rivers (Abell et al. 2016). Globally, approximately 70% of river reaches lack any protection in their upper watersheds, and protection of inland waters as a whole (e.g., wetlands, lakes, rivers, and streams) is difficult to quantify (Abell et al. 2016). However, a recent study by Vanessa Reis and her colleagues (2017) combined data sets relative to threats and protections of inland wetlands with satellite imagery to quantify the proportion of the earth's land surface covered with seasonal inland wetlands and the percentage of these wetlands that are protected by region. Approximately 20% of wetlands are protected in Central America, 18% in South America, 12% in North America, and only 8% in Asia, where human incursions into wetlands are extensive. Protected wetlands reach almost 13% in Africa and approximately 15% each in Europe and Oceania (Reis et al. 2017). Beyond these numbers, however, the level of protection varies and, even in protected areas, human use of wetlands has both direct and indirect impacts. The Nimba otter shrew (*Micropotamogale lamottei*) is protected in the Mount Nimba Strict Nature Reserve, a UNESCO World Heritage Site; however, it is threatened with mining immediately adjacent to the reserve and fire and poaching within the reserve, all of which contribute to its declining population numbers (Stephenson 2016).

To increase the protection of these vital habitats, some ecologists suggest that larger freshwater species that are considered "charismatic" be used as **flagship species** to raise awareness and create support for protection of their freshwater habitats (He et al. 2017). Larger species, including otters and hippopotami, require more space and, therefore, could serve as **umbrella species**, thereby resulting in the protection of other species within the same area. As larger species, their distributions also tend to cover larger geographic areas. Protection of the latitudinal extents of present, and possibly future, distributions would also allow for north-south movements to accommodate for habitat changes caused by climate change and associated wetland loss.

Connectivity is an important consideration for protected area networks. As seen with European water voles (*Arvicola amphibius*), sustaining viable populations is aided by the creation or maintenance of connections among the various habitat patches used by these metapopulations (MacPherson and Bright 2011). For other semi-aquatic mammals, the rapid increase in construction of hydroelectric and other forms of dams since the middle of the twentieth century has dramatically reduced water flows and aquatic connectivity for several species (Lehner et al. 2011). Large hydropower projects are either planned or underway in some of the world's major river basins, including the Amazon, Mekong, Congo, and Ganges (He et al. 2017), all of which support numerous semi-aquatic mammals. The IUCN Red List species reports cite dams as notable threats to seven species of otters, the European mink (*Mustela lutreola*), Musso's fish-eating rat (*Neusticomys mussoi*), and the Japanese water shrew (*Chimarrogale platycephalus*).

Managing seasonal or even daily water releases can help mitigate some of the impacts of these developments. In northern climates, preventing unnecessary releases of water when rivers are covered in ice also would prevent inundation of beaver lodges at a time of year when beavers are unable to escape from under the ice. In hopes of restoring riverine function and biodiversity, some countries, including Denmark, France, and the United States, are now actively removing dams and restoring rivers to their natural state (Bednarek 2001). The Japanese water shrew is mainly affected by small dams; therefore, these types of programs would have an immediate and positive impact on species persistence.

Freshwater and Riparian Habitat Conservation and Restoration

Loss of wetlands and riparian areas is central to the dramatic declines of many semi-aquatic mammal populations. The dual habitats of these species must be considered whenever conservation and restoration actions are contemplated. The loss of one habitat negates the suitability of the other. Despite historic wetland losses in North America, wetland restoration efforts have increased the total area of wetlands in the United States (Junk et al. 2013). Similar activities have occurred in Canada, which has 20% of the world's freshwater, much of which is stored in glaciers and as groundwater. The restoration of freshwater habitats is welcome after losses of approximately 80% of wetlands to wetland drainage for agricultural expansion throughout North America (Mitsch and Gosselink 2007). Anthropogenic wetlands (e.g., rice paddy plantations and fish ponds) can support some species of semi-aquatic mammals, for example, the Nimba otter shrew (*Micropotamogale lamottei*), Coues's rice rat (*Oryzomys couesi*), the Asian small-clawed otter (*Aonyx cinereus*), the smooth-coated otter (*Lutrogale perspicillata*), and the Chinese water deer (*Hydropotes inermis*). However, other species, such as Kemp's grass mouse (*Deltamys kempi*) and the hairy-nosed otter (*Lutra sumatrana*), suffer critical habitat loss because of rice plantations (IUCN 2019).

In both Europe and North American, the use of beavers as a "wetland restoration tool" is increasing (Petro, Taylor, and Sanchez. 2015; Law et al. 2017). The release of two Eurasian beavers (*Castor fiber*) onto a fenced private estate in eastern Scotland in 2002 resulted in a site that was "almost unrecognizable from its initial state, with a mosaic of aquatic, semi-aquatic and terrestrial habitat patches having formed" (Law et al. 2017: 1028). Prior to the release of the beavers, the site had been predominately agricultural in nature, with beavers absent since the sixteenth century. Twelve years postrelease, average plant species richness increased by 46% per plot, and the average cumulative number of species increased by 148%. The study determined that as "part of the tool kit for restoring degraded wetlands or generating new habitats, habitat engineering by beaver is a passive method requiring little continuous management and is a potentially low-cost option" (Law et al. 2017: 1028). Similar "beaver restoration projects" are occurring in North America and throughout Europe, with some early success.

One cost-effective tool to mitigate human-beaver conflicts once populations become well-established are pond levelers (Fig. 13.1), which, once placed in the beaver dam, maintain beaver ponds at a constant level and prevent flooding on adjacent roads and trails (Hood, Manaloor, and Dzioba 2018). The beavers are then able to remain in their ponds, rather than facing an immediate management response of removing the beaver colony, only for the pond to be recolonized by new beavers in a short period of time. These flow devices prevent the regular draining of biologically valuable ponds and the associated disruption of their ecological communities.

Beavers are directly impacted by deforestation of riparian habitats, as are European mink that use bramble patches adjacent to water for denning and protection from predators (Zabala et al. 2003). From 1998 to 2000, the global deforestation rate averaged around 14.6 million hectares annually (Sweeney et al. 2004). Riparian deforestation, in particular, reduces wildlife habitat and movement corridors, and it also degrades stream quality. The reduction of wood inputs, leaf litter, and other sources of carbon changes water biochemistry and has a notable impact on a stream's biological communities (Sweeney et al. 2004). Decreased shading results in higher water temperatures that can stress fish populations, and fish are key prey items for otters and mink. Fewer trees result in a reduced buffer for overland pollutants that would otherwise

Figure 13.1. Pond leveler installation to mitigate flooding by beavers. G. A. HOOD

be intercepted by riparian vegetation. In addition, forested streams have wider channels and more natural configurations (Sweeney et al. 2004). To adequately protect freshwater ecosystems, the entire watershed must be considered, including the upstream drainage network, surrounding lands, the **riparian zone**, and downstream reaches necessary for dispersal or foraging (Dudgeon et al. 2006). Such an approach requires interjurisdictional cooperation and planning.

The Bonn Challenge represents a global commitment by forty-eight countries, and several additional forestry and private organizations, to restore 150 million hectares of deforested and degraded land by 2020, and an additional 200 million hectares by 2030 (IUCN 2016). It is an ambitious project that reflects the cooperative nature of large-scale habitat restoration, and it requires meaningful engagement with local communities and landowners (Holl 2017). These types of initiatives are critical to the long-term survival of many species, including the otter civet (*Cynogale bennettii*) of Southeast Asia.

As well as depending on peat swamps, the otter civet depends on primary forest habitat, which has declined by ~20% in the past decade (Ross et al. 2015). Deforestation is especially concerning in the lowland

Sundaic region and has likely reduced otter civet populations in part because of siltation and water pollution. Peat swamp forests are also being converted to oil palm plantations, which are a rapidly increasing threat to wetland habitats (Ross et al. 2015). Conservation of these endangered mammals requires dedicated protection of forest and riverine habitats (Veron et al. 2006). The IUCN identifies at least thirty species of semi-aquatic mammals that are threaten by deforestation (IUCN 2019).

Invasive Species Management

Just as the physical removal of a wetland or riparian forest results in the loss of biodiversity, so do biological invasions. American mink (*Neovison vison*) constitute one of worst invasive predator species in Europe and southern South America (Bonesi and Palazon 2007). After a thirteen-week live-trapping program to remove American mink from the Butron river basin in Northern Spain, trapping success varied from 44% to 89% (Zuberogoitia et al. 2010). Live trapping was necessary to ensure that any accidently caught European mink could be released immediately upon capture. The research team calculated the costs for capturing the thirty-one mink during the study and four mink detected via field signs after the study (€24,713). They also extrapolated what it would cost to trap the estimated population of mink given their territorial behavior and available habitat (49 mink, €58,300), as well as costs to trap the estimated population if there was more than one individual per territory (70 mink, €172,500). These types of eradication programs are difficult to sustain, not only financially, but logistically.

The researchers noted that trapping efficiency relies on species behavior and ecology, weather, habitat type, and experience of the trappers. As well, wildlife can learn to avoid traps. Regardless, the logistics of such a project are complicated by public and agency support and, in the case of the American mink, what appears to be a constant source of replacement animals from adjacent populations. Ongoing trapping over several months would be required every one to two years to reduce American mink populations to minimum population levels in the Butron river basin (Bonesi and Palazon 2007). Unless a population is isolated on island habitats, such as Hiiumaa Island, Estonia (Maran 2003), full eradication is not possible, but regulating mink farms and mitigating range expansion might be attainable.

Similar issues exist with coypus (*Myocastor coypus*), muskrats (*Ondatra zibethicus*), and North American beavers. In some situations, like with the North American beaver in South America, active eradication programs have had mixed success, and invasive species continue to expand their range (Choi 2008). The primary methods for control of coypu include trapping, poisoning, and shooting, with methods being most effective when combined (Carter and Leonard 2002). As with any eradication program, there can be unforeseen consequences. Several species of semi-aquatic mammals have suffered from secondary poisonings from other pest-control efforts, for example, Himalayan water shrews (*Chimarrogale himalayica*), Pyrenean desmans (*Galemys pyrenaicus*), flat-headed cats (*Prionailurus planiceps*), fishing cats (*P. viverrinus*), and European mink.

The most successful eradication program for coypus was in England beginning in the 1940s, with increased efforts to trap them in the 1960s and full eradication in 1981 (Norris 1967; Gosling and Baker 1989). This program combined an incentive bonus for trappers, population modeling in advance of the campaign, and ongoing participation and advice from biologists. To encourage the trappers to fully eradicate the coypu, authorities restricted funding to a ten-year timeframe and promised trappers a bonus equivalent to three times their annual salaries if coypus were completely eradicated (Gosling and Baker 1989). The bonus, however, gradually declined after six years to promote an early end to the trapping campaign. It worked.

Muskrats were eradicated from Britain before coypus, in part because of previous experience with muskrat damage in Europe. Efforts began in 1932; the last capture of muskrats in England was in 1935 and in Scotland in 1937 (Gosling and Baker 1989). Persistence, preplanning, and an island environment all aided the eradication of these species in Britain. In some areas, however, eradication can be short-lived. Although coypus were eradicated in California by 1978 (Carter and Leonard 2002), they have since recolonized and are once again a pest species in the state.

Invasive plants and fish are extremely difficult to control and almost impossible to eradicate, given their high reproductive outputs, and they can be detrimental to native species. Spotted-necked otters (*Hydrictis maculicollis*) in Lake Victoria in Africa are negatively impacted by introduced Nile perch (*Lates niloticus*) and water hyacinth (*Eichhornia crassipes*). Introduced into Lake Victoria in the 1950s, the Nile perch is directly linked to the extinction of native fish (i.e., cichlids), a key food source for the otters. In North America, marsh rabbits are also negatively affected by introduced aquatic plants (e.g., common water hyacinth). Cats, dogs, red foxes, and exotic snakes are among other species that are difficult to control because of repeated releases into natural habitats by humans and because of their high reproductive rates.

Rüdiger Wittenberg and Matthew Cock (2001) provide a "toolkit" for the prevention and management of invasive alien species. They present four major steps for addressing introduced species: (1) prevention, (2) early detection, (3) eradication, and (4) control. As seen with American mink, this can be easier said than done; however, successful eradication programs for muskrats and nutria in the United Kingdom followed these steps and managed to avoid the fourth one, control. As expected, costs increase with each step. Early detection is key. Species-specific site surveys require people with a knowledge of species ecology and behavior, and people with the political and financial support to conduct the surveys and then follow up if an introduced species is detected. In most places, control of invasive species and mitigation of their impacts are the most reasonable goals.

Ex Situ Conservation

Ex situ conservation (conservation of a species outside its natural range) can complement conservation efforts within a species' natural habitat (**in situ conservation**). Zoos, wildlife parks, captive breeding programs, wildlife rescue centers, and even private wildlife collections can all play a role in the maintenance of a species in a captive setting. Many zoos combine ex situ conservation with education campaigns to gain public and political support for habitat restoration and possible reintroduction efforts (Kasso and Balakrishnan 2013). To prevent the loss of genetic diversity, organizations, specifically the Association of Zoos and Aquariums (AZA), maintain studbooks that document the lineage and demographic history of individual animals living in AZA member

Figure 13.2. Sulawesi babirusa in the San Diego Zoo. D. PATRIQUIN, USED WITH PERMISSION

institutions. Animals kept in captivity, however, are prone to decreased genetic diversity and habituation to people. When captive animals are used to augment or re-establish wild populations, there can be difficulties with low success rates, despite high financial commitments (Canessa et al. 2016). Although many species are now raised and bred in captivity, capturing wild individuals to establish or supplement ex situ breeding programs can result in unforeseen consequences.

A group of British and Indonesian researchers present a poignant case study involving the rare Sulawesi babirusa (*Babyrousa celebensis*) (Clayton et al. 2000). Despite being fully protected under Indonesian law, the Sulawesi babirusa was facing local extirpations because of poaching for the trade in wild meat. Because of these population declines, there has been an effort to keep and breed the Sulawesi babirusa in captivity for conservation purposes (Fig. 13.2). A conservation program was developed to capture wild individuals for an international captive breeding program. On the surface, the program seemed well planned; it included a population and habitat viability analysis workshop in 1996, at which the group recommended the capture of live babirusas for captive breeding (Clayton et al. 2000). Surprisingly, an unprecedented increase in trapping and trade in live babirusas began to occur. There was also an increase in the amount of illegal babirusa meat in the Sulawesi markets, despite a previous drop in trade following recent visits from government inspectors.

Apparently a local person at the international workshop, after hearing about the plan to obtain live babirusa, assumed that there was a lucrative national and international market for the live animals that was officially sanctioned. Thereafter, he issued an illegal letter that "authorized" traders to transport live babirusas. Trappers used leg snares to catch the animals and transported them in ordinary vehicles, often among live dogs, thereby contributing to several deaths of the rare animals. Fortunately, on noticing the increase in trade of live babirusas and babirusa meat in the markets and illegal trade networks, the Indonesian government was able to quickly stop the live trade activity. What appeared to be a well-planned conservation initiative, involving veterinarians and an international team of conservationists, initiated activities that further imperiled the species in the wild. Regardless of the contribution of captive breeding programs to species conservation, the steady decline of Sulawesi babirusas in the wild because of poaching requires an integrated strategy that engages the local community, government, and nongovernmental organizations to promote education and awareness in meaningful ways (Onibala and Laatung 2009).

As discussed previously, Père David's deer (*Elaphurus davidianus*) is an example of how captivity of a species on another continent can inadvertently lead to the species' repatriation much later. Lord Robin Tavistock, fourteenth Duke of Bedford, was instrumental in returning Père David's deer to China from his estate in England almost a hundred years after they were brought to the estate by his great-grandfather (Maddison, Zhigang, and Boyd 2012), and several hundred years after they were first brought into captivity in China. Neil Maddison and his colleagues (2012: 29) call it "a staggering success story; a testimony to several individuals' vision and commitment to safeguarding a species and how, without the support of zoological collections, the world could have lost one of its most spectacular animals." Through the use of fencing, there has been some progress mitigating unforeseen conflicts with farmers growing crops in the region where the deer were released. Following some inadvertent escapes, however, there are 350 Père David's deer now living in the wilds of Hubei and Hunan. Père David's deer represent a tentative success story that is still unfolding. The remaining deer in captivity in various facilities throughout the world now act as reinforcements for the first of its species to live in the wild in nearly a century.

Research and Monitoring

At the heart of species conservation is the research and monitoring required to establish even the most basic information: distributions, habitat needs, demographics, anatomy, and physiology. Several species of semi-aquatic mammals listed as data deficient still require the most basic of research to allow for focused conservation and management plans. While reviewing over 3,500 peer-reviewed studies with a direct focus on the semi-aquatic mammals described in this book, I realized that 62% of the research was disproportionately focused on larger, more charismatic species. Studies on the Eurasian otter topped the list, with 421 articles (10.4%), followed by the North American beaver (396 articles, 9.7%), the American mink (378 articles, 9.3%), the Eurasian beaver (323 articles 7.9%), the platypus, *Ornithorhynchus anatinus* (215 articles, 5.3%), the coypu (202 articles, 5%), the North American river otter (183, 3.7%), and the capybara, *Hydrochoerus hydrochaeris* (148 articles, 3.7%). Articles on the muskrat (130 articles, 3.2%) and the common hippopotamus, *Hippopotamus amphibius* (105 articles, 2.6%) followed closely behind. Almost all are hunted species, and none are listed as endangered, although one is vulnerable (common hippopotamus) and two are near threatened (Eurasian otter and platypus).

Thirty-seven species only appeared in multispecies articles, which primarily assessed closely related species to determine their taxonomic classification, phylogeny, or genetics. These articles included almost all of the *Dasymys* species (a genus that is still a focus of taxonomic debate), most of the *Chimarrogale*, *Sorex*, and *Hydromys* species, and almost half of the *Neusticomys* species. Other species in this category were also small mammals, including the web-footed marsh rat (*Holochilus brasiliensis*), common water rat (*Nectomys rattus*), Mexican water mouse (*Rheomys mexicanus*), Alfaro's rice water rat (*Sigmodontomys alfari*), northern bog lemming (*Synaptomys borealis*), Cape marsh rat (*Dasymys capensis*), Tanzanian vlei rat (*Otomys lacustris*), grooved-toothed swamp rats (*Pelomys* spp.), and four shrews—Kahuzi swamp shrew (*Crocidura stenocephala*), swamp musk shrew (*C. mariquensis*), elegant water shrew (*Nectogale elegans*), and Transcaucasian water shrew (*Neomys teres*). This list includes endangered, vulnerable, and near threatened species, as well as species listed by the IUCN as least concern and data deficient.

Sixteen of the 140 species covered in this book appeared to be absent from the literature. Many of these sixteen species are listed as at-risk species, including the critically endangered Ethiopian water mouse (*Nilopegamys plumbeus*) and the endangered Togian Islands babirusa (*Babyrousa togeanensis*). Four species are listed as vulnerable—Delany's swamp mouse (*Delanymys brooksi*), Pittier's crab-eating rat (*Ichthyomys pittieri*), Musso's fish-eating rat (*Neusticomys mussoi*), and the Venezuelan fish-eating rat (*N. venezuelae*)—and one is near threatened—the crab-eating rat (*Ichthyomys hydrobates*). Only three species are of least concern—the Rwenzori otter shrew (*Micropotamogale ruwenzorii*), western Amazonian nectomys (*Nectomys apicalis*), and earless water rat (*Crossomys moncktoni*)—while four are considered data deficient—the Las Cajas water mouse (*Chibchanomys orcesi*), Chibchan water mouse (*C. trichotis*), New Britain water rat (*Hydromys neobritannicus*), and Hopkins's groove-toothed swamp rat (*Pelomys hopkinsi*). These species represent some of the smallest semi-aquatic mammals in the world and several are at risk of imminent or possible extinction in the near future.

The top three primary areas of research in the articles were diseases and parasites, foraging and diet, and anatomy. Almost all of these studies were done outside of the animal's habitat, mainly in laboratories. Only 35% of the studies were conducted in the field. Over the past twenty years, the study of diseases and parasites grew dramatically compared to other fields of study, especially in the mid-2000s. The top three secondary areas of research were species distribution, species interactions (often associated with habitat change), and the consequences and management of introduced species. The question, then, is how well does this research inform conservation? On the far side of conservation, the disease studies were focused primarily on zoonotic diseases and their threat to human health, while several others addressed captive animals (primarily in zoos and on fur farms). However, the studies on species distribution and habitat use provide critical insights into areas for focused conservation activities.

It is apparent that areas that require more focus are the impacts of climate change on species persistence and reintroduction biology and what is needed for success. Research is also rare relative to how to restore the biochemical and physical habitats of semi-aquatic mammals with appropriate follow-up, and how to have greater inclusion of human dimensions and indigenous involvement into research planning and implementation. Although there are suggestions that focusing research on charismatic species in freshwater systems might improve their effectiveness as umbrella species (Minin and Moilanen 2014; He et al. 2017), these species already dominate existing studies, while many populations of semi-aquatic mammals continue to decline. However, perhaps these studies have slowed the rate of declines because of increased awareness of semi-aquatic mammals and appreciation of their habitats. They certainly helped in the writing of this book.

Emerging Approaches

Threat assessments have benefited from integrating the "evolutionary distinctiveness and global endangeredness" (EDGE) method into the assessment of conservation priorities (Kuntner, May-Collado, and Agnarsson 2011). This method identifies "one-of-a-kind" species that lack many close relatives. If this method is applied to afrotherians, the Nimba otter shrew (*Micropotamogale lamottei*), currently listed as vulnerable, ranks highest for conservation priorities within the afrotherians; the web-footed tenrec (*Limnogale mergulus*) and giant otter shrew (*Potamogale velox*) are also considered top-ranking species (Kuntner, May-Collado, and Agnarsson 2011). It is just one approach, using phylogenetic isolation to assess conservation priorities, yet similar approaches could be developed to look at ecological uniqueness of a species and its functional importance to inform conservation decisions and species recovery plans.

Many of the species noted above as receiving little to no research attention are likely logistically or financially difficult to study. Many are found in remote areas and are difficult to detect. One of the more exciting developments in aquatic research is the use of environmental DNA (eDNA) methods that allow for the detection of freshwater aquatic species by sampling the water itself and analyzing it for DNA signatures of the species using that waterbody (Ficetola et al. 2008). Even if it is difficult to capture and identify the species in the field, the DNA fragments they shed into the water can still be detected using traditional DNA sequencing and more advanced techniques. For rare species, there are museum specimens that might provide baseline DNA to aid in detection in the field. Museum collections were similarly useful for a morphological study of the elusive otter civet (Veron et al. 2006). Currently, the San Diego Zoo has partnered with Australian scientists to use eDNA to detect platypuses in order to map the distribution of the species in hundreds of waterbodies in southeast Australia. The continued advancement of new technologies and citizen science projects (e.g., iNaturalist) present unexpected opportunities to help monitor and advance research on semi-aquatic mammals and the freshwater systems on which they depend.

Legal Protections and Enforcement

Species declines are not just due to habitat loss and degradation, poaching and the wildlife trade pose significant threats to many species of semi-aquatic mammals. In north Sulawesi, overhunting for subsistence and the wildlife trade have resulted in notable declines in local wildlife (Onibala and Laatung 2009). Market demand for wild meat and a global demand for wildlife parts are significant, and laws to prevent illegal activities can be difficult to enforce because of varying degrees of legislative protection and political commitment to enforcement efforts in the field, as discussed in some detail in Chapter 11. The local trade for the capybara, for example, involves the legal and illegal use of its skin as high-quality material for gloves, belts, shoes, and other leather material (Mones and Ojasti 1986). International treaties and agreements (e.g., CITES) help reduce the import and export of wildlife parts. These treaties are then enforced through federal legislation specific to each signatory country to those agreements. As seen with the Sulawesi babirusa, the presence of government inspectors at the ground level can act as a noticeable deterrent to those involved in the illegal wildlife trade (Clayton et al. 2000). Increasing and re-establishing field-level enforcement officers would be a positive step in curbing the negative impacts of the wildlife trade and illegal hunting.

Wildlife Conservation Organizations

Several organizations contribute invaluable resources and expertise to the conservation of wildlife and their habitats. Data from the IUCN and WWF, among other conservation organizations, have been instrumental in relaying the distributions, status, ecology, and conservation needs of almost all the species described in this book. From those organizations, additional working groups and specialist committees, as well as research, monitoring, and fundraising initiatives, have developed, many of which cooperate on multilateral international conservation efforts. Along with these wide-reaching organizations, there have been multiple generations of scientists who have conducted essential research on all aspects of species biology and management. Naturalists have also made extensive contributions to the field, including Harry Burrell, who wrote a book on the platypus, and Lewis H. Morgan, who wrote a book on North American beavers. Early mammalogist, Alfred Brazier Howell's 1930 book, *Aquatic Mammals*, has been a key resource for almost a century of research on aquatic and semi-aquatic mammals.

Despite very different beginnings, many zoos have embraced their role as sources of research and conservation activities to aid the broader field of wildlife conservation. Their role in environmental education is multifaceted and has varied results. Unfortunately, the attractiveness of a species is still directly related to the interest shown by the public. In a 2010 study at the Chester Zoo in the United Kingdom, Andrew Moss and Maggie Esson determined that visitors were far more interested in mammals than other taxonomic groups, in particular they were interested in larger flagship species. Mean holding (observation) times by visitors ranged between just under twenty-three seconds to almost forty six seconds for zoo species, depending on whether they were housed with a flagship species or not (Moss and Esson 2010). Similar studies have shown that zoo visitors spend less time than expected at zoo exhibits (Marcellini and Jenssen 1988), thus prompting new approaches to environmental education programming and outreach.

In the past thirty years, many zoos have moved toward a conservation model and are increasingly investing in species conservation and animal research (Fraser and Wharton 2007). Education and signage in many zoos now emphasize sustainability and conservation needs for natural habitats and their species, although visitor attention spans are still a challenge. Behind the exhibits and front-line programming, however, a consortium of zoo organizations is now working with IUCN Specialist Groups to integrate ex situ and in situ programs into a more integrative path forward for conservation education (ex situ) and in situ field support (Traylor-Holzer, Leus, and Bauman 2019). The "One Plan Approach" (OPA) to conservation planning aims to produce one comprehensive conservation plan for each threatened species to ensure that in situ and ex situ activities reflect coordinated conservation solutions (Byers et al. 2013). Certainly not all zoos are focused on conservation, and some, especially private zoos, continue to fuel the illegal trade in wild animals (Lavorgna 2014); however, most of the major zoos are now actively contributing to wildlife conservation and awareness in cooperation with other conservation organizations.

The Wechiau Community Hippo Sanctuary in Ghana provides an excellent example of community-based natural resource management in collaboration with zoo conservation initiatives (Sheppard et al. 2010). This form of resource management aims to combine conservation projects with the alleviation of poverty. The sanctuary was founded in 1998 to increase protection of the common hippopotamus in a region with comparatively few protected areas (Sheppard et al. 2010). Even within existing protected areas in the region, wildlife populations are in decline (Brashares et al. 2004). The sanctuary was founded as a community-managed sanctuary, with financial assistance from the Nature Conservation Research Centre (a Ghanaian nonprofit organization) and the Calgary Zoological Society (a not-for-profit organization founded by the Calgary Zoo in 1929). Since its inception, the hippo sanctuary has attracted sustained external investment from various donors and has increasing profits from visitors to the sanctuary (Sheppard et al. 2010). At least fifty members of the local community gain regular income from the sanctuary, and it supports another hundred people with occasional employment.

As a reflection of the sanctuary's success, an average of fourteen hippopotami were seen during censuses between 2000 and 2009, and their population is stable, unlike nearby populations outside the sanctuary. In 2008, the Wechiau Community Hippo Sanctuary received the United Nations' Equator Prize for its "outstanding community efforts to reduce poverty through the conservation and sustainable use of biodiversity" (www.equatorinitiative.org). These types of initiatives, done in partnership with local leaders and community members, represent a sense of hope for conservation of semi-aquatic mammals by using innovative and effective collaborations at the local, regional, and international level.

Future Considerations

As species that have an obligatory relationship with both freshwater and terrestrial habitats, many semi-aquatic mammals face an uncertain future in a world with warming climates and ever-increasing human populations. It is important to learn from conservation success stories, yet it is equally important to learn from those attempts that failed. Despite habitat loss and, for many, unprecedented overharvesting, the persistence of semi-aquatic mammals has been remarkable. However, human intervention is required to counter the problems they face today. In the paper "World Scientists' Warning to Humanity: A Second Notice" (Ripple et al. 2017), there is a list of actions that would increase global ecological sustainability for all species.

Increased protection of wild spaces that encompass all ecosystems, including freshwater and coastal habitats, would have immediate and positive impacts on semi-aquatic mammals. Such protection must go hand in hand with preventing the conversion of natural habitats into ecologically nonproductive systems. Restoration of disturbed areas must take hold to enhance ecological connectivity and genetic flow at all geographic scales, and reintroductions could follow once habitat quality is suitable. There are, of course, other suggestions provided by Dr. Ripple and his fellow scientists, all of which are worth heeding. Throughout all of these steps, there is a need to not only enforce existing legislation and protections but to also provide engaging and impactful educational opportunities for nontraditional audiences. As with the common hippopotamus sanctuary in Ghana, community-based education and citizen science programs for fishing cats in Southeast Asia, and even the release of Eurasian beavers onto a private estate in Scotland, conservation of semi-aquatic mammals and the habitats on which they depend requires a global community of support. Research is critical, as are concrete actions and the solutions they support.

Taxonomic List of Semi-aquatic and Riparian-Dependent Mammals

Extant semi-aquatic and riparian-dependent mammals (n = 140) in taxonomic order by family, following taxonomy used by Feldhamer et al. 2015 and updates in Burgin et al. 2018. The first listed common name is the one used in this book.

Order	Family	Species	Common name
Monotremata	Ornithorhynchidae	*Ornithorhynchus anatinus*	platypus, duck-billed platypus
Didelphimorphia	Didelphidae	*Chironectes minimus*	water opossum, yapok
	Didelphidae	*Lutreolina crassicaudata*	little water opossum, thick-talled opossum, lutrino opossum
Afrosoricida	Tenrecidae (Oryzorictinae)	*Limnogale mergulus*	web-footed tenrec, aquatic tenrec
	Tenrecidae (Potamogalinae)	*Micropotamogale lamottei*	Nimba otter shrew, pygmy otter-shrew
	Tenrecidae (Potamogalinae)	*Micropotamogale ruwenzorii*	Rwenzori otter shrew, Ruwenzori otter shrew
	Tenrecidae (Potamogalinae)	*Potamogale velox*	giant otter shrew, giant otter-shrew
Lagomorpha	Leporidae	*Sylvilagus aquaticus*	swamp rabbit, cane cutter, cane Jake, swamper
	Leporidae	*Sylvilagus palustris*	marsh rabbit, Lower Keys rabbit, Lower Keys marsh rabbit, Key rabbit
Rodentia	Castoridae	*Castor canadensis*	North American beaver, American beaver

Order	Family	Species	Common name
Rodentia, continued	Castoridae	*Castor fiber*	Eurasian beaver
	Thryonomyidae	*Thryonomys swinderianus*	greater cane rat
	Capromyidae	*Mesocapromys angelcabrerai*	Cabrera's hutia
	Caviidae	*Hydrochoerus hydrochaeris*	capybara
	Caviidae	*Hydrochoerus isthmius*	lesser capybara
	Myocastoridae	*Myocastor coypus*	coypu, nutria
	Cricetidae	*Amphinectomys savamis*	Ucayali water rat
	Cricetidae	*Anotomys leander*	Ecuador fish-eating rat, Ecuadoran ichthyomyine, aquatic rat, fish-eating rat
	Cricetidae	*Arvicola amphibius*	European water vole, water vole, Eurasian water vole
	Cricetidae	*Arvicola sapidus*	southern water vole, southwestern water vole
	Cricetidae	*Chibchanomys orcesi*	Las Cajas water mouse, Las Cajas Ichthyomyine
	Cricetidae	*Chibchanomys trichotis*	Chibchan water mouse, Chibchan Ichthyomyine
	Cricetidae	*Deltamys kempi*	Kemp's grass mouse
	Cricetidae	*Holochilus brasiliensis*	web-footed marsh rat
	Cricetidae	*Holochilus chacarius*	Chaco marsh rat
	Cricetidae	*Holochilus lagigliai*	Lagiglia's marsh rat
	Cricetidae	*Holochilus sciureus*	marsh rat
	Cricetidae	*Ichthyomys hydrobates*	crab-eating rat
	Cricetidae	*Ichthyomys pittieri*	Pittier's crab-eating rat
	Cricetidae	*Ichthyomys stolzmanni*	Stolzmann's crab-eating rat
	Cricetidae	*Ichthyomys tweedii*	Tweedy's crab-eating rat

Order	Family	Species	Common name
Rodentia, *continued*	Cricetidae	*Lundomys molitor*	Lund's amphibious rat, greater marsh rat
	Cricetidae	*Microtus richardsoni*	North American water vole, water vole
	Cricetidae	*Nectomys apicalis*	western Amazonian nectomys
	Cricetidae	*Nectomys palmipes*	Trinidad water rat
	Cricetidae	*Nectomys rattus*	common water rat
	Cricetidae	*Nectomys squamipes*	South American water rat
	Cricetidae	*Neofiber alleni*	round-tailed muskrat
	Cricetidae	*Neotomys ebriosus*	Andean swamp rat
	Cricetidae	*Neusticomys ferreirai*	Ferreira's fish-eating rat
	Cricetidae	*Neusticomys monticolus*	montane fish-eating rat
	Cricetidae	*Neusticomys mussoi*	Musso's fish-eating rat
	Cricetidae	*Neusticomys oyapocki*	Oyapock's fish-eating rat
	Cricetidae	*Neusticomys peruviensis*	Peruvian fish-eating rat
	Cricetidae	*Neusticomys venezuelae*	Venezuelan fish-eating rat
	Cricetidae	*Neusticomys vossi*	Voss fish-eating rat
	Cricetidae	*Ondatra zibethicus*	muskrat
	Cricetidae	*Oryzomys couesi*	Coues's rice rat
	Cricetidae	*Oryzomys palustris*	marsh rice rat
	Cricetidae	*Rheomys mexicanus*	Mexican water mouse, Mexican fishing mouse, Goodwin's water mouse
	Cricetidae	*Rheomys raptor*	Goldman's water mouse
	Cricetidae	*Rheomys thomasi*	Thomas's water mouse
	Cricetidae	*Rheomys underwoodi*	Underwood's water mouse
	Cricetidae	*Scapteromys aquaticus*	Argentine swamp rat

Order	Family	Species	Common name
Rodentia, *continued*	Cricetidae	*Scapteromys meridionalis*	plateau swamp rat
	Cricetidae	*Scapteromys tumidus*	Waterhouse's swamp rat, swamp rat
	Cricetidae	*Sigmodontomys alfari*	Alfaro's rice water rat, short-tailed sigmodontomys
	Cricetidae	*Synaptomys borealis*	northern bog lemming, Okanagan bog lemming
	Cricetidae	*Synaptomys cooperi*	southern bog lemming, Kansas bog lemming
	Muridae	*Baiyankamys habbema*	mountain water rat, New Guinea mountain water rat
	Muridae	*Baiyankamys shawmayeri*	Shaw Mayer's water rat
	Muridae	*Colomys goslingi*	African wading rat, African water rat
	Muridae	*Crossomys moncktoni*	earless water rat
	Muridae	*Dasymys capensis*	Cape marsh rat
	Muridae	*Dasymys foxi*	Fox's shaggy rat, Fox's marsh rat, Fox's water rat
	Muridae	*Dasymys griseifrons*	
	Muridae	*Dasymys incomtus*	African marsh rat, common Dasymys
	Muridae	*Dasymys longipilosus*	
	Muridae	*Dasymys medius*	
	Muridae	*Dasymys montanus*	montane shaggy rat
	Muridae	*Dasymys nudipes*	Angolan marsh rat, Angolan shaggy rat, Angolan water rat
	Muridae	*Dasymys robertsii*	Robert's shaggy rat
	Muridae	*Dasymys rufulus*	West African shaggy rat
	Muridae	*Dasymys shortridgei*	
	Muridae	*Deomys ferrugineus*	Congo forest mouse

Order	Family	Species	Common name
Rodentia, continued	Muridae	*Hydromys chrysogaster*	rakali, rabe, water-rat, common water rat, golden-bellied water rat
	Muridae	*Hydromys hussoni*	western water rat
	Muridae	*Hydromys neobritannicus*	New Britain water rat
	Muridae	*Hydromys ziegleri*	Ziegler's water rat
	Muridae	*Nesokia bunnii*	Bunn's short-tailed bandicoot rat, long-tailed bandicoot rat, long-tailed nesokia, red pest rat
	Muridae	*Nilopegamys plumbeus*	Ethiopian amphibious rat, Ethiopian water mouse
	Muridae	*Otomys lacustris*	Tanzanian vlei rat
	Muridae	*Parahydromys asper*	waterside rat, coarse-haired water rat
	Muridae	*Pelomys fallax*	creek groove-toothed swamp rat
	Muridae	*Pelomys hopkinsi*	Hopkins's groove-toothed swamp rat
	Muridae	*Pelomys isseli*	Issel's groove-toothed swamp rat
	Muridae	*Rattus lutreolus*	Australian swamp rat
	Muridae	*Waiomys mamasae*	Sulawesi water rat
	Muridae	*Xeromys myoides*	false water rat, water mouse, false swamp rat, false water-rat
	Nesomyidae	*Delanymys brooksi*	Delany's swamp mouse, Delany's mouse
Eulipotyphla	Soricidae (Crocidurinae)	*Crocidura mariquensis*	swamp musk shrew
	Soricidae (Crocidurinae)	*Crocidura stenocephala*	Kahuzi swamp shrew, narrow-headed shrew
	Soricidae	*Chimarrogale hantu*	Malayan water shrew, hantu water shrew, Asiatic water shrew
	Soricidae	*Chimarrogale himalayica*	Himalayan water shrew, elegant water shrew

Order	Family	Species	Common name
Eulipotyphla, *continued*	Soricidae	*Chimarrogale phaeura*	Bornean water shrew, Borneo water shrew, Sunda water shrew
	Soricidae	*Chimarrogale platycephalus*	Japanese water shrew, flat-headed water shrew
	Soricidae	*Chimarrogale styani*	Chinese water shrew, Styan's water shrew
	Soricidae	*Chimarrogale sumatrana*	Sumatran water shrew, Sumatra water shrew
	Soricidae	*Nectogale elegans*	elegant water shrew, web-footed water shrew
	Soricidae	*Neomys anomalus*	southern water shrew, Mediterranean water shrew, Miller's water shrew
	Soricidae	*Neomys fodiens*	Eurasian water shrew, water shrew, northern water shrew
	Soricidae	*Neomys teres*	Transcaucasian water shrew
	Soricidae	*Sorex alaskanus*	Glacier Bay water shrew
	Soricidae	*Sorex bendirii*	Pacific water shrew, marsh shrew
	Soricidae	*Sorex palustris*	American water shrew, water shrew
	Talpidae	*Condylura cristata*	star-nosed mole
	Talpidae	*Desmana moschata*	Russian desman
	Talpidae	*Galemys pyrenaicus*	Pyrenean desman
Carnivora	Felidae	*Prionailurus planiceps*	flat-headed cat
	Felidae	*Prionailurus viverrinus*	fishing cat
	Herpestidae	*Atilax paludinosus*	marsh mongoose, water mongoose
	Viverridae	*Cynogale bennettii*	otter civet, otter-civet, Sunda otter civet
	Viverridae	*Genetta piscivora*	aquatic genet
	Mustelidae	*Mustela lutreola*	European mink
	Mustelidae	*Neovison vison*	American mink

Order	Family	Species	Common name
Carnivora, continued	Mustelidae	Lutra lutra	Eurasian otter, European otter, European river otter, Old World otter, common otter
	Mustelidae	Lutra sumatrana	hairy-nosed otter
	Mustelidae	Hydrictis maculicollis	spotted-necked otter, speckle-throated otter, spot-necked otter
	Mustelidae	Lutrogale perspicillata	smooth-coated otter, Indian smooth-coated otter
	Mustelidae	Lontra canadensis	North American river otter, Northern river otter, North American otter
	Mustelidae	Lontra felina	marine otter, sea cat
	Mustelidae	Lontra longicaudis	Neotropical otter, La Plata otter, Neotropical river otter, South American river otter, long-tailed otter
	Mustelidae	Lontra provocax	southern river otter, huillin
	Mustelidae	Pteronura brasiliensis	giant otter, giant Brazilian otter
	Mustelidae	Aonyx capensis	African clawless otter, Cape clawless otter
	Mustelidae	Aonyx cinereus	Asian small-clawed otter, small-clawed otter, Oriental small-clawed otter
	Mustelidae	Aonyx congicus	Congo clawless otter, Zaire clawless otter, small-clawed otter, small-toothed clawless otter, Cameroon clawless otter
	Procyonidae	Procyon cancrivorus	crab-eating raccoon
Artiodactyla	Suidae	Babyrousa babyrussa	hairy babirusa, babiroussa, babirusa, Buru babirusa, deer hog, golden babirusa, Moluccan babirusa
	Suidae	Babyrousa celebensis	Sulawesi babirusa, North Sulawesi babirusa
	Suidae	Babyrousa togeanensis	Togian Islands babirusa, Malenge babirusa, Togian babirusa

Order	Family	Species	Common name
Artiodactyla, continued	Hippopotamidae	*Choeropsis liberiensis*	pygmy hippopotamus
	Hippopotamidae	*Hippopotamus amphibius*	common hippopotamus, hippopotamus, large hippo
	Tragulidae	*Hyemoschus aquaticus*	water chevrotain
	Cervidae	*Blastocerus dichotomus*	marsh deer
	Cervidae	*Elaphurus davidianus*	Père David's deer, Pere David's deer
	Cervidae	*Hydropotes inermis*	water deer
	Bovidae	*Kobus leche*	southern lechwe
	Bovidae	*Kobus megaceros*	Nile lechwe
	Bovidae	*Tragelaphus spekii*	sitatunga, marshbuck

Online Resources

American Society of Mammalogists
https://www.mammalogy.org

Animal Diversity Web (ADW)
https://animaldiversity.org

Fossilworks and the Paleobiology Database
http://fossilworks.org

International Union for Conservation of Nature:
The IUCN Red List of Threatened Species
https://www.iucnredlist.org

Society for Conservation Biology
https://conbio.org

Society of Wetland Scientists
https://www.sws.org

The Wildlife Society
https://wildlife.org

abduction. Spread out, in reference to digits on an appendage.

adaptation. A trait with a functional role that has evolved through natural selection and aids an individual's evolutionary fitness.

adduction. Drawing together, in reference to digits on an appendage.

alimentary system. The gastrointestinal system; a pathway by which food enters, is processed, and then solid waste is expelled by the body.

allochthonous. Materials brought into a system (e.g., branches brought into ponds by beavers).

allochthonous endemic. A species that dispersed from another location but is now the sole survivor of the original population and is now unique to a specific area.

allometric regression. A statistical formula that quantifies the relationship between body size and an animal's anatomy, physiology, and/or behavior.

alternate pelvic paddling. Alternating movements of hind limbs in the vertical plane while swimming.

alternate pelvic rowing. Alternating movements of hind limbs in the horizontal plane while swimming.

altricial. Having young unable to move around or fend for themselves soon after birth and requiring parental care for more than a few days, and sometimes for more than a year.

amensalism. Species interaction where one organism is harmed while the other is unaffected.

amniotes. Tetrapod vertebrates that have a membrane that surrounds the fetus during development.

anastomosing vessels. Connections that open between blood vessels; arteriovenous anastomoses open when body gets too hot.

Archimedes' principle. A physical law of buoyancy that states that an object completely or partially submerged in a fluid is countered with an upward (buoyant) force equal to the weight of the fluid displaced.

assisted colonization. The translocation of a species outside its natural range to prevent its extinction within its original range.

autochthonous. Materials from within the system (e.g., carbon produced by aquatic bacteria).

autochthonous endemic. A species, unique to an area, that evolved in geographic isolation.

basal metabolic rate. The lowest metabolic rate when animal is at rest, BMR.

basal temperature. Lowest body temperature while an organism is at rest, T_b.

Bering land bridge. A past land connection between what is now Alaska, USA, and Chukotka, Russia.

biodiversity hotspots. Terrestrial ecoregions with high biodiversity and endemism that are a high priority for conservation.

biogeographic relicts. Sole descendants of a previously widely distributed and diverse group.

bradycardia. Slower than normal heart rate.

brown adipose tissue. "Brown fat"; a highly vascularized form of fat, dense with mitochondria.

carangiform. A form of swimming with flexion limited to the posterior region (e.g., tail) and body bending into a partial wave.

carnivore. Consumes terrestrial vertebrates (Withers et al. 2016); consumes other animals.

carrying capacity. Maximum population size of a species that the habitat can support indefinitely.

cathemeral. Active at any time of the day or night.

chromatophores. Skin cells that contain pigment that can reflect light.

cloaca. Common external opening in vertebrates for intestinal, urinary, and reproductive outputs.

commensalism. A species interaction where one species benefits and the other has neutral gain.

competition. A species interaction where both species lose to some degree over resource competition.

competitive exclusion. A species interaction where a superior competitor excludes an inferior competitor from an area through resource competition.

competitive exploitation. An indirect species interaction involving resource competition where a superior competitor alters the use of habitat resources of an inferior competitor but does not exclude it.

conduction. Transfer of heat energy from one object to another; also called diffusion.

conservation introduction. Release of a species into an area outside its natural range for conservation purposes.

conservation translocation. Translocations to achieve deliberate conservation benefits.

convection. Transfer of heat energy from one place to another via movement of fluids past an object.

convergent evolution. Occurs when unrelated organisms independently evolve similar adaptations for similar environments or ecological niches.

cooperative breeding. Young receive care from nonparental group members.

coprophagous. The eating of one's own feces or that of another animal. In herbivores it allows for additional nutrient gain from partially digested feces.

cost of transport. Metabolic cost of moving a unit of mass a given distance.

countercurrent heat exchange. A crossover of heat exchange from spatially close arterioles and venules whose close contact allows heat from arterial blood to transfer to venous blood.

crepuscular. Active at twilight.

crustacivore/clam-eater. Consumes crabs and clams (less mobile prey).

delayed implantation. When the embryo remains in a dormant state and does not immediately implant in the uterus.

diastema. Gap between the incisors and premolars.

diphyletic. Derived from two different ancestral lines.

diurnal. Active during the day.

drag. Resistive force from friction.

drag-based propulsion. Forward movement that is countered by a counterforce that causes resistance.

ecological footprint. A measure of consumption of natural resources that can be computed from the individual to global scale.

ecological replacement. The translocation of a species outside its natural range to have it perform an ecological function in its new location.

ecomorphology. The ecological relationship between the ecological role and morphological adaptations of an organism.

ecosystem engineer. An organism that physically modifies its environment in ecologically impactful ways.

ectoparasite. A parasitic organism that lives on the outside of its host.

ectothermy. Generation of heat from the environment, external to the body.

endoparasite. A parasitic organism that lives inside its host.

endothermy. The physiological generation and regulation of heat through metabolism from within the body.

estrus. The period of sexual receptivity and fertility in the sexual cycle of many species of female mammals.

euryphagous. Having a broad range of foods.

eutherian. A placental mammal that lacks epipubic bones in the pelvis (unlike monotremes and marsupials).

evaporation. Vaporization of sweat or water on the body that promotes heat loss away from the body.

exaptation. A trait that is used for a function that differs from the one for which it originally evolved.

ex situ conservation. Conservation and maintenance of a species in which it is housed outside of its natural habitat (e.g., zoos, off-site wildlife refuges).

extirpated. Extinct from a part of its original distribution but still present elsewhere.

facultatively monogamous. When a male stays in a pair bond with a female because there are no other mating opportunities available to him.

fecundity. The reproductive potential of an organism or population (e.g., number of eggs per female).

fibrillae. Fibrous hairs.

fimbriation. Fringed with short thick hairs that increase surface area.

fitness. Ability to survive to a reproductive age and produce reproductively viable offspring.

flagship species. A species, usually considered charismatic, that is used to raise awareness and support for conservation goals.

foregut fermenter. Herbivores with a multichambered stomach (e.g., ruminants and pseudoruminants) in which microbes aid the breakdown and fermentation of plant materials.

fossa. A depression or hollow in a bone.

frugivore/granivore. Consumes reproductive parts of plants, nuts, and seeds (Withers et al. 2016).

frugivore/omnivore. Consumes reproductive parts of plants (e.g., fruit), some seeds, small invertebrates, and small vertebrates (Withers et al. 2016).

fundamental niche. The entire set of conditions that could support a species or population if interaction with other species were not limiting.

furuncle. A tender dome-shaped skin lesion ("boil").

fusiform. A body form that tapers at the head and tail.

genetic monogamy. A sexually monogamous relationship in which there is genetic evidence of exclusive paternity.

giving-up density. Prey density in a patch at which the animal moves on to other patches.

giving-up time. The length of time from when the animal last fed to the time it leaves the patch for another.

Grande Coupure. A mass extinction at the end of the Eocene (~33.9 Ma).

hard release. The release of an animal into a new environment without any preconditioning to its surroundings.

herbivore. An animal that eats plants.

herbivore/browser. Consumes plant stems, twigs, buds, and leaves (Withers et al. 2016).

herbivore/frugivore. Consumes fruit, reproductive parts of plants, storage roots, fungi, and leaves (Withers et al. 2016).

herbivore/granivore. Consumes reproductive parts of plants, nuts, and seeds (Withers et al. 2016).

herbivore/grazer. Consumes grasses, some browse (Withers et al. 2016).

herptile. A reptile or amphibian.

hibernation. A deep level of dormancy in which body temperatures remain at 2°C to 5°C for several weeks in winter.

hindgut fermenter. Herbivores with a single stomach that rely on a large cecum, the large intestine, or both organs for microbial fermentation.

homeostasis. Maintenance of a relatively steady biological state, often physiological and chemical in nature.

homeothermy. Regulation of a constant internal body temperature via physiological processes.

hydrophilic. Water loving, in particular in reference to wetland plants.

hydrophobic. Water repellent, nonwettable.

ideal free distribution. A theory to understand group foraging that predicts that animals distribute themselves in patches based on prey quality of patches and will choose the most profitable patches to maximize energy intake based in the number of other individuals foraging in the patch.

insectivore/omnivore. Consumes insects, arthropods, mollusks (e.g., snails), some small vertebrates, and fruit (Withers et al. 2016).

in situ conservation. Conservation and maintenance of a species in its natural habitat.

integumentary system. The system of the body composed of skin, hair, nails, glands, and nerves.

intermediate host. The organism that "hosts" a parasite for a short period while the parasite reaches the next developmental stage, but not the adult stage.

introduction. Releasing a species (captive or wild) into an area outside its natural range either through accidental or deliberate means.

keystone species. A species whose activities have a disproportionate effect on the ecosystem (particularly food webs), over and above its own needs.

lentic. Still water, such as a wetland, pond, or lake.

lift-based propulsion. Propulsion that is aided by a perpendicular force to the direction of oncoming flow that causes upward lift that then counters gravity.

light attenuation. Reduction of intensity of light as it travels deeper into the water.

lotic. Running water, such as in a river or stream.

lower critical temperature. The lowest body temperature that can be toler-

ated by an organism before generating heat to counter heat loss, LCT.

macroparasite. A multicellular parasite that often can be seen with the human eye (e.g., flatworms, roundworms, ticks, fleas, lice, flies, and mites).

marginal value theorem. A theory based on patch selection theory that predicts that a foraging animal will leave a patch when capture rates drop to the average capture rate for the patch.

marsupial. A pouched mammal, metatherian.

metabolic rate. Measure of the how quickly chemical processes in the cells produce energy.

metabolism. Chemical processes within an organism that sustain life.

metapopulation. A population of several spatially separated populations in which there is some interaction among patches of occupied and currently unoccupied habitat.

microparasite. A unicellular parasite that is too small to be seen with the human eye (e.g., virus, bacteria, some fungi, and protozoan).

minerotrophic. Water comes from streams or groundwater.

monogamy. A pair bond of two individuals of the same species, usually of the opposite sex.

monomorphic. A species in which the sexes are similar in size and appearance.

monophyletic. A group or clade containing all descendants of a common ancestor.

monotreme. An egg-laying mammal, prototherian.

monotypic. The only species within a taxonomic family.

morphology. Outward appearance (e.g., shape, structure, size).

mutualism. A species interaction in which both species benefit.

niche. The role a species or population plays and relationship it has within its environment.

nictitating membrane. A clear membrane that acts as a third eyelid to protect the eyes underwater.

nocturnal. Active at night.

non-peatlands. Wetlands that form on mineral-based soils (e.g., marshes, open-water ponds).

nonshivering thermogenesis. Activating enzyme systems to metabolize fats.

ombrotrophic. Water comes from streams or groundwater.

omnivore. Eats a broad diet of plant and animal material and consumes prey from more than one trophic level.

optimal foraging theory. An ecological model that predicts how an animal behaves while foraging relative to energetic gains and losses.

oviparous. Producing young from eggs laid outside the mother's body prior to their birth.

ovoviviparous. Producing eggs that develop within the mother's body, rather than externally.

Pangea. Supercontinent during the Late Paleozoic and Early Mesozoic.

parasagittal plane. A plane that divides the body into left and right, but not directly through the midline.

parasite. An organism that lives inside or on another species and obtains nutrients from it.

parasitoid. An insect that lays eggs on a host, upon which the larvae later feed.

parturition. Giving birth to young.

patch selection theory. A behavioral theory that describes animals that forage in patches that are separated by resource-free space that requires travel time from patch to patch.

peatlands. Wetlands that form from thick accumulations of partly decayed plant material (e.g., fens, bogs).

pectoral rowing. Alternating movements of forelimbs in the horizontal plane while swimming.

piloerection. Involuntary constriction of cutaneous muscles that causes hairs to stand upright.

pinnae. The projecting structure, often cartilaginous, of the external ear.

piscivore/squid-eater. Consumes larger fish, and larger invertebrates (e.g., squid) (Withers et al. 2016).

polyandry. A mating system in which one female mates with several males that only mate with one individual in a breeding season.

polyestrous. Coming into estrus more than once per year.

polygynandry. A mating system in which several females and males mate with multiple other breeding partners in a breeding season and have social ties.

polygyny. A mating system in which one male mates with several other females that only mate with one individual in a breeding season.

polytypic genera. Having more than one species within the genus.

population bottleneck. Loss of genetic variation in a population because of reduced population sizes for at least one generation.

positive buoyancy. Floating; an object is lighter than the fluid it displaces.

precocial. Having young born in an advanced state and able to fend for themselves quickly after birth.

predation. A species interaction that involves one organism killing and eating another organism.

prey-switching. A form of frequency-dependent predation by which a predator switches from a formerly dominant prey item to another prey item now more common than the original prey item.

primary consumer. An animal that eats primary producers (e.g., plants).

primary host. The host species in which a parasite reaches maturity and often reproduces. Also called the definitive host.

primary producer. Organism that produces energy on its own (e.g., plants).

promiscuity. A mating system in which several females and males mate with multiple other breeding partners in a breeding season and lack social ties.

radiation. Loss of heat energy generated in a body to the surrounding atmosphere; also received from a radiant heat source.

radiations. The rapid evolution/diversification of new species from an ancestral species.

realized niche. The actual space and resources used by an organism due to limiting external forces.

recruitment. The introduction of new individuals into the adult population.

regional heterothermy. Ability to maintain different temperatures in different zones of the body via physiological processes.

reinforcement translocation. Release of a species into an existing area with an existing population of conspecifics.

reintroduction. Release of a species into its former range where it had been extirpated.

relicts. Species that are the only remaining descendants of once diverse groups.

riparian zone. Plant habitats and communities along the margin of bodies of fresh water.

scansorial. Adapted for climbing.

secondary consumer. An animal that eats primary consumers (i.e., herbivores).

sensorimotor cortex. The part of the brain that contains the sensory and motor areas.

shivering thermogenesis. Using involuntary muscle contractions to free heat energy.

social monogamy. Pair bonding between males and females that then produce and raise young together but can occasionally mate outside their pair bond.

soft release. Prerelease and postrelease conditioning of an animal during translocation.

squalene. Oily film produced from lipids from the sebaceous glands at the base of hairs.

stenophagous. Having a very narrow diet.

stratification. Separation of the water column into distinct layers based on temperature and water density.

stratum compactum. A tightly laminated layer of skin cells that provides support to the body.

superfetation. The co-occurrence of more than one embryo in different stages of development in the same mother.

symbiotic. A close, long-term interaction between two organisms within a biological context.

syntopic. The common occurrence of two species living in the same habitat at the same time.

taxonomic relicts. Sole surviving descendants of a previously diverse taxonomic group.

temporal heterothermy. Regulation of differential temperatures in the body at different times via physiological processes.

tertiary consumer. An animal that eats both primary and secondary consumers.

thermal equilibrium. When there is no net flow of heat between two systems.

thermoneutral zone. The temperature range in which the rate of basal heat production equals the rate of heat lost to the environment.

thrust. Propulsive force.

torpor. A period of dormancy when heart rate, breathing, and body temperature are maintained at lower levels but not to same extent as hibernation. Can be shorter than hibernation.

translocation. Human-mediated movement of living organisms from one area to another.

turbidity. Degree of transparency associated with suspended solids in water.

umbrella species. A species whose in situ protection results in the protection of the many other species living within its large habitat.

upper critical temperature. The highest body temperature that can be tolerated by an organism before energy is used to dissipate heat, UCT.

vector. An organism that spreads infection by transferring a pathogen from one host to another host (often through a bite).

vicariance. Geographical separation of a population that results in two or more closely related species.

viviparous. Giving birth to young that have developed within the mother's body.

VO2 max. Maximum amount of oxygen an organism can use during vigorous exercise.

waif dispersal. Dispersing by hitching a ride with another organism or inanimate objects (e.g., rafts of vegetation).

white adipose tissue. "White fat"; each cell of which contains a lipid droplet to store excess energy.

zoonotic disease. Disease that is transferred from other mammals to humans.

Aadrean, A., B. Kanchanasaka, S. Heng, I. Reza Lubis, P. de Silva, and A. Olsson. 2015. *Lutra sumatrana*. IUCN *Red List of Threatened Species* 2015:e.T12421A21936999.

Aadrean, W., Novarino, and Jabang. 2011. A record of small-clawed otters (*Aonyx cinereus*) foraging on an invasive pest species, golden apple snails (*Pomacea canaliculata*) in a West Sumatra rice field. IUCN *Otter Spec. Group Bull.* 28:34–38.

Abell, R., B. Lehner, M. Thieme, and S. Linke. 2016. Looking beyond the fenceline: Assessing protection gaps for the world's rivers. *Conserv. Lett.* 10:384–394.

Abramov, A. V., 2000. A taxonomic review of the genus *Mustela* (Mammalia, Carnivora). *Zoosyst. Ross.* 8:357–364.

Adhya, T. 2015. Habitat use and diet of two sympatric felids—the fishing cat (*Prionailurus viverrinus*) and the jungle cat (*Felis chaus*)—in a human-dominated landscape in suburban Kolkata. MSc thesis, Nat. Cent. Biol. Sci. Tata Inst. Fundam. Res.

Alarcon, G. G., and P. C. Simões-Lopes. 2004. The Neotropical otter *Lontra longicaudis* feeding habits in a marine coastal area, southern Brazil. IUCN *Otter Spec. Group Bull.* 21:24–30.

Alderton, D. 1996. *Rodents of the World*. London, UK: Facts on File, Inc.

Aleksiuk, M. 1970. The function of the tail as a fat storage depot in the beaver (*Castor canadensis*). *J. Mammal.* 51:145–148.

Aleksiuk, M., and A. Frohlinger. 1971. Seasonal metabolic organization in the muskrat (*Ondatra zibethicus*). I. Changes in growth, thyroid activity, brown adipose tissue, and organ weights in nature. *Can. J. Zool.* 49:1143–1154.

Allbrook, D. 2009. The morphology of the subdermal glands of *Hippopotamus amphibius*. *J. Zool.* 139:67–73.

Allen, C. R., S. Demarais, and R. S. Lutz. 1994. Red imported fire ant impact on wildlife: An overview. *Tex. J. Sci.* 46:51–59.

Al-Robaae, K. 1977. Distribution of *Nesokia indica* (Gray 6 Hardwicke, 1830) in Basrah Liwa, south Iraq. With some biological notes. *Säugetierkd. Mitt.* 25:194–197.

Amador, L. I., and N. P. Giannini. 2016. Phylogeny and evolution of body mass in didelphid marsupials (Marsupialia: Didelphimorphia: Didelphidae). *Org. Divers. Evol.* 13:641–657.

Amstislavsky, S., H. Lindeberg, Y. Ternovskaya, E. Zavjalov, G. Zudova, D. Klochkov, and L. Gerlinskaya. 2009. Reproduction in the European mink, *Mustela lutreola*: Oestrous cyclicity and early pregnancy. *Reprod. Domest. Anim.* 44:489–498.

Anderson, C. B., M. V. Lencinas, P. K. Wallem, A. E. J. Valenzuela, M. P. Simanonok, and G. Martínez Pastur. 2014. Engineering by an invasive species alters landscape-level ecosystem function, but does not affect biodiversity in freshwater systems. *Divers. Distrib.* 20:214–222.

Anderson, C. B., G. M. Pastur, M. V. Lencinas, P. K. Wallem, M. C. Moorman, and A. D. Rosemond. 2009. Do introduced North American beavers *Castor canadensis* engineer differently in southern South America? An overview with implications for restoration. *Mamm. Review* 38:33–52.

Anderson, R. M, and R. M. May. 1979. Population biology of infectious diseases: Part I. *Nature* 280:361–367.

Ando, K., and S. Fujiwara. 2016. Farewell to life on land—thoracic strength as a new indicator to determine paleoecology in secondary aquatic mammals. *J. Anat.* 229:768–777.

Araújo, C. V. M., M. Moreira-Santos, J. P. Sousa, V. Ochoa-Herrera, A. C. Encalada, and R. Ribeiro. 2014. Contaminants as habitat disturbers: PAH-driven drift by Andean paramo stream insects. *Ecotoxicol. Environ. Saf.* 108:89–94.

Archer, M., F. A. Jenkins, S. J. Hand, P. Murray, and H. Godthelp. 1992. Description of the skull and non-vestigial dentition of a Miocene plátypus (*Obdurodon dicksoni* n.sp.) from Riversleigh, Australia, and the problem of monotreme origins. In *Platypus and Echidnas*, edited by M. L. Augee, 15–27. Chipping Norton, NSW: Royal Zool. Soc. New South Wales.

Asahara, M., M. Koizumi, T. E. Macrini, S. J. Hand, and M. Archer. 2016. Comparative cranial morphology in living and extinct platypuses: Feeding behavior, electroreception, and loss of teeth. *Sci. Adv.* 2:e1601329.

Ashdown, P. 1969. For the defense: Florida conservationist speaks out in favor of fire ant control. *Agric. Chem.* 24:10–12.

Asher, R. J., and M. Hofreiter. 2006. Tenrec phylogeny and the noninvasive extraction of nuclear DNA. *Syst. Biol.* 55:181–194.

Asher, R. J., J. Meng, J. E. Wible, M. C., McKenna, G. W. Rougier, D. Dashzeveg, and M. J. Novacek. 2005. Stem Lagomorpha and the antiquity of Glires. *Science* 307:1091–1094.

Bagniewska, J. M., T. Hart, L. A. Harrington, and D. W. Macdonald. 2013. Hidden Markov analysis describes dive patterns in semiaquatic mammals. *Behav. Ecol.* 24:659–667.

Bailey, V. 1923. The combing claws of beavers. *J. Mammal.* 4:77–79.

Baker, B. W., H. C. Ducharme, D. C. Mitchell, T. R., Stanley, and H. R. Peinetti. 2005. Interaction of beaver and elk herbivory reduces standing crop of willow. *Ecol. Appl.* 15:110–118.

Baker, C. M. 1992a. *Atilax paludinosus. Mamm. Species* 408:1–6.

———. 1992b. Observations on the postnatal behavioural development in the marsh mongoose (*Atilax paludinosus*). *Z. Säugetierkunde* 57:335–342.

Baker, C. M., and J. D. Ray. 2013. *Atilax paludinosus* marsh mongoose. In *The Mammals of Africa. V. Carnivores, Pangolins, Equids and Rhinoceroses*, edited by J. Kingdon and M. Hoffmann, 298–302. London, UK: Bloomsbury Publishing.

Bakka, S. V., N. Yu Kiseleva, I. I. Pankratov, I. A. Tarasov, and P. M. Shukov. 2018. The story of the creation and monitoring of the Russian desman (*Desmana moschata* L.) population reintroduced of in the Kerzhenets river floodplain in the Nizhny Novgorod region. *IOP Conf. Series: Earth Environ. Sci.* 115:012036. http://doi .org/10.1088/1755-1315/115/1/012036.

Ballard, K. A., J. G. Sivak, and H. C. Howard. 1988. Intraocular muscles of the Canadian river otter and Canadian beaver and their optical function. *Can. J. Zool.* 67:469–474.

Ballinger, A., and P. S. Lake. 2006. Energy and nutrient fluxes from rivers and streams into terrestrial food webs. *Mar. Freshwater Res.* 57:15–28.

Banfield, A. W. F. 1974. *The Mammals of Canada.* Toronto, ON: Univ. Toronto Press.

Barklow, W. E. 2004. Amphibious communication with sound in hippos, *Hippopotamus amphibius*. *Animal Behav.* 68:1125–1132.

Barnes, W. J., and E. Dibble. 1988. The effects of beaver in riverbank forest succession. *Can. J. Bot.* 66:40–44.

Barros, K. S., R. S. Tokumaru, J. P. Pedroza, and S. S. C. Nogueira. 2011. Vocal repertoire of captive capybara (*Hydrochoerus hydrochaeris*): Structure, context and function. *Ethology* 117:83–93.

Baskin, J. A. 2004. *Bassariscus* and *Probassariscus* (Mammalia, Carnivora, Procyonidae) from the early Barstovian (Middle Miocene). *J. Vert. Paleo.* 24:709–720.

Batson, W. G., I. J. Gordon, D. B. Fletcher, and A. D. Manning. 2015. Translocation tactics: A framework to support the IUCN Guidelines for wildlife translocations and improve the quality of applied methods. *J. Appl. Ecol.* 52:1598–1607.

Baudinette, R. V., and P. Gill. 1985. The energetics of "flying" and "paddling" in water: Locomotion in penguins and ducks. *J. Comp. Physiol.* 155:373–380.

Baumgartner, L. L., and F. C. Bellrose. 1943. Determination of sex and age in muskrats. *J. Wildl. Manag.* 7:77–81.

Bazin, R. C., and R. A. MacArthur. 1992. Thermal benefits of huddling in muskrat (*Ondatra zibethicus*). *J. Mammal.* 73:559–564.

BC Ministry of Forests, Lands and Natural Resource Operations. 2014. Nutria (*Myocastor coypus*). BC Prohibited Species Alert. https://www2.gov.bc.ca/assets/gov/environment/plants-animals-and-ecosystems/invasive-species/alerts/nutria_alert.pdf.

Beckel-Kratz, A. 1977. Preliminary observations of the social behaviour of the North American otter (*Lutra canadensis*). *J. Otter Trust* 1977:28–32.

Becker, K. 1972. Muskrats in central Europe and their control. *Proc. 5th Vertebr. Pest Conference.* 6:18–21.

Bednarek, A. T. 2001. Undamming rivers: A review of the ecological impacts of dam removal. *Environ. Manage.* 27:803–814.

Ben-David, M., T. M. Williams, and O. A. Ormseth. 2000. Effects of oiling on exercise physiology and diving behaviour of river otters: A captive study. *Can. J. Zool.* 78:1380–1390.

Bennie, J. J., J. P. Duffy, R. Inger, and K. J. Gaston. 2014. Biogeography of time partitioning in mammals. *Proc. Natl. Acad. Sci. USA* 111:13727–13732.

Benoit, J., P. R. Manger, L. Norton, V. Fernandez, and B. S. Rubidge. 2017. Synchrotron scanning reveals the palaeoneurology of the head-butting *Moschops capensis* (Therapsida, Dinocephalia). *PeerJ* 5:e3496.

Benstead, J. P., K. H. Barnes, and C. M. Pringle. 2001. Diet, activity patterns, foraging movement and responses to deforestation of the aquatic tenrec *Limnogale mergulus* (Lipotyphla: Tenrecidae) in eastern Madagascar. *J. Zool. Lond.* 254:119–129.

Benstead, J. P., M. M. Douglas, and C. M. Pringle. 2003. Relationships of stream invertebrate communities to deforestation in eastern Madagascar. *Ecol. Appl.* 13:1473–1490.

Bertolino, S., M. L. Guichón, and J. Carter. 2012. *Myocastor coypus* Molina (coypu). In *A Handbook of Global Freshwater Invasive Species*, edited by R. A. Francis, 357–368. London: Earthscan.

Bertolino, S., A. Perrone, and L. Gola. 2005. Effectiveness of coypu control in small Italian wetland areas. *Wildl. Soc. Bull.* 33:714–720.

Bethge, P., S. Munks, H. Otley, and S. Nicol. 2009. Activity patterns and sharing of time and space of platypuses, *Ornithorhynchus anatinus*, in a subalpine Tasmanian lake. *J. Mammal.* 90:1350–1356.

Biewener, A. A. 1989. Mammalian terrestrial locomotion and size: Mechanical design principles define limits. *BioScience* 39:776–783.

Bininda-Emonds, O. R. P., J. L. Gittleman, and C. K. Kelly. 2001 Flippers versus feet: Comparative trends in aquatic and non-aquatic carnivores. *J. Anim. Ecol.* 70:386–400.

Birkenholz, D. E. 1972. *Neofiber alleni. Mamm. Species* 14:1–4.

Blair, W. F. 1936. The Florida marsh rabbit. *J. Mammal.* 17:197–207.

Bochkov, A. V., A. Labrzycka, M. Skoracki, and A. P. Saveljev. 2012. Fur mites of the genus *Schizocarpus* Trouessart (Acari: Chirodiscidae) parasitizing the Eurasian beaver *Castor fiber belorussicus* Lavrov (Rodentia: Castoridae) in NE Poland (Suwalki). *Zootaxa* 3410:1–18.

Boisserie, J. 2005. The phylogeny and taxonomy of Hippopotamidae (Mammalia: Artiodactyla): A review based on morphology and cladistic analysis. *Zool. J. Linn. Soc.* 143:1–26.

Bonesi, L., and D. W. Macdonald. 2004. Differential habitat use promotes sustainable coexistence between the specialist otter and the generalist mink. *Oikos* 106:509–519.

Bonesi, L., and S. Palazon. 2007. The American mink in Europe: Status, impacts, and control. *Biol. Conserv.* 134:470–483.

Bonesi, L., and M. Thom. 2012. *Neovison vison* Schreber (American mink). In *A Handbook of Global Freshwater Invasive Species*, edited by R. A. Francis, 369–381. London: Earthscan.

Bosschere, H., S. Roels, N. Lemmens, and E. Vanopdenbosch. 2005. Canine distemper virus in Asian clawless otter (*Aonyx cinereus*) littermates in captivity. *Vlaams Diergeneeskd. Tijdschr.* 74:299–302.

Botha-Brink, J., and K. D. Angielczyk. 2010. Do extraordinarily high growth rates in Permo-Triassic dicynodonts (Therapsida, Anomodontia) explain their success before and after the end-Permian extinction? *Zool. J. Linn. Soc.* 160:341–365.

Bounds, D. L., M. H. Sherfy, and T. A. Mollett. 2003. Nutria. In *Wild Mammals of North America: Biology, Management, and Conservation*, 2nd ed., edited by G. A. Feldhamer, B. C. Thompson, and J. A. Chapman, 1119–1147. Baltimore: Johns Hopkins Univ. Press.

Bowman, J., A. G. Kidd, R. M. Gorman, and A. I. Schulte-Hostedde. 2007. Assessing the potential for impacts by feral mink on wild mink in Canada. *Biol. Conserv.* 139:12–18.

Bowman, J., A. G. Kidd, L. A. Nituch, C. Sadowski, and A. Schulte-Hostedde. 2014. Testing for Aleutian mink disease virus in the river otter (*Lontra canadensis*) in sympatry with infected American mink (*Neovison vison*). *J. Wildl. Dis.* 50:689–693.

Brashares, J. S., P. Arcese, M. K. Sam, P. B. Coppolillo, A. R. E. Sinclair, and A. Balmford. 2004. Bushmeat hunting, wildlife declines, and fish supply in West Africa. *Science* 306:1180–1183.

Bromley, C. K., and G. A. Hood. 2013. Beaver (*Castor canadensis*) facilitate early access by Canada geese (*Branta canadensis*) to nesting habitat and open water in Canada's boreal wetlands. *Mamm. Biol.* 78:73–77.

Brommer, J. E., R. Alakoski, V. Selonen, and K. Kauhala. 2017. Population dynamics of two beaver species in Finland inferred from citizen-science census data. *Ecosphere* 8:e01947.

Brooks, R. P. 1985. Microenvironments and activity patterns of burrow-dwelling muskrats (*Ondatra zibethicus*) in rivers. *Acta Zool. Fenn.* 173:47–49.

Brooks, R. P., and W. E. Dodge. 1981. Identification of muskrat (*Ondatra zibethicus*) habitat in riverine environments. In *Proceedings of the Worldwide Furbearer Conference*, edited by J. A. Chapman and D. Pursley, 113–128. Frostburg, MD: Worldwide Furbearer Conf., Inc.

———. 1986. Estimation of habitat quality and summer population density for muskrats on a watershed basis. *J. Wildl. Manage.* 50:269–273.

Brooks, R. P., T. L. Serfass, M. Triska, and L. Rebelo. 2011. Ramsar protected wetlands of international importance as habitats for otters. *IUCN Otter Spec. Group Bull.* 28:47–63.

Brzeziński, M., J. Romanowski, M. Żmihorski, and K. Karpowicz. 2010. Muskrat (*Ondatra zibethicus*) decline after the expansion of American mink (*Neovison vison*) in Poland. *Eur. J. Wildl. Res.* 56:341–348.

Bullard, R. W., and G. M. Rapp. 1970. Problems of body heat loss in water immersion. *Aerosop. Med.* 41:1269–1277.

Burgin, C. J., J. P. Colella, P. L. Kahn, and N. S. Upham. 2018. How many species of mammals are there? *J. Mammal.* 99:1–14.

Burnett, S., P. Menkhorst, M. Ellis, and M. Denny. 2016. *Rattus lutreolus* (errata version published in 2017). *IUCN Red List of Threatened Species* 2016:e.T19343A115147713.

Burrell, H. J. 1927. *The Platypus: Its Discovery, Zoological Position, Form and Characteristics, Habits, Life History, etc.* Sydney: Angus and Robertson.

Butler, P. J. 1982. Respiratory and cardiovascular control during diving in birds and mammals. *J. Exp. Biol.* 100:195–221.

Butler, P. J., and D. R. Jones. 1997. Physiology of diving birds and mammals. *Physiol. Rev.* 77:837–899.

Byers, O., C. Lees, J. Wilcken, and C. Schwitzer. 2013. The "One Plan Approach": The philosophy and implementation of CBSG's approach to integrated species conservation planning. *WAZA Mag.* 14:2–5.

Cabrera-Guzmán, E., M. R. Crossland, D. Pearson, J. K. Webb, and R. Shine. 2015. Predation on invasive cane toads (*Rhinella marina*) by native Australian rodents. *J. Pest. Sci.* 88:143–153.

Calder, W. A. 1969. Temperature relations and underwater endurance of the smallest homeothermic diver, the water shrew. *Comp. Biochem. Physiol.* 30:1075–1082.

Caldwell, W. H. 1887. The embryology of Monotremata and Marsupialia. *Phil. Trans. R. Soc. Lond. B* 178:463–486.

Caley, M. J. 1987. Dispersal and inbreeding avoidance in muskrats. *Animal Behav.* 35:1225–1233.

Campbell, K. L., and R. A. MacArthur. 1996. Digestibility of animal tissue by muskrats. *J. Mammal.* 77:755–760.

Canessa, S., S. J. Converse, M. West, N. Clemann, G. Gillespie, M. McFadden, A. J. Silla, K. M. Parris, and M. A. McCarthy. 2016. Planning for ex situ conservation in the face of uncertainty. *Conserv. Biol.* 30:599–609.

Cantoni, D. 1993. Social and spatial organization of free-ranging shrews, *Sorex coronatus* and *Neomys fodiens* (Insectivora, Mammalia). *Animal Behav.* 45:975–995.

Carleton, M. D., and S. L. Olson. 1999. Amerigo Vespucci and the rat of Fernando de Noronha: A new genus and species of Rodentia (Muridae, Sigmodontinae) from a volcanic island off Brazil's continental shelf. *Am. Mus. Novit.* 3256:1–59.

Carpenter, S. R., E. H. Stanley, and M. J. Vander Zanden. 2011. State of the world's freshwater ecosystems: Physical, chemical, and biological changes. *Annu. Rev. Environ. Resour.* 36:75–99.

Carrick, F. N., T. R. Grant, and P. D. Temple-Smith. 2008. Platypus, *Ornithorhynchus anatinus*. In *The Mammals of Australia*, 3rd ed., edited by S. Van Dyck and R. Strahan, 32–35. Sydney, AU: Reed New Holland.

Carroll, L. 1872. *Through the Looking Glass and What Alice Found There*. London: Macmillan and Co.

Carson, R. 2002. *Silent Spring*. Boston: Houghton Mifflin.

Carter, J., and B. P. Leonard. 2002. A review of the literature on the worldwide distribution, spread of, and efforts to eradicate the coypu (*Myocastor coypus*). *Wildl. Soc. Bull.* 30:162–175.

Carter, S. K., F. C. W. Rosas, A. B. Cooper, and A. C. Cordeiro-Duarte. 1999. Consumption rate, food preferences and transit time of captive giant otters *Pteronura brasiliensis*: Implications for the study of wild populations. *Aquat. Mamm.* 25:79–90.

Carthey, A. J., and P. B. Banks. 2015. Foraging in groups affects giving-up densities: Solo foragers quit sooner. *Oecologia* 178:707–713.

Cassola, F. 2016. *Sorex bendirii* (errata version published in 2017). *IUCN Red List of Threatened Species* 2016:e.T41389A115183051.

Castellini, M. A. 1988. Visualizing metabolic transitions in aquatic mammals: Does apnea plus swimming equal "diving"? *Can. J. Zool.* 66:40–44.

Catania, K. C. 1999. A nose that looks like a hand and acts like an eye: The unusual mechanosensory system of the star-nosed mole. *J. Comp. Physiol.* 185:367–372.

———. 2013. The neurobiology and behavior of the American water shrew (*Sorex palustris*). *J. Comp. Physiol. A.* 199:545–554.

Catania, K. C., J. F. Hare, and K. L. Campbell. 2008. Water shrews detect movement, shape and smell to find prey underwater. *Proc. Natl. Acad. Sci. USA* 106:571–576.

Chanin, P. 1985. *The Natural History of Otters*. New York: Facts on File.

Chapman, J. A., and G. A. Feldhamer. 1981. *Sylvilagus aquaticus*. *Mamm. Species* 151:1–4.

Chapman, J. A., and G. R. Willner. 1981. *Sylvilagus palustris*. *Mamm. Species* 1531:1–3.

Charnov, E. L. 1976. Optimal foraging: The marginal value theorem. *Theor. Popul. Biol.* 9:129–136.

Cheal, D. C., 1987. The diets and dietary preferences of *Rattus-fuscipes* and *Rattus-lutreolus* at Walkerville in Victoria. *Aust. Wildl. Res.* 14:35–44.

Chen, M., Z.-X. Luo, and G. P. Wilson. 2017. The postcranial skeleton of *Yanoconodon allini* from the Early Cretaceous of Hebei, China, and its implications for locomotor adaptation in eutriconodontan mammals. *J. Vertebr. Paleontol.* 37:e1315425.

Chen, M., and G. P. Wilson. 2015. A multivariate approach to infer locomotor modes in Mesozoic mammals. *Paleobiol.* 41:280–313.

Cherin, M., D. A. Iurino, G. Willemsen, and G. Carnevale. 2016. A new otter from the Early Pleistocene of Pantalla (Italy), with remarks on the evolutionary history of Mediterranean Quaternary Lutrinae (Carnivora, Mustelidae). *Quat. Sci. Rev.* 135:92–102.

Chimento, N., F. Agnolin, and A. G. Martinelli. 2016. Mesozoic mammals from South America: Implications for understanding early mammalian faunas from Gondwana. In *Historia Evolutiva y Paleobiogeográfica de los Vertebrados de América del Sur*, edited by F. L. Agnolin, G. L. Lio, F. Brissón Egli, N. R. Chimento, F. E. Novas, 199–209. Buenos Aires: Museo Argentino de Ciencias Naturales.

Chiozza, F. 2016. *Chimarrogale sumatrana*. *IUCN Red List of Threatened Species* 2016:e.T4649A22282082.

Choi, C. 2008. Tierra del Fuego: The beavers must die. *Nature* 453:968.

Christensen, L. S., L. Gram-Hansen, M. Chriel, and T. H. Jensen. 2011. Diversity and stability of Aleutian mink disease during bottleneck transitions resulting from eradication in domestic mink in Denmark. *Vet. Microbiol.* 149:64–71.

Christoffersen, L. E. 1997. IUCN: A bridge-builder for nature conservation. In *Green Globe Yearbook 1997: Yearbook of International Co-operation on Environment and Development*, edited by H. O. Bergesen and G. Parmann, 59–69. Oxford: Oxford Univ. Press.

Cianfrani, C., O. Broennimann, A. Loy, and A. Guisan. 2018. More than range exposure: Global otter vulnerability to climate change. *Biol. Conserv.* 221:103–113.

Cincotta, R. P., J. Wisnewski, and R. Engelman. 2000. Human population in the biodiversity hotspots. *Nature* 404:990–992.

Clare, S., N. Krogman, L. Foote, and N. Lemphers. 2011. Where is the avoidance in the implementation of wetland law and policy? *Wetl. Ecol. Manag.* 19:165–182.

Clausen, G., and A. Ersland. 1970. Blood O_2 and acid-base changes in the beaver during submersion. *Respir. Physiol.* 11:104–112.

Clauss, M., W. J. Streich, A. Schwarm, S. Ortmann, and J. Hummel. 2007. The relationship of food intake and ingesta passage predicts feeding ecology in two different megaherbivore groups. *Oikos* 116:209–216.

Clayton, L., and D. W. MacDonald. 1999. Social organization of the babirusa (*Babyrousa babyrussa*) and their use of salt licks in Sulawesi, Indonesia. *J. Mammal.* 80:1147–1157.

Clayton, L. M., E. J. Milner-Gulland, D. W. Sinaga, and A. H. Mustari. 2000. Effects of a proposed ex situ conservation program on in situ conservation of the babirusa, and endangered suid. *Conserv. Biol.* 14:382–385.

Clemens, E. T., and G. M. O. Malioy. 1982. The digestive physiology of three East African herbivores: The elephant, rhinoceros and hippopotamus. *J. Zool.* 198:141–156.

Cockburn, A. 2006. Prevalence of different modes of parental care in birds. *Proc. Biol. Sci.* 273:1375–1383.

Collins, K., C. Bragg, and C. Birss. 2019. *Bunolagus monticularis. IUCN Red List of Threatened Species* 2019:e.T3326A45176532.

Colwell, R. K., and T. F. Rangel. 2009. Hutchinson's duality: The once and future niche. *Proc. Natl. Acad. Sci. USA* 106:19651–19658.

Connolly, J. H. 2009. A review of mucormycosis in the platypus (*Ornithorhynchus anatinus*). *Aust. J. Zool.* 57:235–244.

Cook, J. A., C. J. Conroy, and J. D. Herriges. 1997. Northern record of the water shrew, *Sorex palustris*, in Alaska. *Can. Field-Nat.* 111:638–640.

Cornelis, G., C. Vernochet, S. Malicorne, S. Souquere, A. C. Tzika, S. M. Goodman, F. Catzeflis, T. J. Robinson, M. C. Milinkovitch, G. Pierron, O. Heidmann, A. Dupressoir, and T. Heidmann. 2014. Retroviral envelope syncytin capture in an ancestrally diverged mammalian clade for placentation in the primitive Afrotherian tenrecs. *Proc. Natl. Acad. Sci. USA* 111:E4332–E4341.

Costa, D. P., and G. L. Kooyman. 1982. Oxygen consumption, thermoregulation, and the effects of fur oiling and washing on the sea otter, *Enhydra lutris. Can. J. Zool.* 60:2761–2767.

Coughlin, B. L., and F. E. Fish. 2009. Hippopotamus underwater locomotion: Reduced-gravity movements for a massive animal. *J. Mammal.* 90:675–679.

Cove, M. V., B. Gardner, T. R. Simons, and A. F. O'Connell. 2018. Co-occurrence dynamics of endangered Lower Keys marsh rabbits and free-ranging domestic cats: Prey responses to an exotic predator removal program. *Ecol. Evol.* 8:4042–4052.

Cox, J. M., S. L. Marinier, and A. J. Alexander. 1988. Auditory communication in the cane rat (*Thryonomys swinderianus*). *J. Zool.* 216:141–167.

Cozzuol, M. A., F. Goin, M. Reyes, and A. Ranzi. 2006. The oldest species of *Didelphis* (Mammalia, Marsupialia, Didelphidae), from the late Miocene of Amazonia. *J. Mammal.* 87:663–667.

Crandall, L. S. 1964. *The Management of Wild Mammals in Captivity*. Chicago: Univ. Chicago Press.

Crawford, J. C., Z. Liu, T. A. Nelson, C. K. Nielsen, and C. K. Bloomquist. 2008. Microsatellite analysis of mating and kinship in beavers (*Castor canadensis*). *J. Mammal.* 89:575–581.

Curtis, P. G., C. M. Slay, N. L. Harris, A. Tyukavina, and M. C. Hansen. 2018. Classifying drivers of global forest loss. *Science* 361:1108–1111.

Cutright, W. J., and T. McKean. 1979. Countercurrent blood vessel arrangement in beaver (*Castor canadensis*). *J. Morphol.* 161:169–176.

Czech-Damal, N. U., G. Dehnhardt, P. Manger, and W. Hanke. 2013. Passive electroreception in aquatic mammals. *J. Comp. Physiol. A.* 199:555–563.

Dagg, A. I., and D. E. Windsor. 1972. Swimming in northern terrestrial mammals. *Can. J. Zool.* 50:117–130.

Dalerum, F. 2007. Phylogenetic reconstruction of carnivore social organizations. *J. Zool.* 273:90–97.

Dalsgaard, N. J. 2002. Prion diseases. An overview. *APMIS* 110:3–13.

Danell, K. 1996. Introductions of aquatic rodents: Lessons of the muskrat *Ondatra zibethicus* invasion. *Wildl. Biol.* 2:213–220.

Danilov, P., V. Kanshiev, and F. Fyodorov. 2011. Characteristics of North American and Eurasian beaver ecology in Karelia. In *Restoring the European Beaver: 50 Years of Experience*, edited by G. Sjöberg and J. P. Ball, 55–72. Sofia, BG: Pensoft Publishers.

Darwin, C., 1859. *On the Origin of Species by Means of Natural Selection, or Preservation of Favoured Races in the Struggle for Life*. London: John Murray.

Davis, J. A. 1978. A classification of otters. In *Otters: Proceedings of the First Meeting of the Otter Specialist Group*, edited by N. Duplaix, 14–33. Morges, FR: IUCN.

Davison, A., H. I. Griffiths, R. C. Brookes, T. Maran, D. W. Macdonald, V. E. Sidorovich, A. C. Kitchener, I. Irizar, I. Villate, J. González-Esteban, J. C. Ceña, A. Ceña, I Moya, and S. Palazón Miñano. 2000. Mitochondrial DNA and palaeontological evidence for the origins of endangered European mink, *Mustela lutreola*. *Anim. Conserv.* 4:345–355.

Dawson, T. J., and P. D. Fanning. 1981. Thermal and energetic problems of semi-aquatic mammals: A study of the Australian water rat, including comparisons with the platypus. *Physiol. Zool.* 54:285–296.

Delany, M. J. 1975. *The Rodents of Uganda*. London: British Mus. Nat Hist.

Deocampo, D. M. 2002. Sedimentary structures generated by *Hippopotamus amphibius* in a lake-margin wetland, Ngorongoro Crater, Tanzania. *Palaios* 17:212–217.

Derrickson, E. M. 1992. Comparative reproductive strategies of altricial and precocial eutherian mammals. *Funct. Ecol.* 6:57–65.

de Silva, P., W. A. Khan, B. Kanchanasaka, I. Reza Lubis, M. M. Feeroz, and O. F. Al-Sheikhly. 2015. *Lutrogale perspicillata*. *IUCN Red List of Threatened Species* 2015:e.T12427A21934884.

Devictor, V., J. Clavel, R. Julliard, S. Lavergne, D. Mouillot, W. Thuiller, P. Venail, S. Villeger, and N. Mouquet. 2010. Defining and measuring ecological specialization. *J. Appl. Ecol.* 47:15–25.

de Witte, G. F., and S. Frechkop. 1955. Sur une espèce encore inconnue de mammifère Africain, *Potamogale ruwenzorii* sp. N. *Bull. Institut Royal Sci. Nat. Belgique* 31:1–11.

Di Minin, E., and T. Toivonen. 2015. Global protected area expansion: Creating more than paper parks. *BioScience* 65:637–638.

Doby, J. 2015. A systematic review of the Soricimorph Eulipotyphla (Soricidae: Mammalia) from the Gray fossil site (Hemphillian). MSc thesis, East Tennessee State Univ.

Dodds, W. K, and M. R. Whiles. 2010. *Freshwater ecology: Concepts and environmental applications of limnology.* 2nd ed. London: Academic Press.

Döll, P., B. Jiménez-Cisneros, T. Oki, N. W. Arnell, G. Benito, J. G. Cogley, T. Jiang, Z. W. Kundzewicz, S. Mwakalila, and A. Nishijima. 2015. Integrating risks of climate change into water management. *Hydrolog. Sci. J.* 60:4–13.

Dong, T., M. Tep, S. Lim, S. Soun, and T. Chrin. 2010. Distribution of otters in the Tropeang Roung, Koh Kong Province, Cambodia. *IUCN Otter Spec. Group Bull.* 27:62–74.

Donkor, N. T., and J. M. Fryxell. 1999. Impact of beaver foraging on structure of lowland boreal forests of Algonquin Provincial Park, Ontario. *For. Ecol. Manage.* 118:83–92.

Doronina, L., A. Matzke, G. Churakov, M. Stoll, A. Huge, and J. Schmitz. 2017. The beaver's phylogenetic lineage illuminated by retroposon reads. *Sci. Rep-UK.* 7:43562.

Dorward, L. J. 2015. New record of cannibalism in the common hippo, *Hippopotamus amphibius* (Linnaeus, 1758). *Afr. J. Ecol.* 53:385–387.

Douady, C. J., and E. J. P. Douzery. 2003. Molecular estimation of eulipotyphlan divergence times and the evolution of "Insectivora." *Mol. Phylogenet. Evol.* 28:285–296.

Dozier, H. L. 1942. Identification of sex in live muskrats. *J. Wildl. Manag.* 6:292–293.

———. 1952. The present status and future of nutria in the southeast states. *Proc. Annual Conference Southeast. Assoc. Game Fish Comm.* 6:368–373.

Dubost, G. 1984. Comparison of the diets of frugivorous forest ruminants of Gabon. *J. Mammal.* 65:298–316.

Duckworth, J. W., and D. M. Hills. 2008. A specimen of hairy-nosed otter *Lutra sumatrana* from far northern Myanmar. *IUCN Otter Spec. Group Bull.* 25:60–67.

Dudgeon, D., A. H. Arthington, M. O. Gessner, Z. Kawabata, D. J. Knowler, C. Lévêque, R. J. Naiman, A. H. Prieur-Richard, D. Soto, M. L. J. Stiassny, and C. A. Sullivan. 2006. Freshwater biodiversity: Importance, threats, status and conservation challenges. *Biol. Rev.* 81:163–182.

Dudley, J. P., B. M. Hang'ombe, F. H. Leendertz, L. J. Dorward, J. de Castro, A. L. Subalusky, and M. Clauss. 2016. Carnivory in the common hippopotamus *Hippopotamus amphibius*: Implications for the ecology and epidemiology of anthrax in African landscapes. *Mamm. Rev.* 46:191–203.

Dunstone, N. 1978. The fishing strategy of mink (*Mustela vison*); time-budgeting of hunting effort? *Behaviour* 67:157–177.

———. 1993. *The mink.* London: T and A. D. Poyser Ltd.

———. 1998. Adaptations to the semi-aquatic habit and habitat, In *Behaviour and Ecology of Riparian Mammals*, edited by N. Dunstone and M. C. Gorman, 1–16. Cambridge: Cambridge Univ. Press.

Dunstone, N., and M. C. Gorman, eds. 1998. *Behaviour and Ecology of Riparian Mammals.* Cambridge: Cambridge Univ. Press.

Dunstone, N., and R. J. O'Connor. 1979. Optimal foraging in an amphibious mammal. I. The aqualung effect. *Anim. Behav.* 27:1182–1194.

Duplaix, N. 1980. Observations on the ecology and behavior of the giant river otter *Pteronura brasiliensis* in Suriname. *Rev. Ecol. Terre Vie* 34:495–620.

Duplaix, N., and M. Savage. 2018. *The Global Otter Conservation Strategy.* Salem, OR: IUCN/SSC Otter Specialist Group.

Dyck, A. P., and R. A. MacArthur. 1992. Seasonal patterns of body temperature and activity in free-ranging beaver (*Castor canadensis*). *Can. J. Zool.* 70:1668–1672.

Eadie, W. R., and W. J. Hamilton. 1956. Notes on reproduction in the star-nosed mole. *J. Mammal.* 37:223–231.

Eberhardt, R. T. 1973. Some aspects of mink-waterfowl relationships of prairie wetlands. *Prairie Nat.* 5:17–19.

Edwards, M. J., and J. E. Deakin. 2013. The marsupial pouch: Implications for reproductive success and mammalian evolution. *Aust. J. Zool.* 61:41–47.

Eisenberg, J. F. 1981. *The Mammalian Radiations: An Analysis of Trends in Evolution, Adaptation, and Behavior.* Chicago: Univ. Chicago Press.

Elton, C. S. 1927. *Animal Ecology.* London: Sidgwick and Jackson.

Eltringham, S. 2002. *The Hippos.* London: Bloomsbury Publishing PLC.

Enders, R. K. 1937. Panniculus carnosus and formation of the pouch in didelphids. *J. Morphol.* 61:1–26.

Engelhart, A., and D. Müller-Schwarze. 1995. Responses of beaver (*Castor canadensis* Kuhl) to predator chemicals. *J. Chem. Ecol.* 21:1349–1364.

Ermala, A., S. Lahti, and P. Vikberg. 1999. Bäverstammen ökar fortfarande—fangsten redan närmare 2500 bävrar. *Jägaren* 4:28–31.

Errington, P. L. 1939. Observations on young muskrats in Iowa. *J. Mammal.* 20:465–478.

———. 1946. *Special Report on Muskrat Disease.* Iowa Coop. Wildl. Res. Unit, Quart. Rep.

———. 1963. *Muskrat Populations.* Ames: Iowa State Univ. Press.

Errington, P. L., R. J. Siglin, and R. C. Clark. 1963. The decline of a muskrat population. *J. Wildl. Manage.* 27:1–8.

Essbauer, S., S. Hartnack, K. Misztela, J. Kiessling-Tsalos, W. Baeumler, and M. Pfeffer. 2009. Patterns of orthopox virus wild rodent hosts in South Germany. *Vector-Borne Zoonotic Dis.* 9:301–311.

Estes, J. A. 1986. Marine otters and their environment. *Ambio* 15:181–183.

———. 1989. Adaptations for aquatic living by carnivores. In *Carnivore Behavior, Ecology, and Evolution*, Part 1, edited by J. L. Gittleman, 242–282. Ithaca, NY: Cornell Univ. Press.

Estes, R. D. 1991. *The Behavior Guide to African Mammals, Including Hoofed Mammals, Carnivores and Primates.* Berkeley: Univ. California Press.

Evans, B. K., D. R. Jones, J. Baldwin, and G. R. J. Gabbott. 1994. Diving ability of the platypus. *Austr. J. Zool.* 42:17–27.

Everson, K. M., V. Soarimalala, S. M. Goodman, and L. E. Olson. 2016. Multiple loci and complete taxonomic sampling resolve the phylogeny and biogeographic history of tenrecs (Mammalia: Tenrecidae) and reveal higher speciation rates in Madagascar's humid forests. *Syst. Biol.* 65:890–909.

Fabre, P., L. Hautier, D. Dimitrov, and E. J. P. Douzery. 2012. A glimpse on the pattern of rodent diversification: A phylogenetic approach. *BMC Evol. Biol.* 12:88.

Fabre, P., J. T. Vilstrup, M. Raghavan, C. Der Sarkissian, E. Willerslev, E. J. P. Douzery, and L. Orlando. 2014. Rodents of the Caribbean: Origin and diversification of hutias unravelled by next-generation museomics. *Biol. Lett.* 10:20140266.

Fabre, P., N. S. Upham, L. H. Emmons, F. Justy, Y. L. R. Leite, A. C. Loss, L. Orlando, M. Tilak, B. D. Patterson, and E. J. P. Douzery. 2017. Mitogenomic phylogeny, diversification, and biogeography of South American spiny rats. *Mol. Biol. Evol.* 34:613–633.

Falcão, C. B., I. L. Lima, J. M. Duarte, J. R. de Oliveira, R. A. Torres, A. M. Wanderley, J. E. Gomes da Cunha, and J. E. Garcia. 2017. Are Brazilian cervids at risk of prion diseases? *Prion* 11:65–70.

Fanning, F. D., and T. J. Dawson. 1980. Body temperature variability in the Australian water rat, *Hydromys chrysogaster*, in air and water. *Aust. J. Zool.* 28:229–238.

Fasola, F., J. Muzio, C. Chehébar, M. Cassini, and D. W. Macdonald. 2011. Range expansion and prey use of American mink in Argentinean Patagonia: Dilemmas for conservation. *Eur. J. Wildl. Res.* 57:283–294.

Faulhaber, C. A., N. D. Perry, N. J. Silvy, R. R. Lopez, P. A. Frank, and M. J. Peterson. 2006. Reintroduction of Lower Keys marsh rabbits. *Wild. Soc. Bull.* 34:1198–1202.

Feldhamer, G. A., L. C. Drickamer, S. H. Vessey, J. F. Merritt, and C. Krajewski. 2007. *Mammalogy: Adaptation, Diversity, Ecology.* 3rd ed. Baltimore: Johns Hopkins Univ. Press.

———. 2015. *Mammalogy: Adaptation, Diversity, Ecology.* 4th ed. Baltimore: Johns Hopkins Univ. Press.

Feldhamer, G. A., B. C. Thompson, and J. A. Chapman, eds. 2003. *Wild Mammals of North America: Biology, Management, and Conservation.* 2nd ed. Baltimore: Johns Hopkins Univ. Press.

Fernández, F. J., J. Torres, M. Tammone, J. M. López, and U. F. J. Pardiñas. 2017. New data on the endemic cricetid rodent *Holochilus lagigliai* from central-western Argentina: Fossil record and potential distribution. *Mammalia.* 81:621–625.

Ficetola, G. F., C. Miaud, F. Pompanon, and P. Taberlet. 2008. Species detection using environmental DNA from water samples. *Biol. Lett.* 40:423–425.

Field, C. R. 1970. A study of the feeding habits of the hippopotamus (*Hippopotamus amphibius* Linn.) in the Queen Elizabeth National Park, Uganda, with some management implications. *Zool. Afr.* 5:71–86.

Figueira, C. J. M., J. S. R. Pires, A. Andriolo, M. J. R. P. Costa, and J. M. B. Duarte. 2005. Marsh deer (*Blastocerus dichotomus*) reintroduction in the Jataí Ecological Station (Luís Antônio, SP): Spatial preferences. *Braz. J. Biol.* 65:263–270.

Finn, F. 1929. *Sterndale's Mammalia of India.* Calcutta: Thacker, Spink and Co.

Fischer, J., and D. B. Lindenmayer. 2000. An assessment of the published results of animal translocations. *Biol. Conserv.* 96:1–11.

Fish, F. E. 1979. Thermoregulation in the muskrat (*Ondatra zibethicus*): The use of regional heterothermia. *Comp. Biochem. Physiol.* 64:391–397.

———. 1982a. Aerobic energetics of surface swimming in the muskrat *Ondatra zibethicus.* *Physiol. Zool.* 55:180–189.

———. 1982b. Function of the compressed tail of surface swimming muskrats (*Ondatra zibethicus*). *J Mammal.* 63:591–597.

———. 1992. Aquatic locomotion. In *Mammalian Energetics: Interdisciplinary Views of Metabolism and Reproduction*, edited by T. E. Tomasi and T. H. Horton, 34–63, 276. Ithaca, NY: Cornell Univ. Press.

———. 1993. Influence of hydrodynamic design and propulsive mode on mammalian swimming energetics. *Aust. J. Zool* 42:79–101.

———. 1994. Association of propulsive swimming mode with behavior in river otters (*Lutra canadensis*). *J. Mammal.* 75:989–997.

———. 1996. Transitions from drag-based to lift-based propulsion in mammalian swimming. *Amer. Zool.* 36:628–641.

———. 2000. Biomechanics and energetics in aquatic and semiaquatic mammals: Platypus to whale. *Physiol. Biochemic. Zool.* 73:683–698.

Fish, F. E., R. V. Baudinette, P. B. Frappell, and M. P. Sarre. 1997. Energetics of swimming by the platypus *Ornithorhynchus anatinus*: Metabolic effort associated with rowing. *J. Exp. Biol.* 200:2647–2652.

Fish, F. E., J. Smelstoys, R. V. Baudinette, and P. S. Reynolds. 2002. Fur does not fly, it floats: Buoyancy of pelage in semi-aquatic mammals. *Aquat. Mamm.* 28.2:103–112.

Fisher, R. E., K. M. Scott, and V. L. Naples. 2007. Forelimb myology of the pygmy hippopotamus (*Choeropsis liberiensis*). *Anat. Rec.* 290:673–693.

Flannery, T. F. 1995. *The Mammals of New Guinea.* 2nd ed. Sydney, AU: Reed Books.

Flynn, J. J., and G. D. Wesley-Hunt. 2005. Carnivora. In *The Rise of Placental Mammals: Origins and Relationships of the Major Extant Clades,* edited by K. D. Rose and J. D. Archibold, 175–198. Baltimore: Johns Hopkins Univ. Press.

Fortes, F. S., L. C. Santos, Z. S. Cubas, I. R. Barros-Filho, A. W. Biondo, I. Silveira, M. B. Labruna, and M. B. Molento. 2011. Anti-*Rickettsia* spp. antibodies in free-ranging and captive capybaras from southern Brazil. *Pesqui. Vet. Bras.* 31:1014–1018.

Forys, E. A., C. R. Allen, and D. P. Wojcik. 2002. Influence of the proximity and amount of human development and roads on the occurrence of the red imported fire ant in the lower Florida Keys. *Biol. Conserv.* 108:27–33.

Foster-Turley, P. 1992. Conservation ecology of sympatric Asian otters *Aonyx cinerea* and *Lutra perspicillata.* PhD diss., Univ. Florida.

Frafjord, K. 2016. Influence of reproductive status: Home range size in water voles (*Arvicola amphibius*). *PLoS One* 11:e0154338.

France, R. L. 1997. The importance of beaver lodges in structuring littoral communities in boreal headwater lakes. *Can. J. Zool.* 75:1009–1013.

Frantz, L., E. Meijaard, J. Gongora, J. Haile, M. A. Groenen, and G. Larson. 2016. The evolution of Suidae. *Annu. Rev. Anim. Biosci.* 4:61–85.

Fraser, J., and D. Wharton. 2007. The future of zoos: A new model for cultural institutions. *Curator* 50:41–54.

Fretwell, S. D., and H. L. Lucas. 1970. On territorial behavior and other factors influencing habitat distribution in birds. I. Theoretical Development. *Acta Biotheor.* 19:16–36.

Furlan, E., J. Stoklosa, J. Griffiths, N. Gust, R. Ellis, R. M. Huggins, and A. R. Weeks. 2012. Small population size and extremely low levels of genetic diversity in island populations of the platypus, *Ornithorhynchus anatinus.* *Ecol. Evol.* 2:844–857.

Gallant, D., C. H. Berube, E. Tremblay, and L. Vasseur. 2004. An extensive study of the foraging ecology of beavers (*Castor canadensis*) in relation to habitat quality. *Can. J. Zool.* 82:922–933.

Galliez, M., M. de Souza Leite, T. Lopes Queiroz, and F. A. dos Santos Fernandez. 2009. Ecology of the water opossum *Chironectes minimus* in Atlantic forest streams of southeastern Brazil. *J. Mammal.* 90:93–103.

Games, I. 1983. Observations on the sitatunga *Tragelaphus spekei selousi* in the Okavango Delta of Botswana. *Biol. Conserv.* 27:157–170.

Garland, E. 2008. The elephant in the room: Confronting the colonial character of wildlife conservation in Africa. *Afri. Stud. Rev.* 51:51–74.

Gatti, A., R. Bianchi, C. R. Xavier Rosa, and S. Lucena Mendes. 2006. Diet of two sympatric carnivores, *Cerdocyon thous* and *Procyon cancrivorus,* in a restinga area of Espirito Santo State, Brazil. *J. Trop. Ecol.* 22:227–230.

Gaubert, P., C. A. Fernandes, M. W. Bruford, and G. Veron. 2004. Genets (Carnivora, Viverridae) in Africa: An evolutionary synthesis based on cytochrome *b* sequences and morphological characters. *Biol. J. Linn. Soc.* 81:589–610.

Gaywood, M. J. 2018. Reintroducing the Eurasian beaver *Castor fiber* to Scotland. *Mamm. Rev.* 48:48–61.

Ge, D., Z. Wen, L. Xia, Z. Zhang, M. Erbajeva, C. Huang, and Q. Yang, O. 2013. Evolutionary history of lagomorphs in response to global environmental change. *PLoS One* 8:e59668.

Geist, V. 1999. *Deer of the World: Their Evolution, Behaviour, and Ecology.* Shrewsbury, UK: Swan Hill Press.

Gelling, M., D. W. Macdonald, S. Telfer, T. Jones, K. Bown, R. Birtles, and F. Mathews.

2012. Parasites and pathogens in wild populations of water voles (*Arvicola amphibius*) in the UK. *Eur. J. Wildl. Res.* 58:615–619.

Gerell, R. 1967. Dispersal and acclimatization of the mink (*Mustela vison* Schreb.) in Sweden. *Viltrevy* 5:1–38.

———. 1968. Food habits of the mink, *Mustela vison* Schreber, in Sweden. *Viltrevy* 5:119–121.

Gherman, C. M., A. D. Sándor, Z. Kalmár, M. Marinov, and A. D. Mihalca. 2012. First report of *Borrelia burgdorferi sensu lato* in two threatened carnivores: The marbled polecat, *Vormela peregusna* and the European mink, *Mustela lutreola* (Mammalia: Mustelidae). *BMC Vet. Res.* 8:137.

Gingerich, P. D. 2003. Land-to-sea transition in early whales: Evolution of Eocene Archaeoceti (Cetacea) in relation to skeletal proportions and locomotion of living semiaquatic mammals. *Paleobiology* 29:429–454.

Ginsburg, L. 1999. Order Carnivora. In *The Miocene: Land Mammals of Europe*, edited by G. E. Rössner and K. Heissig, 109–148. München: Pfeil.

Gittleman, J. L. 1985. Carnivore body size: Ecological and taxonomic correlates. *Oecologia* 67:540–544.

Gleason, J. S., R. A. Hoffman, and J. M. Wendland. 2005. Beavers, *Castor canadensis*, feeding on salmon carcasses: Opportunistic use of a seasonally superabundant food source. *Can. Field-Nat.* 119:591–593.

Gomez, L., and J. Bouhuys. 2018. *Illegal Otter Trade in Southeast Asia*. Petaling Jaya, MY: TRAFFIC.

Gomez, L., B. T. C. Leupen, M. Theng, K. Fernandez, and M. Savage. 2016. *Illegal Otter Trade: An Analysis of Seizures in Selected Asian Countries (1980–2015)*. Petaling Jaya, MY: TRAFFIC.

Gong, B., and G. Zhang. 2014. Interactions between plants and herbivores: A review of plant defense. *Acta Ecol. Sin.* 34:325–336.

González, E. M., and U. G. J. Pardiñas. 2002. *Deltamys kempi. Mamm. Species* 711:1–4.

Goodwin, G. G. 1959. Descriptions of some new mammals. *Am. Mus. Novit.* 1967:1–8.

Gorman, M. L., D. Jenkins, and R. J. Harper. 1978. The anal scent sacs of the otter (*Lutra Intra*). *J. Zool.* 186:463–471.

Gosling, L. M., and S. J. Baker. 1989. The eradication of muskrats and coypus from Britain. *Biol. J. Linn. Soc.* 38:39–51.

Gould, S. J., and E. S. Vrba. 1982. Exaptation—a missing term in the science of form. *Paleobiology* 8:4–15.

Graells, G., D. Corcoran, and J. C. Aravena. 2015. Invasion of North American beaver (*Castor canadensis*) in the province of Magallanes, Southern Chile: Comparison between dating sites through interviews with the local community and dendrochronology. *Rev. Chil. Hist. Nat.* 88:UNSP 3.

Graf, P. M., R. P. Wilson, L. C. Sanchez, K. Hackländer, and F. Rosell. 2018. Diving behavior in a free-living, semi-aquatic herbivore, the Eurasian beaver *Castor fiber*. *Ecol. Evol.* 8:997–1008.

Grant, T. R., and T. J. Dawson. 1978a. Temperature regulation in the platypus, *Ornithorhynchus anatinus*: Production and loss of metabolic heat in air and water. *Physiol. Zool.* 51:315–332.

———. 1978b. Temperature regulation in the platypus, *Ornithorhynchus anatinus*: Maintenance of body temperature in air and water. *Physiol. Zool.* 51:1–6.

Grant, T. R., and P. D. Temple-Smith. 1998. Field biology of the platypus (*Ornithorhynchus anatinus*): Historical and current perspectives. *Philos. Trans. R. Soc. Lond. Ser. B Biol. Sci.* 353:1081–1091.

————. 2003. Conservation of the platypus, *Ornithorhynchus anatinus*: Threats and challenges. *Aquat. Ecosyst. Health Manag.* 6:5–18.

Green, J. 1977. Sensory perception in hunting otters, *Lutra lutra* L. *J. Otter Trust* 1977:13–16.

Greer, K. R. 1955. Yearly food habits of the river otter in the Thompson Lakes region, northwestern Montana, as indicated by scat analyses. *Am. Midl. Nat.* 54:299–313.

Griffiths, M., M. A. Elliott, R. M. C. Leckie, and G. I. Schoefl. 1973. Observations of comparative anatomy and ultrastructure of mammary glands and on the fatty acids of the triglycerides in platypus and echidna milk fats. *J. Zool.* 169:255–279.

Grinnell, J. 1917. The niche relationships of the California thrasher. *Auk* 34:427–433.

Groenendijk, J., N. Duplaix, M. Marmontel, P. Van Damme, and C. Schenck. 2015. *Pteronura brasiliensis*. *IUCN Red List of Threatened Species* 2015:e.T18711A21938411.

Groves, C. 2016. Systematics of the Artiodactyla of China in the 21(st) century. *Zool. Res.* 37:119–125.

Grützner, F., T. Ashley, D. M. Rowell, and J. A. Marshall Graves. 2006. How did the platypus get its sex chromosome chain? A comparison of meiotic multiples and sex chromosomes in plants and animals. *Chromosoma* 115:75–88.

Guichón, M. L., M. Borgnia, C. Fernández Righi, G. H. Cassini, and M. H. Cassini. 2003. Social behavior and group formation in the coypu (*Myocastor coypus*) in the Argentinean pampas. *J. Mammal.* 84:254–262.

Guilbride, P. D. L., T. J. Coyle, E. G. McAnulty, L. Barber, and G. D. Lomax. 1962. Some pathogenic agents found in hippopotamus in Uganda. *J. Comp. Path. Therap.* 72:137–141.

Guilday, J. E., P. S. Martin, and A. D. McCrady. 1964. New Paris No 4: A Pleistocene cave deposit in Bedford County, Pennsylvania. *Bull. Natl. Speleol. Soc.* 26:121–194.

Gunnell, G. F., T. M. Bown, J. H. Hutchinson, and J. I. Bloch. 2008. Lipotyphla. In *Evolution of Tertiary Mammals of North America: Small Mammals, Xenarthrans, and Marine Mammals*, vol. 2, edited by C. M. Janis, G. F. Gunnell, and M. D. Uhen, 89–125. Cambridge: Cambridge Univ. Press.

Guo, G., and E. Zhang. 2005. Diet of the Chinese water deer (*Hydropotes inermis*) in Zhoushan Archipelago, China. *Acta Theriol. Sinica* 25:122–130.

Gust, N., and J. Griffiths. 2009. Platypus mucormycosis and its conservation implications. *Austr. Mycol.* 28:1–8.

Guth, C., H. Heim de Balsac, and M. Lamotte. 1959. Recherches sur la morphologie de *Micropotamogale lamottei* et l'évolution des Potamogalinae: Écologie, denture, anatomie crânienne. *Mammalia* 23:423–447.

Hadlow, W. J., and L. Karstad. 1968. Transmissible encephalopathy of mink in Ontario. *Can. Vet. J.* 9:193–196.

Hagerman, A. E., and C. T. Robbins. 1993. Specificity of tannin-binding salivary proteins relative to diet selection by mammals. *Can. J. Zool.* 71:628–633.

Halanych, K. M., and T. J. Robinson. 1997. Phylogenetic relationships of cottontails (*Sylvilagus*, Lagomorpha): Congruence of 12S rDNA and cytogenetic data. *Mol. Phylogenet. Evol.* 7:294–302.

Haldane, J. B. S. 1949. *What is Life?* London: Lindsay Drummond.

Hall, B. K. 1999. The paradoxical platypus. *BioScience* 49:211–218.

Halley, D. J., and F. Rosell. 2002. The beaver's reconquest of Eurasia: Status population development and management of a conservation success. *Mamm. Rev.* 32:153–178.

Hamilton, W. D, and M. Zuk. 1982. Heritable true fitness and bright birds: A role for parasites? *Science* 218:384–387.

Hamilton, W. J. 1931. Habits of the star-nosed mole, *Condylura cristata*. *J. Mammal.* 12:345–355.

Hamilton, W. J., and W. R. Eadie. 1964. Reproduction in the otter, *Lutra canadensis*. *J. Mammal.* 45:242–252.

Hamrick, M. W. 2001. Development and evolution of the mammalian limb: Adaptive diversification of nails, hooves, and claws. *Evol. Dev.* 3:355–363.

Hanke, W., and G. Dehnhardt. 2013. Sensory biology of aquatic mammals. *J. Comp. Physiol. A.* 199:417–420.

Hanson, J. D., G. D'Elía, S. B. Ayers, S. B. Cox, S. F. Burneo, and T. E. Lee. 2015. A new species of fish-eating rat, genus Neusticomys (Sigmodontinae), from Ecuador. *Zool. Stud.* 54:49.

Haque, N. M., and V. Vijayan. 1993. Food habits of the fishing cat (*Felis viverrina*) in Keoladeo National Park Bharatpur, Rajasthan. *J. Bombay Nat. Hist. Soc.* 90:498–500.

———. 1995. Food habits of the smooth Indian otter (*Lutra perspicillata*) in Keoladeo National Park, Bharatpur, Rajasthan, India. *Mammalia* 59:345–348.

Harding, L. E., and F. A. Smith. 2009. *Mustela* or *Vison*? Evidence for the taxonomic status of the American mink and a distinct biogeographic radiation of American weasels. *Mol. Phylogenet. Evol.* 52:632–642.

Härkönen, S. 1999. Management of the North American beaver (*Castor canadensis*) on the South-Savo game management district, Finland (1983–1997). In *Beaver Protection, Management, and Utilization in Europe and North America*, edited by P. E. Busher and R. M. Dzięciołowski, 7–14. New York: Kluwer Academic/Plenum.

Harlow, H. J. 1984. The influence of Hardarian gland removal and fur lipid removal on heat loss and water flux to and from the skin of muskrats, *Ondatra zibethicus*. *Physiol. Zool.* 57:349–356.

Harrington, L. A., G. C. Hays, L. Fasola, A. L. Harrington, D. Righton, and D. W. Macdonald. 2012. Dive performance in a small-bodied, semi-aquatic mammal in the wild. *J. Mammal.* 93:198–210.

Hart, J. A., and R. M. Timm. 1978. Observations on the aquatic genet in Zaire. *Carnivore* 1:130–132.

Hashimoto, K., Y. Saikawa, and M. Nakata. 2007. Studies on the red sweat of the *Hippopotamus amphibius*. *Pure Appl. Chem.* 79:507–517.

Hassanin, A., F. Delsuc, A. Ropiquet, C. Hammer, B. Jansen van Vuuren, C. Matthee, M. Ruiz-Garcia, F. Catzeflis, V. Areskoug, T. T. Nguyen, and A. Couloux. 2012. Pattern and timing of diversification of Cetartiodactyla (Mammalia, Laurasiatheria), as revealed by a comprehensive analysis of mitochondrial genomes. *C. R. Biol.* 335:32–50.

Hawkins, M., and A. Battaglia. 2009. Breeding behaviour of the platypus (*Ornithorhynchus anatinus*) in captivity. *Aust. J. Zool.* 57:283–293.

Hayes, R. A., H. F. Nahrung, and J. C. Wilson. 2006. The response of native Australian rodents to predator odours varies seasonally: A by-product of life history variation? *Anim. Behav.* 71:1307–1314.

Hayssen, V., and T. J. Orr. 2017. *Reproduction in Mammals: The Female Perspective*. Baltimore: Johns Hopkins Univ. Press.

Hayward, J. S., and P. A. Lisson. 1992. Evolution of brown fat: Its absence in marsupials and monotremes. *Can. J. Zool.* 70:171–179.

Hayward, M. W., and G. I. H. Kerley. 2005. Prey preferences of the lion (*Panthera leo*). *J. Zool. Lond.* 267:309–322.

He, F., C. Zarfl, V. Bremerich, A. Henshaw, W. Darwall, K. Tockner, and S. C. Jähnig. 2017. Disappearing giants: A review of threats to freshwater megafauna. *Wiley Interdiscip. Rev.—Water* 2017 4:UNSP e1208.

He, K., Y. Li, M. C. Brandley, L. Lin, Y. Wang, Y. Zhang, and X. Jiang. 2010. A multi-locus phylogeny of Nectogalini shrews and influences of the paleoclimate on speciation and evolution. *Mol. Phylogenet. Evol.* 56:734–746.

Heim de Balsac, H. 1954. Un genre inédit et inattendu de Mammifère (Insectivore Tenrecidae) d'Afrique Occidentale. *C. R. Acad. Sei. Paris.* 239:102–104.

Helbing, H. 1927: Une genette miocène trouvée dans les argiles de Capitieux (Gironde). *Verh. Naturforsch. Ges. Basel* 38:305–315.

Helder, J., and H. K. Andrade. 1997. Food and feeding habits of the Neotropical river otter *Lontra longicaudis* (Carnivora, Mustelidae). *Mammalia* 61:193–203.

Helgen, K. M. 2005. The amphibious murines of New Guinea (Rodentia, Muridae): The generic status of *Baiyankamys* and description of a new species of *Hydromys*. *Zootaxa.* 913:1–20.

Helm, S. R., and R H. Chabreck. 2006. Notes on food habits of swamp rabbits in the Atchafalaya Basin, Louisiana. *J. Miss. Acad. Sci.* 51:129–133.

Hemmer, H. 1976. Fossil history of living Felidae. In *The World's Cats*, edited by R. L. Eaton, 1–14. Seattle, WA: Carnivore Res. Inst. Burke Museum.

Heng, S., T. Dong, N. Hon, and A. Olsson. 2016. The hairy-nosed otter *Lutra sumatrana* in Cambodia: Distribution and notes on ecology and conservation. *Cambodian J. Nat. Hist.* 2016:102–110.

Hentschke, J., H. Meyer, U. Wittstatt, A. Ochs, S. Burkhardt, and A. Aue. 1999. An outbreak of cowpox in beavers (*Castor fiber canadensis*) and bearcats (*Ailurus fulgens*) in Berlin zoo. *Tieraerztl. Umsch.* 54:311–316.

Herrera, E. A. 2013. Capybara social behavior and use of space: Patterns and processes. In *Capybara: Biology, Use and Conservation of an Exceptional Neotropical Species*, edited by J. R. Moreira, K. M. P. M. B. Ferraz, E. A. Herrera, and D. W. Macdonald, 195–208. New York: Springer.

Herrera, E. A., and D. W. Macdonald. 1989. Resource utilization and territoriality in group-living capybaras (*Hydrochoerus hydrochaeris*). *J. Anim. Ecol.* 58:667–679.

Hershkovitz, P. 1969. The recent mammals of the Neotropical region; a zoogeographie and ecological review. *Q. Rev. Biol.* 44:1–70.

Hertel, H. 1966. *Structure, Form, Movement.* New York: Reinhold.

Hickman, G. C. 1984. Swimming ability of talpid moles, with particular reference to the semi-aquatic *Condylura cristata*. *Mammalia* 48:505–513.

Hill, A., R. Drake, L. Tauxe, M. Monaghan, J. C. Barry, A. K. Behrensmeyer, G. Curtis, B. Fine Jacobs, L. Jacobs, N. Johnson, and D. Pilbeam. 1985. Neogene palaeontology and geochronology of the Baringo Basin, Kenya. *J. Hum. Evol.* 14:759–773.

Hocking, D. P., F. G. Marx, T. Park, E. M. G. Fitzgerald, and A. R. Evans. 2017. A behavioural framework for the evolution of feeding in predatory aquatic mammals. *Proc. R. Soc. Lond. B.* 284:20162750.

Holl, K. D. 2017. Restoring tropical forests from the bottom up. *Science* 355:455–456.

Holland, N., and S. M. Jackson. 2002. Reproductive behaviour and food consumption associated with the captive breeding of platypus (*Ornithorhynchus anatinus*). *J. Zool. Lond.* 256:279–288.

Holler, N. R., and H. M. Marsden. 1970. Onset of evening activity of swamp rabbits and cottontails in relation to sunset. *J. Wildl. Manage.* 34:349–353.

Holt, B. G., J. Lessard, M. K. Borregaard, S. A. Fritz, M. B. Araújo, D. Dimitrov, P. Fabre, C. H. Graham, G. R. Graves, K. A. Jønsson, D. Nogués-Bravo, Z. Wang, R. J. Whittaker, J. Fjeldså, and C. Rahbek. 2013. An update of Wallace's zoogeographic regions of the world. *Science* 339:74–78.

Honeycutt, R. L., L. J. Frabotta, and D. L. Rowe. 2007. Rodent evolution, phylogenetics, and biogeography. In *Rodent Societies: An Ecological and Evolutionary Perspective*, edited by J. Wolfe and P. Sherman, 8–27. Chicago: *Univ. Chicago Press* Chicago.

Hood, G. A. 2011. *The Beaver Manifesto*. Calgary, AB: Rocky Mountain Books.

———. 2015. Post-release survival of beavers exposed to bitumen. Technical Report, Canadian Natural Resources Ltd. Camrose, AB: Univ. of Alberta.

Hood, G. A., and S. E. Bayley. 2008a. The effects of high ungulate densities on foraging choices by beaver (*Castor canadensis*) in the mixed-wood boreal forest. *Can. J. Zool.* 86:484–496.

———. 2008b. Beaver (*Castor canadensis*) mitigate the effects of climate on the area of open water in boreal wetlands in western Canada. *Biol. Conserv.* 141:556–567.

———. 2009. A comparison of riparian plant community response to herbivory by beaver (*Castor canadensis*) and ungulates in Canada's boreal mixed-wood forest. *For. Ecol. Manage.* 258:1979–1989.

Hood, G. A., and D. G. Larson. 2015. Ecological engineering and aquatic connectivity: A new perspective from beaver-modified wetlands. *Freshwat. Biol.* 60:198–208.

Hood, G. A., V. Manaloor, and B. Dzioba. 2018. Mitigating infrastructure loss from beaver flooding: A cost-benefit analysis. *Hum. Dimens. Wildl.* 23:146–159.

Horn, S., W. Durka, R. Wolf, A. Ermala, A. Stubbe, M. Stubbe, and M. Hofreiter. 2011. Mitochondrial genomes reveal slow rates of molecular evolution and the timing of speciation in beavers (Castor), one of the largest rodent species. *PLoS One* 6:e14622.

Horovitz, I., T. Martin, J. Bloch, S., Ladevèze, C. Kurz, and M. R. Sánchez-Villagra. 2009. Cranial anatomy of the earliest marsupials and the origin of opossums. *PLoS One* 4:e8278.

Howell, A. B. 1930. *Aquatic Mammals: Their Adaptation to Life in the Water*. Springfield, IL: Charles C. Thomas.

Hugueney, M. 1976. Un stade primitive dans l'évolution des Soricinae (Mammalia, Insectivora): Srinitium marteli nov. gen. nov. sp. de Oligocéne moyen de Saint-Martin-de-Castillon (Vaucluse). *C. R. Acad. Sci. D* 282:981–984.

Hugueney, M., and F. Escuillié. 1995. K strategy and adaptive specialization in *Steneofiber* from Montaigu-le-Blin (dept. allier, France; Lower Miocene, MN 2a, ſſ23 Ma): First evidence of fossil life-history strategies in castorid rodents. *Paleogeogr. Paleoclimatol. Paleoecol.* 113:217–225.

Hulbert, R. C., A. Kerner, and G. S. Morgan. 2014. Taxonomy of the Pleistocene giant beaver *Castoroides* (Rodentia: Castoridae) from the southeastern United States. *Bull. Fla. Mus. Nat. Hist.* 53:26–43.

Hung, N., and C. J. Law. 2016. Lutra lutra (Carnivora: Mustelidae). *Mamm. Species* 48:109–122.

Hunsaker, D. 1977. *The Biology of Marsupials*. New York: Academic Press.

Hunt, R. M. 1996. Biogeography of the Order Carnivora. In *Carnivore Behavior, Ecology and Evolution*, edited by J. L. Gittleman, 485–541. Ithaca, NY: Cornell Univ. Press.

Hunt, T. P. 1959. Breeding habits of the swamp rabbit with notes on its life history. *J. Mammal.* 40:82–91.

Hussain, S. A. 1996. Group size, group structure and breeding in smooth-coated otter *Lutra perspicillata*, Geoffroy (Carnivora, Mustelidae) in National Chambal Sanctuary, India. *Mammalia* 60: 289–297.

Hussain, S. A., S. K. Gupta, and P. K. de Silva. 2011. Biology and ecology of Asian small-clawed otter *Aonyx cinereus* (Illiger, 1815): A review. *IUCN Otter Spec. Group Bull.* 28:63–75.

Hutchinson, G. E. 1957. Concluding remarks. *Cold Spring Harbor Symp. Quant. Biol.* 22:415–427.

Hutterer, R., B. Kryštufek, N. Yigit, G. Mitsain, H. Meinig, S. Bertolino, and L. J. Palomo. 2016a. *Neomys anomalus* (errata version published in 2017). *IUCN Red List of Threatened Species* 2016:e.T29657A115169785.

Hutterer, R., H. Meinig, S. Bertolino, B. Kryštufek, B. Sheftel, M. Stubbe, R. Samiya, J. Ariunbold, V. Buuveibaatar, S. Dorjderem, Ts. Monkhzul, M. Otgonbaatar, and M. Tsogbadrakh. 2016b. *Neomys fodiens* (errata version published in 2017). *IUCN Red List of Threatened Species* 2016:e.T29658A115170106.

Hwang, Y. T., and S. Larivière. 2005. *Lutrogale perspicillata. Mamm. Species* 786:1–4.

Hyvärinen, H. 1994. Brown fat and the wintering of shrews, In *Advances in the Biology of Shrews*, edited by J. F. Merritt, G. L. Kikland, and R. K. Rose, 259–266. Carnegie Museum of Natural History spec. pub. no. 16. Pittsburgh, PA: Carnegie Mus. Nat. Hist.

Hyvärinen, H., A. Palviainen, U. Strandberg, and I. J. Holopainen 2009. Aquatic environment and differentiation of vibrissae: Comparison of sinus hair systems of ringed seal, otter and pole cat. *Brain Behav. Evol.* 74:268–279.

IJzerman, H., J. A. Coan, F. M. A., Wagermans, M. A. Missler, I. van Beest, S. Lindenberg, and M. Tops. 2015. A theory of social thermoregulation in human primates. *Front. Psychol.* 6:UNSP 464.

Innis, H. A. 1930. *The Fur Trade in Canada: An Introduction to Canadian Economic History.* New Haven, CT: Yale Univ. Press.

Irving, L. 1939. Respiration in diving mammals. *Physiol. Rev.* 19:112–134.

———. 1973. Aquatic mammals. In *Comparative Physiology of Thermoregulation: Special Aspects of Thermoregulation*, Vol. 3, edited by G. C. Whittow, 47–96. London: Academic Press.

Irving, L., and M. D. Orr. 1935. The diving habits of the beaver. *Science* 82:569.

IUCN. 2016. *Bonn Challenge.* https://www.bonnchallenge.org.

IUCN. 2017. *Guidelines for Using the IUCN Red List Categories and Criteria. Version 13.* Prepared by the Standards and Petitions Subcommittee.

IUCN. 2019. *IUCN Red List of Threatened Species. Version 2019-1.* https://www.iucnredlist .org.

IUCN/SSC. 2013. *Guidelines for Reintroductions and Other Conservation Translocations. Version 1.0.* Gland, Switzerland: IUCN Species Survival Commission.

Iversen, J. A. 1972. Basal energy metabolism of mustelids. *J. Comp. Physiol.* 81:341–344.

Ivlev, Y. F., M. V. Rutovskaya, and O. S. Luchkina. 2013. The use of olfaction by the Russian desman (*Desmana moschata* L.) during underwater swimming. *Dokl. Biol. Sci.* 452:280–283.

Jacob, J., and E. von Lehmann. 1976. Chemical composition of the nasal gland secretion from the marsh deer *Odocoileus* (*Dorcelaphus*) *dichotomus* (Illiger). *Z. Naturforsch C.* 31:496–498.

Jacques, H., N. Duplaix, and G. Chapron. 2004. The Congo clawless otter: State of knowledge, and needs for further research. *IUCN Otter Spec. Group* Bull. 21A:2004.

Jacques, H., J. Reed-Smith, C. Davenport, and M. J. Somers. 2015. *Aonyx congicus. IUCN Red List of Threatened Species* 2015:e.T1794A14164772.

Jaksic, F. M., J. A. Iriarte, J. E. Jiménez, and D. R. Martínez. 2002. Invaders without frontiers: Cross-border invasions of exotic mammals. *Biol. Invasions* 4:157–173.

Jansa, S. A., and R. S. Voss. 2011. Adaptive evolution of the venom-targeted vWF protein in opossums that eat pitvipers. *PLoS One* 6:e20997.

Janzen, D. H. 1963. Observations on populations of adult beaver beetles, *Platypsyllus castoris* (Platypsyllidae: Coleoptera). *Pan-Pac. Entomol.* 34:215–228.

Jenkins, P. D., and A. A. Barnett. 1997. A new species of water mouse, of the genus *Chib-*

chanomys (Rodentia: Muridae: Sigmodontinae) from southern Ecuador. *Bull. Nat. Hist. Mus. Zool.* 63:123–128.

Jenkins, S. H. 1975. Food selection by beavers: A multidimensional contingency table analysis. *Oecologia* 21:157–173.

Ji, Q., Z.-X., Luo, C.-X. Yuan, and A. R. Tabrum. 2006. A swimming mammaliaform from the Middle Jurassic and ecomorphological diversification of early mammals. *Science* 311:1123–1127.

Jiang, Z., and R. B. Harris. 2016. *Elaphurus davidianus.* IUCN *Red List of Threatened Species* 2016:e.T7121A22159785.

Johansen, K. 1962a. Buoyancy and insulation in the muskrat. *J. Mammal.* 43:64–68.

———. 1962b. Heat exchange through the muskrat tail. Evidence for vasodilator nerves to the skin. *Acta Physiol. Scand.* 55:160–169.

Johnson, W. E., E. Eizirik, J. Pecon-Slattery, W. J. Murphy, A. Antunes, E. Teeling, and S. J. O'Brien. 2006. The late Miocene radiation of modern Felidae: A genetic assessment. *Science* 311:73–77.

Jones, C. G., J. H. Lawton, and M. Shachak. 1994. Organisms as ecosystem engineers. *Oikos* 69:373–386.

José, H., and H. K. De Andrade. 1997. Food and feeding habits of the Neotropical river otter *Lontra longicaudis* (Carnivora, Mustelidae). *Mammalia* 61:193–203.

Jung, J., Y. Shimizu, K. Omasa, S. Kim, and S. Lee. 2016. Developing and testing a habitat suitability index model for Korean water deer (*Hydropotes inermis argyropus*) and its potential for landscape management decisions in Korea. *Anim. Cells Syst.* 20:218–227.

Junk, W. J., S. An, C. M. Finlayson, B. Gopal, J. Květ, S. A. Mitchell, W. J. Mitsch, and R. D. Robarts. 2013. Current state of knowledge regarding the world's wetlands and their future under global climate change: A synthesis. *Aquat. Sci.* 75:151–167.

Käkelä, R., and H. Hyvärinen. 1996a. Fatty acids in extremity tissues of Finnish beavers (*Castor canadensis* and *Castor fiber*) and muskrat (*Ondatra zibethicus*). *Comp. Biochem. Physiol.* 113:113–124.

———. 1996b. Site-specific fatty acid composition adipose tissues of several northern aquatic and terrestrial mammals. *Comp. Biochem. Physiol.* 115:501–514.

Kanchanasaka, B. K. 2007. Food Habitats of the hairy-nosed otter (*Lutra sumatrana*) and the small clawed otter (*Amblonyx cinerea*) in Pru Toa Daeng Peat Swamp Forest, Southern Thailand. Xth International Otter Colloquium, Hwacheon, South Korea.

Kanga, E. M., J. O. Ogutu, H. Olff, and P. Santema. 2011. Population trend and distribution of the vulnerable common hippopotamus *Hippopotamus amphibius* in the Mara Region of Kenya. *Oryx* 45:20–27.

Kasso, M., and M. Balakrishnan. 2013. *Ex situ* conservation of biodiversity with particular emphasis to Ethiopia. ISRN *Biodiver.* 2013:985037.

Kemp, T. S. 2017. *Mammals: A Very Short Introduction.* Oxford: Oxford Univ. Press.

Kennerley, R., and S. T. Turvey. 2016. *Desmana moschata.* IUCN *Red List of Threatened Species* 2016:e.T6506A22321477.

Kenyon, K. W. 1969. The sea otter in the eastern Pacific Ocean. *N. Amer. Fauna* 68:1–352.

Kerbis Peterhans, J., and L. Lavrenchenko. 2008. *Nilopegamys plumbeus.* IUCN *Red List of Threatened Species* 2008:e.T40766A10363474.

Khlyap, L. A., V. V. Bobrov, and A. A. Warshavsky. 2010. Biological invasions on Russian territory: Mammals. *Russ. J. Biol. Invasions.* 1:127–140.

Kim, B. J., N. S. Lee, and S. D. Lee. 2011. Feeding diets of the Korean water deer (*Hydro-*

potes inermis argyropus) based on a 202 bp rbcL sequence analysis. *Conserv. Genet.*
12:851–856.

Kim, J., J, Y. Lee, T. Han, K. Han, S. S. Kang, C. S. Bae, and S. H. Choi. 2005. A case of
malocccluded incisor teeth in a beaver (*Castor canadensis*). *J. Vet. Sci.* 6:173–175.

King, G. M. 1990. The aquatic *Lystrosaurus*: A palaeontological myth. *Hist. Biol.*
4:285–321.

Kingdon, J. 1974. *East African mammals*. Vol. 2. New York: Academic Press.

Kleiman, D. G. 1977. Monogamy in mammals. *Q. Rev. Biol.* 52:39–69.

———. 1989. Reintroduction of captive mammals for conservation: Guidelines for reintroducing endangered species into the wild. *BioScience* 39:152–161.

Klug, H. 2018. Why monogamy? A review of potential drivers. *Front. Ecol. Evol.* 6:30.

Knox, R. 1823. Observations on the anatomy of the duck-billed animal of New South
Wales, the *Ornithorynchus paradoxus* of naturalists. *Mem. Wernerian Nat. Hist. Soc.*
5:26–41.

Koepfli, K.-P., K. A. Deere, G. J. Slater, C. Begg, K. Begg, L. Grassman, M. Lucherini, G.
Veron, and R. K. Wayne. 2008a. Multigene phylogeny of the Mustelidae: Resolving
relationships, tempo and biogeographic history of a mammalian adaptive radiation.
BMC Biol. 6:10.

Koepfli, K.-P., M. E. Gompper, E. Eizirik, C. Ho, L. Linden, J. E. Maldonado, and R. K.
Wayne. 2007. Phylogeny of the Procyonidae (Mammalia: Carnivora): Molecules,
morphology and the Great American Interchange. *Mol. Phylogenetics Evol.* 43:1076–
1095.

Koepfli, K.-P., B. Kanchanasaka, H. Sasaki, H. Jacques, K. D. Y. Louie, T. Hoai, N. X.
Dang, E. Geffen, A. Gutleb, S. Han, T. M. Heggberget, L. LaFontaine, H. Lee, R.
Melisch, J. Ruiz-Olmo, M. Santos-Reis, V. E. Sidorovich, M. Stubbe, and R. K.
Wayne. 2008b. Establishing the foundation for an applied molecular taxonomy of
otters in Southeast Asia. *Conserv. Genet.* 9:1589–1604.

Koepfli, K.-P., and R. K. Wayne. 1998. Phylogenetic relationships of otters (Carnivora:
Mustelidae) based on mitochondrial cytochrome *b* sequences. *J. Zool. Lond.*
246:401–416.

Köhler, D. 1991. Notes on the diving behaviour of the water shrew, *Neomys fodiens*
(Mammalia, Soricidae). *Zool. Anz.* 227:218–228.

Kowalski, K., P. Marciniak, G. Rosiński, and L. Rychlik. 2017. Evaluation of the physiological activity of venom from the Eurasian water shrew *Neomys fodiens*. *Front. Zool.*
14:46.

Koyama, S., H. Wu, T. Easwaran, S. Thopady, and J. Foley. 2013. The nipple: A simple
intersection of mammary gland and integument, but focal point of organ function.
J. Mammary Gland Biol. Neoplasia 18:121–131.

Krause, W. J. 1974. Intestinal mucosa of the platypus, *Ornithorhynchus anatinus*. *Anat.*
Rec. 181:251–266.

Kruuk, H. 2006. *Otters: Ecology, Behaviour and Conservation*. Oxford: Oxford Univ. Press.

Kruuk, H., and D. Balharry. 1990. Effects of sea water on thermal insulation of the otter,
Lutra. *J. Zool. Lond.* 220:405–415.

Kruuk, H., E. Balharry, and P. T. Taylor. 1994. Oxygen consumption of the Eurasian otter
Lutra lutra in relation to water temperature. *Physiol. Zool.* 67:1174–1185.

Kruuk, H., and P. C. Goudswaard. 1990. Effects of changes in fish populations in Lake
Victoria on the food of otters (*Lutra maculicollis* Schinz and *Aonyx capensis* Lichtenstein). *Afr. J. Ecol.* 28:322–329.

Kruuk, H., B. Kanchanasaka, S. O'Sullivan, and S. Wanghongsa. 1994. Niche separation in three sympatric otters *Lutra perspicillata*, *L. lutra* and *Aonyx cinerea* in Huai Kha Khaeng, Thailand. *Biol. Conserv.* 69:115–120.

Kruuk, H., and A. Moorhouse. 1991. The spatial organization of otters *Lutra lutra* L. in Shetland. *J. Zool. Lond.* 224:41–57.

Kruuk, H., D. Wansink, and A. Moorhouse. 1990. Feeding patches and diving success of otters, *Lutra lutra*, in Shetland. *Oikos* 57:68–72.

Kryštufek, B., and A. Bukhnikashvili. 2016. *Neomys teres*. *IUCN Red List of Threatened Species* 2016:e.T29659A22282493.

Kuhn, H. J. 1964. Zur Kenntnis von *Micropotamogale lamottei* (Heim de Balsac). *Z. Säugetierkd.* 29:152–173.

Kuhn, R. A., and W. Meyer. 2010. Comparative hair structure in the Lutrinae (Carnivora: Mustelidae). *Mammalia* 74:291–303.

Kuntner, M., L. J. May-Collado, and I. Agnarsson. 2011. Phylogeny and conservation priorities of afrotherian mammals (Afrotheria, Mammalia). *Zool. Scr.* 40:1–15.

Lacher, T. E., W. J. Murphy, J. Rogan, A. T. Smith, and N. S. Upham. 2016. Evolution, phylogeny, ecology and conservation of the Clade Glires: Lagomorpha and Rodentia. In *Handbook of the Mammals of the World*, Vol. 6, edited by D. E. Wilson, T. E. Lacher, and R. A. Mittermeier, 15–26. Barcelona: Lynx Ediciones.

Lambert, C. T., A. C. Sabol, and N. G. Solomon. 2018. Genetic monogamy in socially monogamous mammals is primarily predicted by multiple life history factors: A meta-analysis. *Front. Ecol. Evol.* 6:10.3389/fevo.2018.00139.

Lambert, W. D. 1997. The osteology and paleoecology of the giant otter *Enhydritherium terraenovae*. *J. Vertebr. Paleontol.* 17:738–749.

Langer, P. 1975. Macroscopic anatomy of the stomach of the Hippopotamidae. *Anat. Histol. Embryol.* 4:334–359.

Larivière, S. 1998. *Lontra felina*. *Mamm. Species* 575:1–5.

———. 1999. *Lontra longicaudis*. *Mamm. Species* 609:1–5.

———. 2001. *Aonyx capensis*. *Mamm. Species* 671:1–6.

———. 2003. *Amblonyx cinereus*. *Mamm. Species* 720:1–5.

Lavorgna, A. 2014. Wildlife trafficking in the Internet age. *Crime Sci* 3:5.

Lavrenchenko, L. A., and A. Bekele. 2017. Diversity and conservation of Ethiopian mammals: What have we learned in 30 years? *Ethiop. J. Biol. Sci.* 16:1–20.

Law, A., M. J. Gaywood, K. C. Jones, P. Ramsay, and N. J. Willby. 2017. Using ecosystem engineers as tools in habitat restoration and rewilding: Beaver and wetlands. *Sci. Total Environ.* 605:1021–1030.

Laws, R. M. 1984. Hippopotamuses. In *The Encyclopedia of Mammals*, edited by D. W. Macdonald, 506–511. New York: Facts on File.

Lee, D., M. Lee, Y. Kim, I. R. Kim, H. K. Kim, D. G. Jeong, J. R. Lee, and J. H. Kim. 2018. Complete mitochondrial genome of the invasive semi-aquatic mammal, nutria *Myocastor coypus* (Rodentia; Myocastoridae). *Conservation Genet. Resour.* 10:613–616.

Lehner, B., and P. Döll. 2004. Development and validation of a global database of lakes, reservoirs and wetlands. *J. Hydrol.* 296:1–22.

Lehner, B., C. R. Liermann, C. Revenga, C. Vorosmarty, B. Fekete, P. Crouzet, P. Doll, M. Endejan, K. Frenken, J. Magome, C. Nilsson, J. C. Robertson, R. Rödel, N. Sindorf, and D. Wisser. 2011. High-resolution mapping of the world's reservoirs and dams for sustainable river-flow management. *Front. Ecol. Environ.* 9:494–502.

Leighton, A. H. 1933. Notes on the relations of beavers to one another and to the muskrat. *J. Mammal.* 14:27–35.

Leite, R. N., S. Kolokotronis, F. C. Almeida, F. P. Werneck, D. S. Rogers, and M. Weksler.

2014. In the wake of invasion: Tracing the historical biogeography of the South American Cricetid radiation (Rodentia, Sigmodontinae). *PLoS One* 9:e100687.

Leopold, A. 1949. *A Sand County Almanac, and Sketches Here and There*. New York: Oxford Univ. Press.

Leus, K., A. Macdonald, J. Burton, and I. Rejek. 2016. *Babyrousa celebensis*. IUCN Red List of Threatened Species 2016:e.T136446A44142964.

Levins, R. 1969. Some demographic and genetic consequences of environmental heterogeneity for biological control. *Bull. Entomol. Soc. Am.* 15:237–240.

Lewison, R. L., and J. Carter. 2004. Exploring behavior of an unusual megaherbivore: A spatially explicit foraging model of the hippopotamus. *Ecol. Model.* 171:127–138.

Li, Z., G. Beauchamp, and M. S. Mooring. 2014. Relaxed selection for tick-defense grooming in Père David's deer? *Biol. Conserv.* 178:12–18.

Liat, L. B., D. M. Belabut, and R. Hashim. 2013. Ecological study of the Malaysian water shrew. *Raffles Bull. Zool.* 29:155–159.

Lima, D. O., G. M. Pinho, and F. A. S. Fernandez. 2016. Spatial patterns of the semi-aquatic rodent *Nectomys squamipes* in Atlantic forest streams. *J. Nat. Hist.* 50:497–511.

Lindroth, R. L., and G. O. Batzli. 1984. Plant phenolics as chemical defenses: Effects of natural phenolics on survival and growth of prairie voles (*Microtus ochrogaster*). *J. Chem. Ecol.* 10:229–244.

Ling, J. K. 1970. Pelage and molting in wild mammals with special reference to aquatic forms. *Q. Rev. Biol.* 45:16–54.

Liu, A. G. S. C., E. R. Seiffert, and E. L. Simons. 2008. Stable isotope evidence for an amphibious phase in early proboscidean evolution. *Proc. Natl. Acad. Sci. USA* 105:5786–5791.

Liwanag, H. E. M., A. Berta, D. P. Costa, M. Abney, and T. M. Williams. 2012. Morphological and thermal properties of mammalian insulation: The evolution of fur for aquatic living. *Biol. J. Linn. Soc.* 106:926–939.

Lodé, T., J. P. Cormier, and D. Le Jacques. 2001. Decline in endangered species as an indicator of anthropic pressures: The case of European mink *Mustela lutreola* western population. *Environ. Manage.* 28:727–735.

Lopatin, A. V. 2002. The earliest shrew (Soricidae, Mammalia) from the Middle Eocene of Mongolia. *Paleontol. J.* 36:650–659.

López-Antoñanzas, R., S. Sen, and P. Mein. 2004. Systematics and phylogeny of the cane rats (Rodentia: Thryonomyidae). *Zool. J. Linn. Soc.* 142:423–444.

Lovegrove, B. G. 2012. The evolution of endothermy in Cenozoic mammals: A plesiomorphic-apomorphic continuum. *Biol. Rev.* 87:128–162.

Lowe, C. E. 1958. Ecology of the swamp rabbit in Georgia. *J. Mammal.* 39:116–127.

Lowery, G. H. 1974. *The Mammals of Louisiana and Its Adjacent Waters*. Baton Rouge: Louisiana State Univ. Press.

Lubis, R. 2005. First recent record of hairy-nosed otter in Sumatra, Indonesia. IUCN Otter Spec. Group Bull. 22:14–20.

Lucas, J. R., P. M. Waser, and S. R. Creel. 1994. Death and disappearance: Estimating mortality risks associated with philopatry and dispersal. *Behav. Ecol.* 5:135–141.

Luck, C. P., and P. G. Wright. 1964. Aspects of the anatomy and physiology of the skin of the hippopotamus (*H. amphibius*). *Q. J. Exp. Physiol. Cogn. Med. Sci.* 49:1–14.

Ludwig, D. R. 1984. *Microtus richardsoni*. *Mamm. Species* 223:1–6.

———. 1988. Reproduction and population dynamics of the water vole, *Microtus richardsoni*. *J. Mammal.* 69:532–541.

Lukas, D., and T. Clutton-Brock. 2012. Cooperative breeding and monogamy in mammalian societies. *Proc. R. Soc. B* 279:2151–2156.

———. 2013. The evolution of social monogamy in mammals. *Science* 341:526–530.

Lunney, D. 2008. Swamp rat, *Rattus lutreolus*. In *The Mammals of Australia*, 3rd ed., edited by S. Van Dyck and R. Strahan, 690–692. Sydney, AU: Reed New Holland.

Luo, Z. 2007. Transformation and diversification in early mammal evolution. *Nature* 450:1011–1019.

MacArthur, R. A. 1977. Behavioral and physiological aspects of temperature regulation in the muskrat (*Ondatra zibethicus*). PhD diss., Winnipeg: Univ. of Manitoba.

———. 1979. Seasonal patterns of body temperature and activity in free-ranging muskrats (*Ondatra zibethicus*). *Can. J. Zool.* 57:25–33.

———. 1984. Aquatic thermoregulation in the muskrat (*Ondatra zibethicus*): Energy demands of swimming and diving. *Can. J. Zool.* 62:241–248.

———. 1986. Brown fat and aquatic temperature regulation in muskrats, *Ondatra zibethicus*. *Physiol. Zool.* 59:306–331.

———. 1989a. Aquatic mammals in cold. In *Advances in Comparative and Environmental Physiology*, Vol. 4, edited by L. C. H. Wang, 289–325. Berlin: Springer-Verlag.

———. 1989b. Energy metabolism and thermoregulation of beaver (*Castor canadensis*). *Can. J. Zool.* 67:651–657.

———. 1992. Gas bubble release by muskrats diving under ice: Lost gas or potential oxygen pool? *J. Zool.* 226:151–164.

MacArthur, R. A., and M. Aleksiuk. 1979. Seasonal microenvironments of the muskrat (*Ondatra zibethicus*) in a northern marsh. *J. Mammal.* 60:146–154.

MacArthur, R. A., and A. P. Dyck. 1990. Aquatic thermoregulation of captive and free-ranging beavers (*Castor canadensis*). *Can. J. Zool.* 68:2409–2416.

MacArthur, R. A., M. M. Humphries, G. A. Fines, and K. L. Campbell. 2001. Body oxygen stores, aerobic dive limits and the diving abilities of juvenile and adult muskrats (*Ondatra zibethicus*). *Physiol. Biochem. Zool.* 74:178–190.

MacArthur, R. A., M. M. Humphries, and D. Jeske. 1997. Huddling behavior and the foraging efficiency of muskrats. *J. Mammal.* 78:850–858.

MacArthur, R. H., and E. R. Pianka. 1966. On the optimal use of a patchy environment. *Am. Nat.* 100:603–609.

Macdonald, D. W. 1981. Dwindling resources and the social behaviour of capybaras, *Hydrochoerus hydrochaeris*. *J. Zool.* 194:371–391.

Macdonald, D. W., and L. A. Harrington. 2003. The American mink: The triumph and tragedy of adaptation out of context. *N. Z. J. Zool.* 30:421–441.

Macdonald, A., K. Leus, I. Masaaki, and J. Burton, J. 2016. *Babyrousa togeanensis*. IUCN *Red List of Threatened Species* 2016:e.T136472A44143172.

MacPherson, J. L., and P. W. Bright. 2011. Metapopulation dynamics and a landscape approach to conservation of lowland water voles (*Arvicola amphibius*). *Landscape Ecol.* 26:1395–1404.

Maddison, N. R., J. Zhigang, and M. Boyd. 2012. The Père David's deer: Clinging on by a cloven hoof. WAZA *Magazine* 13:29–32.

Maran, T. 2003. European mink: Setting of goal for conservation and the Estonian case study. *Galemys* 15:1–11.

Maran, T., H. Kruuk, D. W. Macdonald, and M. Polma. 1998. Diet of two species of mink in Estonia: Displacement of *Mustela lutreola* by *M. vison*. *J. Zool.* 245:218–222.

Maran, T., D. Skumatov, A. Gomez, M. Põdra, A. V. Abramov, and V. Dinets. 2016. *Mustela lutreola*. IUCN *Red List of Threatened Species* 2016:e.T14018A45199861.

Marcellini, D. L., and T. A. Jenssen. 1988. Visitor behavior in the National Zoo's reptile house. *Zoo Biol.* 7:329–338.

Marchand, P. J. 2013. *Life in the Cold: An Introduction to Winter Ecology.* 4th ed. Hanover: Univ. Press of New England.

Marcus, M. J. 1985. Feeding associations between capybaras and jacanas: A case of inter-specific grooming and possibly mutualism. *Ibis* 127:240–243.

Marín, C. D., and C. Sánchez-Giraldo. 2017. Far away from the endemism area: First record of the Ecuador fish-eating rat *Anotomys leander* (Cricetidae: Sigmodontinae) in the Colombian Andes. *Mammalia* 81:627–633.

Marinelli, L., and F. Messier. 1995. Strategies of parental care among muskrats in a female-biased population. *Can. J. Zool.* 73:1503–1510.

Marinelli, L., F. Messier, and Y. Plante. 1997. Consequences of following a mixed repro-ductive strategy in muskrats. *J. Mammal.* 78:163–172.

Marino, L. 2007. Cetacean brains: How aquatic are they? *Anat. Rec.* 290:694–700.

Marshall, L. G. 1977. First Pliocene record of the water opossum *Chironectes minimus* (Didelphidae, Marsupialia). *J. Mammal.* 58:434–436.

———. 1978a. *Chironectes minimus. Mamm. Species* 109:1–6.

———. 1978b. *Lutreolina crassicaudata. Mamm. Species* 91:1–4.

Martin, J. E. 2017. A rare occurrence of the fossil water mole *Gaillardia* (Desmanini, Talpidae) from the Neogene in North America. *Proc. South Dak. Acad. Sci.* 96:94–98.

Martin, T. 2005. Postcranial anatomy of *Haldanodon exspectatus* (Mammalia, Docodonta) from the Late Jurassic (Kimmeridgian) of Portugal and its bearing for mammalian evolution. *Zool. J. Linn. Soc.* 145:219–248.

———. 2006. Paleontology. Early mammalian evolutionary experiments. *Science* 311:1109–1110.

Martiínez-Lanfranco, J. A., D. Flores, J. P. Jayat, and G. D'Elía. 2014. A new species of lutrine opossum, genus *Lutreolina* Thomas (Didelphidae), from the South American Yungas. *J. Mammal.* 95:225–240.

Martino, P. E., N. Radman, E. Parrado, E. Bautista, C. Cisterna, M. P. Silvestrini, and S. Corba. 2012. Note on the occurrence of parasites of the wild nutria (*Myocastor coypus*, Molina, 1782) *Helminthologia* 49:164–168.

Martinsen, G. D., E. M. Driebe, and T. G. Whitham. 1998. Indirect interactions mediated by changing plant chemistry: Beaver browsing benefits beetles. *Ecology* 79:192–200.

Mason, C. F., and S. M. Macdonald. 1993. Impact of organochlorine pesticide residues and PCBs on otters (*Lutra lutra*): A study from western Britain. *Sci. Total Environ.* 138:127–145.

Mason, M. J. 2016. Internally coupled ears in living mammals. *Biol. Cybern.* 110:345–358.

Maté, I., and J. Barrull. 2012. First documented attack of southern water vole *Arvicola sapidus* Miller, 1908 on a viperine snake *Natrix maura* (Linnaeus, 1758), on the Mont-sant River (NE Iberian Peninsula). *Galemys* 24:91–92.

Mate, I., J. Barrull, J. Gosálbez, J. Ruiz-Olmo, and M. Salicrú. 2015. The role of the southern water vole Arvicola sapidus in the diet of predators: A review. *Mammal Rev.* 45:30–40.

Mate, I., J. Barrull, M. Salicrú, J. Ruiz-Olmo, and J. Gosálbez. 2013. Habitat selection by southern water vole (*Arvicola sapidus*) in riparian environments of Mediterranean mountain areas: A conservation tool for the species. *Acta Theriol.* 58:25–37.

Matthew, W. D. 1932. New fossil mammals from the Snake Creek quarries. *Am. Mus. Novit.* 540:1–8.

Matthews, L. H. 1952. *British Mammals.* London: Collins.

May, J., and R. Lindholm. 2013. *Tragelaphus spekii*. In *The Mammals of Africa*, Vol. 6, edited by J. Kingdon, D. Happold, T. Butynski, M. Hoffman, M. Happold, and J. Kalina, 172–178. London: Bloomsbury Natural History.

Mayer, M., A. Zedrosser, and F. Rosell. 2017a. When to leave: The timing of natal dispersal in a large, monogamous rodent, the Eurasian beaver. *Anim. Behav.* 123:375–382.

———. 2017b. Couch potatoes do better: Delayed dispersal and territory size affect the duration of territory occupancy in a monogamous mammal. *Ecol. Evol.* 7:4347–4356.

Mayer, M., F. Künzel, A. Zedrosser, and F. Rosell. 2017. The 7-year itch: Non-adaptive mate change in the Eurasian beaver. *Behav. Ecol. Sociobiol.* 71:UNSP 32.

McCarthy, T. S., W. N. Ellery, and A. Bloem. 1998. Some observations on the geomorphological impact of hippopotamus (*Hippopotamus amphibius* L.) in the Okavango Delta, Botswana. *Afr. J. Ecol.* 36:44–56.

McCleery, R. A., A. Sovie, R. N. Reed, M. W. Cunningham, M. E. Hunter, and K. M. Hart. 2015. Marsh rabbit mortalities tie pythons to the precipitous decline of mammals in the Everglades. *Proc. R. Soc. B* 282:20150120.

McEvoy, J., D. L. Sinn, and E. Wapstra. 2008. Know thy enemy: Behavioural response of a native mammal (*Rattus lutreolus velutinus*) to predators of different coexistence histories. *Austral Ecol.* 33:922–931.

McGinn, S. M. 2010. Weather and climate patterns in Canada's prairie grasslands. In *Arthropods of Canadian Grasslands*, Vol. 1, edited by J. D. Shorthouse and K. D. Floate, 105–119. Charlottetown, PE: Biological Survey of Canada.

McIntyre, I. W. 2000. Diving energetics and temperature regulation of the star-nosed mole, *Condylura cristata*, with comparisons to non-aquatic talpids and the water shrew, *Sorex palustris*. M.Sc. thesis, Univ. of Manitoba.

McIntyre, I. W., K. L. Campbell, and R. A. MacArthur. 2002. Body oxygen stores, aerobic dive limits and diving behaviour of the star-nosed mole (*Condylura cristata*) and comparisons with non-aquatic talpids. *J. Exp. Biol.* 205:45–54.

McKinstry, M. C., and S. H. Anderson. 2002. Survival, fates, and success of transplanted beavers, *Castor canadensis*, in Wyoming. *Can. Field-Nat.* 116:60–68.

McNab, B. K. 1978. The comparative energetics of Neotropical marsupials. *J. Comp. Physiol.* 125:115–128.

———. 2008. An analysis of the factors that influence the level and scaling of mammalian BMR. *Comp. Biochem. Physiol. Part A Mol. Integr. Physiol.* 151:5–28.

Medina, G. 1998. Seasonal variations and changes in the diet of southern river otter in different freshwater habitats in Chile. *Acta Theriol.* 43:285–292.

Medina-Vogel, G., and C. Gonzalez-Lagos. 2008. Habitat use and diet of endangered southern river otter *Lontra provocax* in a predominantly palustrine wetland in Chile. *Wildl. Biol.* 14:211–220.

Medina-Vogel, G., V. S. Kaufman, R. Monsalve, and V. Gomez. 2003. The influence of riparian vegetation, woody debris, stream morphology and human activity on the use of rivers by southern river otters in *Lontra provocax* in Chile. *Oryx* 37:422–430.

Meehl, G. A., T. F. Stocker, W. D. Collins, P. Friedlingstein, A. T. Gaye, J. M. Gregory, A. Kitoh, R. Knutti, J. M. Murphy, A. Noda, S. C. B. Raper, I. G. Watterson, A. J. Weaver, and Z.-C. Zhao. 2007. Global climate projections. In *Climate Change 2007: The Physical Science Basis. Contribution of Working Group I to the fourth assessment report of the Intergovernmental Panel on Climate Change*, edited by S. Solomon, D. Qin, M. Manning, Z. Chen, M. Marquis, K. B. Averyt, M. Tignor, and H. L. Miller, 749–845. Cambridge, UK: Cambridge Univ. Press.

Meena, V. 2002. Otter poaching in Palni Hills. *Zoos' Print J.* 17:696–698.

Melero, Y., M. Plaza, G. Santulli, D. Saavedra, J. Gosàlbez, J. Ruiz-Olmo, and S. Palazón.

2012. Evaluating the effect of American mink, an alien invasive species, on the abundance of a native community: Is coexistence possible? *Biodivers. Conserv.* 21:1795–1809.

Melquist, W. E., and A. E. Dronkert. 1987. River otter. In *Wild Furbearer Management and Conservation in North America*, edited by M. Novak, J. Baker, M. E. Obbard, and B. Malloch, 627–641. Toronto: Ontario Trappers Association.

Melquist, W. E., and M. G. Hornocker. 1983. Ecology of river otters in west central Idaho. *Wildl. Monogr.* 83:1–60.

Melquist, W. E., P. J. Polechla, and D. Toweill. 2003. River otter *Lontra canadensis*. In *Wild Mammals of North America: Biology, Management, and Conservation*, 2nd ed., edited by G. A. Feldhamer, B. C. Thompson, and J. A. Chapman, 708–734. Baltimore: Johns Hopkins Univ. Press.

Mendes-Soares, H., and L. Rychlik. 2009. Differences in swimming and diving abilities between two sympatric species of water shrews: *Neomys anomalus* and *Neomys fodiens*. *J. Ethol.* 27:317–325.

Menvielle, M. F., M. Funes, L. Malmierca., D. Ramadori, B. Saavedra, A. Schiavini, and N. Soto Volkart. 2010. American beaver eradication in the southern tip of South America: Main challenges of an ambitious project. In *Aliens: The Invasive Species Bulletin*, no. 29, edited by P. Genovesi and R. Scalera, 9–16. Aliens: IUCN/SSC Invasive Species Specialist Group.

Merritt, J. F. 2010. *The Biology of Small Mammals*. Baltimore: Johns Hopkins Univ. Press.

Messier, F., and J. A. Virgil. 1992. Differential use of bank burrows and lodges by muskrats (*Ondatra zibethicus*) in a northern marsh environment. *Can. J. Zool.* 70:1180–1184.

Meylan, A. 1977. Fossorial forms of the water vole, *Arvicola terrestris* (L.) in Europe. EPO *Bull.* 7:209–221.

Michaux, J. R., O. J. Hardy, F. Justy, P. Fournier, A. Kranz, M. Cabria, A. Davison, R. Rosoux, and R. Libois. 2005. Conservation genetics and population history of the threatened European mink *Mustela lutreola*, with an emphasis on the west European population. *Mol. Ecol.* 14:2373–2388.

Minin, E., and A. Moilanen. 2014. Improving the surrogacy effectiveness of charismatic megafauna with well- surveyed taxonomic groups and habitat types. *J. Appl. Ecol.* 51:281–288.

Mitsch, W. J., and J. G. Gosselink. 2007. *Wetlands*. 4th ed. Hoboken, NJ: Wiley.

Moncunill-Solé, B., X. Jordana, and M. Köhler. 2016. How common is gigantism in insular fossil shrews? Examining the "Island Rule" in soricids (Mammalia: Soricomorpha) from Mediterranean Islands using new body mass estimation models. *Zool. J. Linn. Soc.* 178:163–182.

Mones, A., and J. Ojasti. 1986. *Hydrochoerus hydrochaeris*. *Mamm. Species* 264:1–7.

Morgan, L. H. 1868. *The American Beaver and His Works*. Philadelphia: Lippincott and Co.

Morgan, S. M. D., C. E. Pouliott, R. J. Rudd, and A. D. Davis. 2015. Antigen detection, rabies virus isolation, and Q-PCR in the quantification of viral load in a natural infection of the North American beaver (*Castor canadensis*). *J. Wildl. Dis.* 51:287–289.

Mori, K., S. Suzuki, D. Koyabu, J. Kimura, S. Han, and H. Endo. 2015. Comparative functional anatomy of hindlimb muscles and bones with reference to aquatic adaptation of the sea otter. *J. Vet. Med. Sci.* 77:571–578.

Mortola, J. P. 2015. The heart rate-breathing rate relationship in aquatic mammals: A comparative analysis with terrestrial species. *Curr. Zool.* 61:569–577.

Mosepele, K., P. B. Moyle, G. S. Merron, D. R. Purkey, and B. Mosepele. 2009. Fish, floods, and ecosystem engineers: Aquatic conservation in the Okavango Delta, Botswana. *BioScience* 59:53–64.

Moss, A., and M. Esson. 2010. Visitor interest in zoo animals and the implications for collection planning and zoo education programmes. *Zoo Biol.* 29:715–731.

Mott, C. L., C. K. Bloomquist, and C. K. Nielsen. 2013. Within-lodge interactions between two ecosystem engineers, beavers (*Castor canadensis*) and muskrats (*Ondatra zibethicus*). *Behaviour* 150:1325–1344.

Moutou, F. 1997. Mammifères aquatiques et semi-aquatiques introduits en France. Risques et consequences. *Bull. Fr. Pêche Piscic.* 344/345:133–139.

Müller-Schwarze, D. 2011. *The Beaver: Its Life and Impact.* 2nd ed. Ithaca, NY: Comstock Publishing Associates.

Mullin, S. K., N. Pillay, and P. J. Taylor. 2005. The distribution of the water rat *Dasymys* (Muridae) in Africa: A review. *S. Afr. J. Sci.* 101:117–124.

Musser, G. D., and M. D. Carleton. 1993. Family Muridae. In *Mammal Species of the World: A Taxonomic and Geographic Reference*, edited by D. E. Wilson and D. M. Reeder, 501–755. Washington, DC: Smithsonian Institution Press.

Muul, I., and B. Lim. 1970. Ecological and morphological observations of *Felis planiceps*. *J. Mammal.* 51:806–808.

Nagl, A., N. Kneidinger, K. Kiik, H. Lindeberg, T. Maran, and F. Schwarzenberger. 2015. Noninvasive monitoring of female reproductive hormone metabolites in the endangered European mink (*Mustela lutreola*). *Theriogenology* 84:1472–1481.

Nagorsen, D. 1996. *Opossums, Shrews of British Columbia.* Victoria, BC: Royal British Columbia Museum.

Nagulu, V. 1992. News from India. *IUCN Otter Spec. Group Bull.* 7:41–42.

Ndawula, J., M. Tweheyo, D. M. Tumusiime, and G. Eilu. 2011. Understanding sitatunga (*Tragelaphus spekii*) habitats through diet analysis in Rushebeya-Kanyabaha wetland, Uganda. *Afr. J. Ecol.* 49:481–489.

Nefdt, R. J. C., and S. J. Thirgood. 1997. Lekking, resource defense, and harassment in two subspecies of lechwe antelope. *Behav. Ecol.* 8:1–9.

Nel, J. A. J., and M. J. Somers. 2007. Distribution and habitat choice of Cape clawless otters, *Aonyx capensis*, in South Africa. *S. Afr. J. Wildl. Res.* 37:61–70.

Nguyen, X. D. 2005. Current status of otters (Mammalia: Lutrinae) in Viet Nam with conservation implications. *Tiger Paper* 33:8–14.

Nguyen, X. D., T. A. Pham, and H. T. Le. 2001. New Information about the hairy-nosed otter (*Lutra sumatrana*) in Vietnam. *IUCN Otter Spec. Group Bull.* 18:64–75.

Nicol, S. C. 2017. Energy homeostasis in monotremes. *Front. Neurosci.* 11:195.

Nicoll, M. 1985. The biology of the giant otter shrew *Potamogale velox. Nat. Geogr. Soc. Res.* 21:331–337.

Nicoll, M. E., and G. B. Rathbun. 1990. *African Insectivora and Elephant-Shrews. An Action Plan for Their Conservation.* Gland, CH: IUCN.

Nimje, P. S., 2018. The effect of social organization on genetic estimates of fitness in Eurasian beavers. PhD diss., Univ. College of Southeast Norway.

Nituch, L. A., J. Bowman, K. B. Beauclerc, and A. I. Schulte-Hostedde. 2011. Mink farms predict Aleutian disease exposure in wild American mink. *PLoS One* 6:e21693.

Nolet, B. A., and F. Rosell. 1998. Comeback of the beaver *Castor fiber*: An overview of old and new conservation problems. *Biol. Conserv.* 83:165–173.

Norris, J. D. 1967. A campaign against feral coypus (*Myocastor coypus* Molina) in Britain. *J. Appl. Ecol.* 4:191–199.

Novakowski, N. S. 1967. The winter bioenergetics of a beaver population in northern latitudes. *Can. J. Zool.* 45:1107–1118.

Nowak, R. M., 1999. *Walker's Mammals of the World*. 6th ed., Vols. 1 and 2. Baltimore: Johns Hopkins Univ. Press.

Nowak, R. M., and J. Paradiso. 1983. *Walker's Mammals of the World*: 4th ed. Baltimore: Johns Hopkins Univ. Press.

Nowell, K., and P. Jackson. 1996. *Wild Cats. Status Survey and Conservation Action Plan*. Gland, CH: IUCN/SSC Cat Specialist Group.

Ojasti, J. 1978. The relation between population and production in the capybara. PhD diss., Univ. of Georgia, Athens.

Oliveira, E. J. F., J. E. Garcia, E. P. B. Contel, and J. M. B. Duarte. 2005. Genetic structure of *Blastocerus dichotomus* populations in the Paraná River Basin (Brazil) based on protein variability. *Biochem. Genet.* 43:211–222.

Oliveira, G. C, J. F. M. Barcellos, S. M. Lazzarini, and F. C. W. Rosas. 2011. Gross anatomy and histology of giant otter (*Pteronura brasiliensis*) and Neotropical otter (*Lontra longicaudis*) testes. *Anim. Biol.* 61:175–183.

Oliver, L. R. 1976. The management of yapoks (*Chironectes minimus*) at Jersey Zoo, with observations on their behavior. *Jersey Wildl. Preserv. Trust* 13:32–36.

Olson, L. E., 2013. Tenrecs. *Curr. Biol.* 23:R5–R8.

Onibala, J. S. I. T., and S. Laatung. 2009. Bushmeat hunting in north Sulawesi and related conservation strategies (A case study at the Tangkoko Nature Reserve). *J. Agr. Rural Dev. Trop. Subtrop.* 90:110–116.

Orians, G. H., and N. E. Pearson. 1979. On the theory of central place foraging. In *Analysis of Ecological Systems*, edited by D. J. Horn, R. D. Mitchell, and G. R. Stairs, 154–177. Columbus: Ohio State Univ. Press.

Orr, T. J., and M. Zuk. 2014. Reproductive delays in mammals: An unexplored avenue for post-copulatory sexual selection. *Biol. Rev. Camb. Philos. Soc.* 89:889–912.

Osborn, D. J., 1955. Techniques of sexing beaver, *Castor canadensis*. *J. Mammal.* 36:141–143.

Ostende, L. W. V. 1995. Insectivore faunas from the lower Miocene of Anatolia. 3. Dimylidae. *Proc. K. Ned. Akad. Wet.* 98:19–38.

Ostfeld, R. S., L. Ebensperger, L. L. Klosterman, and J. C. Castilla. 1989. Foraging, activity budget, and social behavior of the South American marine otter *Lutra felina* (Molina 1782). *Natl. Geogr. Res.* 5:422–438.

Pacheco, V., H. Zeballos, and E. Vivar. 2008. *Amphinectomys savamis*. *IUCN Red List of Threatened Species* 2008:e.T136723A4332353.

Pacini, N., and D. M. Harper. 2008. Aquatic, semi-aquatic and riparian vertebrates. In *Tropical Stream Ecology*, edited by D. Dudgeon, 147–197. Cambridge, MA: Academic Press.

Paine, R. T. 1966. Food web complexity and species diversity. *Am. Nat.* 100:65–75.

———. 1969. A note on trophic complexity and community stability. *Am. Nat.* 103:91–93.

Palazón, S., and Y. Melero. 2014. Status, threats and management actions on the European mink *Mustela lutreola* (Linnaeus, 1761) in Spain: A review of the studies performed since 1992. *Munibe Monogr. Nat. Ser.* 3:109–118.

Pardiñas, U. F. J. 2008. A new genus of oryzomyine rodent (Cricetidae: Sigmodontinae) from the Pleistocene of Argentina. *J. Mammal.* 89:1270–1278.

Pardiñas, U. F. J., and P. Teta. 2011. Fossil history of the marsh rats of the genus *Holochilus* and *Lundomys* (Cricetidae, Sigmodontinae) in southern South America. *Estud. Geol-Madrid.* 67:111–129.

Parker, G. A., and J. M. Smith. 1990. Optimality theory in evolutionary biology. *Nature* 348:27–33.

Parker, H., P. Nummi, G. Hartman, and F. Rosell. 2012. Invasive North American beaver *Castor canadensis* in Eurasia: A review of potential consequences and a strategy for eradication. *Wild. Biol.* 18:354–365.

Pascual, R., M. Archer, E. O. Juareguizar, J. L. Prado, H. Godthelp, and S. J. Hand. 1992. First discovery of monotremes in South America. *Nature* 356:704–706.

Pasitschniak-Arts, M., and L. Marinelli. 1998. *Ornithorhynchus anatinus*. *Mamm. Species* 585:1–9.

Patou, M., P. A. Mclenachan, C. G. Morley, A. Couloux, A. P. Jennings, and G. Veron. 2009. Molecular phylogeny of the Herpestidae (Mammalia, Carnivora) with a special emphasis on the Asian *Herpestes*. *Mol. Phylogenet. Evol.* 53:69–80.

Pattie, D. 1973. *Sorex bendirii*. *Mamm. Species*. 27:1–2.

Payne, A. P. 1994. The harderian gland: A tercentennial review. *J. Anat.* 185:1–49.

Payne, N. F. 1984. Mortality rates of beaver in Newfoundland. *J. Wildl. Manage.* 48:117–126.

Peck, S. B. 2006. Distribution and biology of the ectoparasitic beaver beetle *Platypsyllus castoris* Ritsema in North America (Coleoptera: Leiodidae: Platypsyllinae). *Insecta Mundi* 20:85–94.

Peigné, S., L. de Bonis, A. Likius, H. T. Mackaye, P. Vignaud, and M. Brunet. 2005. The earliest modern mongoose (Carnivora, Herpestidae) from Africa (late Miocene of Chad). *Naturwissenschaften* 92:287–292.

Pérez, M. E., and D. Pol. 2012. Major radiations in the evolution of caviid rodents: Reconciling fossils, ghost lineages, and relaxed molecular clocks. *PLoS One* 7:e48380.

Pérez-Hernandez, R., D. Brito, T. Tarifa, N. Cáceres, D. Lew, and S. Solari. 2016. *Chironectes minimus*. *IUCN Red List of Threatened Species* 2016:e.T4671A22173467.

Persson, S., T. H. Jensen, A. Blomstrom, M. T. Appelberg, and U. Magnusson. 2015. Aleutian mink disease in free-ranging mink from Sweden. *PLoS One* 10: e0122194.

Peters, E., I. Brinkmann, F. Krüger, S. Zwirlein, and I. Klaumann. 2009. Reintroduction of the European mink *Mustela lutreola* in Saarland, Germany. Preliminary data on the use of space and activity as revealed by radio-tracking and live-trapping. *Endang. Species Res.* 10:305–320.

Petersen, K. E., and T. L. Yates. 1980. *Condylura cristata*. *Mamm. Species* 129:1–4.

Petro, V. M., J. D. Taylor, and D. M. Sanchez. 2015. Evaluating landowner-based beaver relocation as a tool to restore salmon habitat. *Glob. Ecol. Conserv.* 3:477–486.

Petzold, D., F. Trillmich, and G. Dehnhardt. 1997. The role of water depth and food types for foraging and diving behaviour in Australian water rats (*Hydromys chrysogaster*). *Adv. Ethol.* 32:187.

Pian, R., Archer, M., and S. J. Hand. 2013. A new, giant platypus, *Obdurodon tharalkooschild*, sp. Nov. (Monotremata, Ornithorhynchidae), from the Riversleigh World Heritage Area, Australia. *J. Verterbr. Paleontol.* 33:1255–1259.

Pierce, G. J., and J. G. Ollason. 1987. Eight reasons why optimal foraging theory is a complete waste of time. *Oikos* 49:111–117.

Pietrek, A. G., and L. Fasola. 2014. Origin and history of the beaver introduction in South America. *Mastozool. Neotrop.* 21:355–359.

Pihlström, H. 2008. Comparative anatomy and physiology of chemical senses in aquatic mammals. In *Sensory Evolution on the Threshold: Adaptations in Secondarily Aquatic Vertebrates*, edited by J. G. M. Thewissen and S. Nummela, 95–105. Berkeley: Univ. California Press.

Pita, R., A. Mira, and P. Beja. 2010. Spatial segregation of two vole species (*Arvicola sapidus* and *Microtus cabrerae*) within habitat patches in a highly fragmented farmland landscape. *Eur. J. Wildl. Res.* 56:651–662.

Platt, W. J., 1974. Metabolic rates of short-tailed shrews. *Physiol. Zool.* 47:75–90.

Pluháček, J., and B. L. Steck. 2015. Different sex allocations in two related species: The case of the extant hippopotamus. *Ethology* 121:462–471.

Pocock, R. I. 1914. On the facial vibrissae of Mammalia. *Proc. Zool. Soc. Lond.* 1914:889–912.

Polo, G., C. M. Acosta, M. B. Labruna, and F. Ferreira. 2017. *PLoS Neglect. Trop. Dis.* 11:e0005613.

Poole, C. M. 2003. The first records of hairy-nosed otter *Lutra sumatrana* from Cambodia with notes on the national status of three other otter species. *Nat. Hist. Bull. Siam Soc.* 51:273–280.

Poole, T. B., and N. Dunstone. 1976. Underwater predatory behavior of the American mink (*Mustela vison*). *J. Zool.* 178:395–412.

Prado, J. L., M. T. Alberdi, B. Azanza, B. Sánchez, and D. Frassinetti. 2005. The Pleistocene Gomphotheriidae (Proboscidea) from South America. *Quat. Int.* 126:21–30.

Prendergast, J. A., and W. E. Jensen. 2012. Consequences of parasitic mite infestation on muskrat (*Ondatra zibethicus*). *West. N. Am. Nat.* 71:516–522.

Prigioni, C., G. Smiroldo, L. Remonti, and A. Balestrieri. 2009. Distribution and diet of reintroduced otters (*Lutra lutra*) on the River Ticino (NW Italy). *Hystrix* 20:45–54.

Pudjihastuti, E., S. P. Pangemanan, and C. L. Kaunang. 2009. A study of carcass and meat chemical composition of babirusa (*Babyrousa babyrussa celebensis* Deniger). *J. Agr. Rural Dev. Trop. Subtrop.* 90:83–93.

Quammen, D. 1997. *Song of the Dodo: Island Biogeography in an Age of Extinctions.* New York: Scribner.

Radford, H. V. 1908. *Artificial Preservation of Timber and History of the Adirondack Beaver.* Albany, NY: J. B. Lyon Company.

Raesly, E. J. 2001. Progress and status of river otter reintroduction projects in the United States. *Wildl. Soc. Bull.* 29:856–862.

Randall, D., W. Burggren, and K. French. 1997. *Eckert Animal Physiology: Mechanisms and Adaptations.* 4th ed. New York: W. H. Freeman and Company.

Ray, S., Chinsamy, A., and S. Bandyopadhyay. 2005. *Lystrosaurus murrayi* (Therapsida; Dicyndontia): Bone histology, growth and lifestyle adaptations. *Palaeontol.* 48:1169–1185.

Reed-Smith, J., I. Oluoch, M. Origa, T. S. Kihedu, M. M. Muhabi, M. M. Yusuf, M. Ogada, A. Lobora, and T. Serfass. 2010. Consumptive uses of and lore pertaining to spotted-necked otters in East Africa—a preliminary report from the Lake Victoria area of Kenya, Tanzania, and Uganda. *IUCN Otter Spec. Group Bull.* 27:85–88.

Reed-Smith, J., T. Serfass, T. S. Kihudu, and M. Mussa. 2014. Preliminary report on the behavior of spotted-necked otter (*Lutra maculicollis*, Lichtenstein, 1835) living in a lentic ecosystem. *Zoo Biol.* 33:121–130.

Regidor, H. A., M. Gorostiague, and S. Sühring. 1999. Reproduction and dental age classes of the little water opossum (*Lutreolina crassicaudata*) in Buenos Aires, Argentina. *Rev. Biol. Trop.* 47:271–272.

Reid, D. G., T. E. Code, A. C. H. Reid, and S. M. Herrero. 1994. Food habits of the river otter in a boreal ecosystem. *Can. J. Zool.* 72:1306–1313.

Reid, D. G., S. M. Herrero, and T. E. Code. 1988. River otters as agents of water loss from beaver ponds. *J. Mammal.* 69:100–107.

Reidenberg, J. S. 2007. Anatomical adaptations of aquatic mammals. *Anat. Rec.* 290:507–513.

Reis, V., V. Hermoso, S. K. Hamilton, D. Ward, E. Fluet-Chouinard, B. Lehner, and S.

Linke. 2017. A global assessment of inland wetland conservation status. *BioScience* 67:523–533.

Reumer, J. W. F. 1989. Speciation and evolution in the Soricidae (Mammalia: Insectivora) in relation with the paleoclimate. *Rev. Suisse Zool.* 96:81–90.

Reynolds, P. S. 1993. Size, shape, and surface area of beaver, *Castor canadensis*, a semiaquatic mammal. *Can. J. Zool.* 71:876–882.

Ribas, C., H. A. Cunha, G. Damasceno, W. E. Magnusson, A. Solé-Cava, and G. Mourão. 2016. More than meets the eye: Kinship and social organization in giant otters (*Pteronura brasiliensis*). *Behav. Ecol. Sociobiol.* 70:61–72.

Ribas, C., G. Damasceno, W. Magnusson, C. Leuchtenberger, and G. Mourão. 2012. Giant otters feeding on caiman: Evidence for an expanded trophic niche of recovering populations. *Stud. Neotrop. Fauna Environ.* 47:19–23.

Richard, P. B. 1973. Capture, transport, and husbandry of the Pyrenean desman *Galemys pyrenaicus*. *Int. Zoo Yearb.* 13:174–177.

Richardson, J. S., E. Taylor, D. Schluter, M. Pearson, and T. Hatfield. 2010. Do riparian zones qualify as critical habitat for endangered freshwater fishes? *Can. J. Fish. Aquat. Sci.* 67:1197–1204.

Ripple, W. J., C. Wolf, T. M. Newsome, M. Galetti, M. Alamgir, E. Crist, M. I. Mahmoud, and W. F. Laurance. 2017. World scientists' warning to humanity: A second notice. *BioScience* 67:1026–1028.

Robertson, S. I., M. T. P. Gilbert, P. F. Campos, F. M. Salleh, S. Tridico, and D. Hills. 2017. Lowe's otter civet *Cynogale lowei* does not exist. *Small Carnivore Conserv.* 55:42–58.

Robinson, P. T. 1970. The status of the pygmy hippopotamus and other wildlife in West Africa. M.Sc. thesis, Michigan State Univ.

Rode-Margono, J., and K. A. Nekaris. 2015. Cabinet of curiosities: Venom systems and their ecological function in mammals, with a focus on primates. *Toxins* 7:2639–2658.

Rodrigues, H. G., L. Marivaux, and M. Vianey-Liaud. 2010. Phylogeny and systematic revision of Eocene Cricetidae (Rodentia, Mammalia) from Central and East Asia: On the origin of cricetid rodents. *J. Zool. Syst. Evol. Res.* 48:259–268.

Román, J. 2007. Historia natural de la rata de agua (*Arvicola sapidus*) en Doñana. PhD diss., Univ. Autonoma de Madrid.

Romashova, N. B. 2016. History of conservation and research activities of the Eurasian beaver (*Castor fiber*) in the Voronezhsky Nature Reserve. *Russ. J. Theriol.* 15:8–19.

Roos, A., A. Loy, P. de Silva, P. Hajkova, and B. Zemanová, B. 2015. *Lutra lutra*. IUCN Red List of Threatened Species 2015:e.T12419A21935287.

Rose, K. D., V. Burke DeLeon, P. Missiaen, R. S. Rana, A. Sahni, L. Singh, and T. Smith. 2007. Early Eocene lagomorph (Mammalia) from Western India and the early diversification of Lagomorpha. *Proc. R. Soc. Lond. B Biol. Sci.* 275:1203–1208.

Rose, K. D., and W. Von Koenigswald. 2005. An exceptionally complete skeleton of *Palaeosinopa* (Mammalia, Cimolesta, Pantolestidae) from the Green River Formation, and other postcranial elements of the Pantolestidae from the Eocene of Wyoming. *Palaeontogr. Abt. A.* 273:55–96.

Rosell, F. 2002. Do Eurasian beavers smear their pelage with castoreum and anal gland secretion? *J. Chem. Ecol.* 28:1697–1701.

Rosell, F., and A. Czech. 2000. Responses of foraging Eurasian beavers *Castor fiber* to predator odours. *Wildl. Biol.* 6:13–21.

Ross, J., A. Wilting, D. Ngoprasert, B. Loken, L. Hedges, J. W. Duckworth, S. Cheyne, J. Brodie, W. Chutipong, A. Hearn, M. Linkie, J. McCarthy, N. Tantipisanuh, and

I. A. Haidir. 2015. *Cynogale bennettii*. IUCN *Red List of Threatened Species* 2015:e. T6082A45197343.

Roth, H., B. Hoppe-Dominik, M. Muhlenberg, B. Steinhauer-Burkart, and F. Fischer. 2004. Distribution and status of the hippopotamids in the Ivory Coast. *Afr. Zool.* 39:211–224.

Rowe, K. C., A. S. Achmadi, and J. A. Esselstyn. 2014. Convergent evolution of aquatic foraging in a new genus and species (Rodentia: Muridae) from Sulawesi Island, Indonesia. *Zootaxa* 3815:541–564.

Rowe, K. C., M. L. Reno, D. M. Richmond, R. M. Adkins, and S. J. Steppan. 2008. Pliocene colonization and adaptive radiations in Australia and New Guinea (Sahul): Multilocus systematics of the old endemic rodents (Muroidea: Murinae). *Mol. Phylogenet. Evol.* 47:84–101.

Rowe, T., T. H. Rich, P. Vickers-Rich, M. Springer, and M. O. Woodburne 2008. The oldest platypus and its bearing on divergence timing of the platypus and echidna clades. *Proc. Natl. Acad. Sci.* 105:1238–1242.

Rowe-Rowe, D. T. 1977a. Food ecology of otters in Natal. *Oikos* 28:210–219.

———. 1977b. Prey capture and feeding behaviour of South African otters. *Lammergeyer* 23:13–21.

Rümke, C. G. 1985. A review of fossil and recent Desmaninae (Talpidae, Insectivora). *Utrecht Micropal. Bull. Sp. Pub.* 4:1–241.

Ryan, J. R., C. S. Apperson, P. E. Orndorff, and J. F. Levine. 2000. Characterization of Lyme disease spirochetes isolated from ticks and vertebrates in North Carolina. *J. Wildl. Dis.* 36:48–55.

Rybczynski, N. 2007. Castorid phylogenetics: Implications for the evolution of swimming and tree-exploitation in beavers. *J. Mammal. Evol.* 14:1–35.

Rybczynski, N., E. M. Ross, J. X. Samuels, and W. W. Korth. 2010. Re-evaluation of *Sinocastor* (Rodentia: Castoridae) with implications on the origin of modern beavers. *PLoS One* 5:e13990.

Rychlik, L., and E. Jancewicz. 2002. Prey size, prey nutrition, and food handling by shrews of different body sizes. *Behav. Ecol.* 13:216–223.

Rzebik-Kowalska, B., and L. I. Rekovets. 2016. New data on Eulipotyphla (Insectivora, Mammalia) from the Late Miocene to the Middle Pleistocene of Ukraine. *Palaeontol. Electron.* 19:1–31.

Safonov, V. G. 1975. Ergebnisse der Wiedereinbürgerung des Flussbibers (*Castor fiber* L.) in der UdSSR. *Beitr. Jagd Wildforschung* 9:397–405.

Samuels, J. X., J. A. Meachen, and S. A. Sakai. 2013. Postcranial morphology and the locomotor habits of living and extinct carnivorans. *J. Morphol.* 274:121–146.

Sandell, M. 1990. The evolution of seasonal delayed implantation. *Q. Rev. Biol.* 65:23–42.

Sansalone, G., T. Kotsakis, and P. Piras. 2016. *Condylura* (Mammalia, Talpidae) reloaded: New insights about the fossil representatives of the genus. *Palaeontol. Electron.* 19:54A.

Santori, R. T., M. V. Vieira, O. Rocha-Barbosa, J. A. Magnan-Neto, and N. Nivar Gobbi. 2008. Water absorption of the fur and swimming behavior of semiaquatic and terrestrial oryzomine rodents. *J. Mammal.* 89:1152–1161.

Santoro, M., N. D'Alessio, A. Cerrone, M. G. Lucibelli, G. Borriello, G. Aloise, C. Auriemma, N. Riccone, and G. Galiero. 2017. The Eurasian otter (*Lutra lutra*) as a potential host for rickettsial pathogens in southern Italy. *PLoS One* 12: e0173556.

Sasaki, H., B. Nor, and B. Kanchanasaka. 2009. Past and present distribution of hairy-nosed otter *Lutra sumatrana* Grey 1985. *Mamm. Study* 34:223–229.

Sato, J. J., T. Hosoda, M. Wolsan, K. Tsuchiya, Y. Yamamoto, and H. Suzuki. 2003. Phylogenetic relationships and divergence times among mustelids (Mammalia: Carnivora) based on nucleotide sequences of the nuclear interphotoreceptor retinoid binding protein and mitochondrial cytochrome b genes. *Zool. Sci.* 20:243–264.

Sato, J. J., M. Wolsan, F. J. Prevosti, G. D'Elía, C. Begg, K. Begg, T. Hosoda, K. L. Campbell, and H. Suzuki. 2012. Evolutionary and biogeographic history of weasel-like carnivorans (Musteloidea). *Mol. Phylogenet. Evol.* 63:745–757.

Savage, R. J. G., and M. R. Long. 1989. *Evolution: An Illustrated Guide.* New York: Facts on File.

Schaller, G. B., and J. M. C. Vasconcelos. 1978. Jaguar predation on capybara. *Z. Säugetierkd.* 43:296–301.

Scheich, H., G. Langner, C. Tidemann, R. B. Coles, and A. Guppy. 1986. Electroreception and electrolocation in platypus. *Nature* 319:401–402.

Schenk, J. J., K. C. Rowe, and S. J. Steppan. 2013. Ecological opportunity and incumbency in the diversification of repeated continental colonizations by muroid rodents. *Syst. Biol.* 62:837–864.

Schindler, D. W., and W. F. Donahue. 2006. An impending water crisis in Canada's western Prairie Provinces. *P. Nat. Acad. Sci. USA.* 103:7210–7216.

Schmidt, P. M., R. A. McCleery, R. R. Lopez, N. J. Silvy, J. A. Schmidt, and N. D. Perry. 2011. Influence of patch, habitat, and landscape characteristics on patterns of Lower Keys marsh rabbit occurrence following Hurricane Wilma. *Landscape Ecol.* 26:1419–1431.

Schmidt-Kittler, N. 1973. *Dimyloides*-Neufunde aus der oberoligozanen Spaltenfüllung "Ehrenstein 4" (Süiddeutschland) und die systematische Stellung der Dimyliden (Insectivora, Mammalia). *Mitt. Bayer. Staatsslg. Paläont. Hist. Geol.* 13:115–139.

Schuster, R. H. 1976. Lekking behavior in Kafue lechwe. *Science* 192:1240–1242.

Schusterman, R. J., and B. Barrett. 1973. Amphibious nature of visual acuity in the Asian "clawless" otter. *Nature* 244:518–519.

Schüttler, E., J. Cárcamo, and R. Rozzi. 2008. Diet of the American mink *Mustela vison* and its potential impact on the native fauna of Navarino Island, Cape Horn Biosphere Reserve, Chile. *Rev. Chil. Hist. Nat.* 81:585–598.

Schwarm, A., S. Ortmann, H. Hofer, W. J. Streich, E. J. Flach, R. Kühne, J. Hummel, J. C. Castell, and M. Clauss. 2006. Digestion studies in captive Hippopotamidae: A group of large ungulates with an unusually low metabolic rate. *J. Anim. Physiol. Anim. Nutr.* 90:300–308.

Sepúlveda, M. A., R. S. Singer, E. A. Silva-Rodriguez, A. Eguren, P. Stowhas, and K. Pelican. 2014. Invasive American mink: Linking pathogen risk between domestic and endangered carnivores. *EcoHealth* 11:409–419.

Serfass, T., S. S. Evans, and P. Polechla. 2015. *Lontra canadensis.* IUCN *Red List of Threatened Species* 2015:e.T12302A21936349.

Serfass, T. L. 1995. Cooperative foraging by North American river otters, *Lutra canadensis. Can. Field-Nat.* 109:458–459.

Serfass, T. L., M. T. Whary, R. L. Peper, R. P. Brooks, T. J. Swimley, W. R. Lawrence, and C. E. Rupprecht. 1995. Rabies in a river otter (*Lutra canadensis*) intended for reintroduction. *J. Zoo. Wildl. Med.* 26:311–314.

Shaw, G. 1799. *The Naturalist's Miscellany.* London: Nodder and Co.

Sheppard, D. J., A. Moehrenschlager, J. M. McPherson, and J. J. Mason. 2010. Ten years of adaptive community-governed conservation: Evaluating biodiversity protection and poverty alleviation in a West African hippopotamus reserve. *Environ. Conserv.* 37:270–282.

Shiklomanov, I. 1993. World fresh water resources. In *Water in Crisis: A Guide to the World's Fresh Water Resources*, edited by P. H. Gleick, 13–24. New York: Oxford Univ. Press.

Shimatani, Y., Y. Fukue, R. Kishimoto, and R. Masuda. 2010a. Genetic variation and population structure of the feral American mink (*Neovison vison*) in Nagano, Japan, revealed by microsatellite analysis. *Mamm. Study* 35:1–7.

Shimatani, Y., T. Takeshita, S. Tatsuzawa, T. Ikeda, and R. Masuda. 2010b. Sex determination and individual identification of American minks (*Neovison vison*) on Hokkaido, Northern Japan, by fecal DNA analysis. *Zool. Sci.* 27:243–247.

Shotwell, J. A. 1970. Pliocene mammals of southeast Oregon and adjacent Idaho. *Mus. Nat. Hist. Oregon Bull.* 17:1–103.

Siemer, W. F., S. A. Jonker, D. J. Decker, and J. F. Organ. 2013. Toward an understanding of beaver management as human and beaver densities increase. *Hum. Wildl. Interact.* 7:114–131.

Simberloff, D., and B. Von Holle. 1999. Positive interactions of nonindigenous species: Invasional meltdown? *Biol. Invasions* 1:21–32.

Sinclair, A. R. E., J. M. Fryxell, and G. Caughley. 2006. *Wildlife Ecology, Conservation, and Management*. 2nd ed. Malden, MA: Blackwell Publishing.

Sjöåsen, T. 1996. Survivorship of captive-bred and wild-caught reintroduced European otters *Lutra lutra* in Sweden. *Biol. Conserv.* 76:161–165.

———. 1997. Movements and establishment of reintroduced European otters *Lutra lutra*. *J. Appl. Ecol.* 34:1070–1080.

Sjöberg, G., and J. P. Ball, eds. 2011. *Restoring the European Beaver: 50 Years of Experience*. Sofia, Bulgaria: Pensoft Publishers.

Skinner, D., 1984. Selection of winter food by beavers at Elk Island National Park. MSc thesis, Univ. of Alberta.

Skinner, J. D., and C. T. Chimimba. 2005. *The Mammals of the Southern African Subregion*. 3rd ed. Cambridge: Cambridge Univ. Press.

Skoczeń, S. 1976. Condylurini Dobson, 1883 (Insectivora, Mammalia) in the Pliocene of Poland. *Acta Zool. Cracov.* 21:291–313.

Skyrienė, G., and A. Paulauskas. 2012. Distribution of invasive muskrats (*Ondatra zibethicus*) and impact on ecosystem. *Ekologija* 58:357–367.

Smith, D. W., and S. H. Jenkins. 1997. Seasonal change in body mass and size of tail of northern beavers. *J. Mammal.* 78:869–876.

Smith, J. M., and R. J. G. Savage. 1956. Some locomotory adaptations in mammals. *J. Linn. Soc. London Zool.* 42:603–622.

Smuts, G. L., and I. J. Whyte. 1981. Relationships between reproduction and environment in the hippopotamus *Hippopotamus amphibius* in the Kruger National Park. *Koedoe.* 24:169–185.

Snyder, G. K. 1983. Respiratory adaptations in diving mammals. *Respir. Physiol.* 54:269–294.

Sokolov, W. 1962. Adaptations of the mammalian skin to the aquatic mode of life. *Nature* 195:464–466.

Somers, M. J. 2000. Foraging behaviour of Cape clawless otters *Aonyx capensis* in a marine habitat. *J. Zool.* 252:473–480.

Somers, M. J., and J. A. J. Nel. 2003. Diet in relation to prey of Cape clawless otters in two rivers in the Western Cape Province, South Africa. *Afr. Zool.* 38:317–326.

———. 2004. Movement patterns and home range of Cape clawless otters (*Aonyx capensis*), affected by high food density patches. *J. Zool.* 262:91–98.

Sorensen, M. F., J. P. Rogers, and T. S. Baskett. 1968. Reproduction and development in confined swamp rabbits. *J. Wildl. Manage.* 32:520–531.

Soy, J., and G. Silva. 2008. *Mesocapromys angelcabrerai. IUCN Red List of Threatened Species* 2008:e.T13215A3420155.

Sparti, A. 1992. Thermogenic capacity of shrews (Mammalia, Soricidae) and its relationship with basal rate of metabolism. *Physiol. Zool.* 65:77–96.

Spealman, C. R. 1946. Body cooling of rats, rabbits and dogs following immersion in water, with a few observations on man. *Am. J. Physiol.* 146:262–266.

Stanhope, M. J., V. G. Waddell, O. Madsen, W. de Jong, S. B. Hedges, G. C. Cleven, D. Kao, and M. S. Springer. 1998. Molecular evidence for multiple origins of Insectivora and for a new order of endemic African insectivore mammals. *Proc. Natal. Acad. Sci. USA* 95:9967–9972.

Starrett, A., and G. F. Fisler. 1970. Aquatic adaptations of the water mouse, *Rheomys underwoodi*. Los Angeles: Co. Nat. Hist. Mus. Contrib. Sci., no. 182.

Steen, I., and J. B. Steen. 1965. Thermoregulatory importance of the beaver's tail. *Comp. Biochem. Physiol.* 15:267–270.

Stegeman, L. C. 1930. Notes on *Synaptomys cooperi cooperi* in Washtenaw County, Michigan. *J. Mamm.* 11:460–466.

Stehlen, H. G. 1910. Remarques sur les faunules de Mammifères des couches eocenes et oligocenes du Bassin de Paris. *B. Soc. Geol. Fr. 4e série, IX:* 488–520.

Stein, B. R. 1988. Morphology and allometry in several genera of semiaquatic rodents (*Ondatra, Nectomys,* and *Oryzomys*). *J. Mammal.* 69:500–511.

———. 1989. Bone density and adaptation in semiaquatic mammals. *J. Mamm.* 70:467–476.

Stephenson, A. B. 1969. Temperatures within a beaver lodge in winter. *J. Mammal.* 50:134–136.

Stephenson, P. J. 1994. Resting metabolic rate and body temperature in the aquatic tenrec *Limnogale mergulus* (Insectivora: Tenrecidae). *Acta Theriol.* 39:89–92.

———. 2016. *Micropotamogale ruwenzorii. IUCN Red List of Threatened Species* 2016:e.T13394A21287768.

Stephenson, R., J. R. Lovvorn, M. R. A. Heieis, D. R. Jones, and R. W. Blake. 1989. A hydromechanical estimate of the power requirements of diving and surface swimming in lesser scaup (*Aythya affinis*). *J. Exp. Biol.* 147:507–518.

Steppan, S. J., R. M. Adkins, and J. Anderson. 2004. Phylogeny and divergence-date estimates of rapid radiations in muroid rodents based on multiple nuclear genes. *Syst. Biol.* 53:533–553.

Steppan, S. J., and J. J. Schenk. 2017. Muroid rodent phylogenetics: 900-species tree reveals increasing diversification rates. *PLoS One* 12:e0183070.

Stevens, S. S., and T. L. Serfass. 2005. Sliding behavior in Nearctic river otters: Locomotion or play? *Northeast. Nat.* 12:241–244.

Stibbe, E. P. 1928. A comparative study of the nictitating membrane of birds and mammals. *J. Anat.* 62:159–176.

Stone, R. D., and M. L. Gorman. 1985. Social organization of the European mole (*Talpa europaea*) and the Pyrenean desman (*Galemys pyrenaicus*). *Mamm. Rev.* 15:35–42.

Straka, J. R., A. Antoine, R. Bruno, D. Campbell, R. Campbell, R. Campbell, J. Cardinal, G. Gibot, Q. Z. Gray, S. Irwin, R. Kindopp, R. Ladouceur, W. Ladouceur, J. Lankshear, B. Maclean, S. Macmillan, F. Marcel, G. Marten, L. Marten, J. McKinnon, L. D. Patterson, C. Voyageur, M. Voyageur, G. S. Whiteknife, and L. Wiltzen. 2018. "We used to say rats fell from the sky after a flood": Temporary recovery of muskrat following ice jams in the Peace-Athabasca Delta. *Arctic* 71:218–228.

Stunkard, H. W. 1924. A new trematode, *Oculotrema hippopotami* n.g., n.sp., from the eye of the hippopotamus. *Parasitology* 16:436–440.

Sun, L., and D. Müller-Schwarze. 1998. Anal gland secretion codes for relatedness in the beaver, *Castor canadensis*. *Ethology* 104:917–927.

Sundqvist, C., L. C. Ellis, and A. Bartke. 1988. Reproductive endocrinology of the mink (*Mustela vison*). *Endocr. Rev.* 9:247–266.

Svendsen, G. E. 1978. Castor and anal glands of the beaver (*Castor canadensis*). *J. Mammal.* 59:618–620.

Svenning, J., and S. Faurby. 2017. Prehistoric and historic baselines for trophic rewilding in the Neotropics. *Perspect. Ecol. Conserv.* 15:282–291.

Sweeney, B. W., T. L. Bott, J. K. Jackson, L. A. Kaplan, J. D. Newbold, L. J. Standley, W. C. Hession, and R. J. Horwitz. 2004. Riparian deforestation, stream narrowing, and loss of stream ecosystem services. *Proc. Natl. Acad. Sci. USA.* 101:14132–14137.

Swinnen, K. R. R., N. K. Hughes, and H. Leirs. 2015. Beaver (*Castor fiber*) activity patterns in a predator-free landscape. What is keeping them in the dark? *Mamm. Biol.* 80:477–483.

Syrůčková, A., A. P. Saveljev, C. Frosch, W. Durka, A. A. Savelyev, and P. Munclinger. 2015. Genetic relationships within colonies suggest genetic monogamy in the Eurasian beaver (*Castor fiber*). *Mammal. Res.* 60:139–147.

Tape, K. D., B. M. Jones, C. D. Arp, I. Nitze, and G. Grosse. 2018. Tundra be dammed: Beaver colonization of the Arctic. *Global Change Biol.* 24:4478–4488.

Tarasoff, F. J. 1974. Anatomical adaptations in the river otter, sea otter and harp seal with reference to thermal regulation. In *Functional Anatomy of Marine Mammals*, Vol. 2, edited by R. H. Harrington, 111–141. New York: Academic Press.

Tarasoff, F. J., A. Bisaillon, J. Piérard, and A. P. Whitt. 1972. Locomotory patterns and external morphology of the river otter, sea otter, and harp seal (Mammalia). *Can. J. Zool.* 50:915–929.

Taylor, M. E. 1989. Locomotor adaptations by carnivores. In *Carnivore Behavior, Ecology, and Evolution*, Part 1, edited by J. L. Gittleman, 382–409. Ithaca, NY: Cornell Univ. Press.

Taylor, M. L., T. A. R. Price, and N. Wedell. 2014. Polyandry in nature: A global analysis. *Trends Ecol. Evol.* 29:376–383.

Taylor, W, P. 1914. The problem of aquatic adaptation in the Carnivora, as illustrated in the osteology and evolution of the sea-otter. *Univ. Calif. Publ. Geol. Dept. Geol.* 7:465–495.

Temple-Smith, P. D. 1973 Seasonal breeding biology of the platypus, *Ornithorhynchus anatinus* Shaw 1799, with special reference to the male. PhD diss., Australian National Univ.

Terrel, T. L. 1972. The swamp rabbit (*Sylvilagus aquaticus*) in Indiana. *Amer. Midland Nat.* 87:283–295.

Thewissen, J. G. M., S. T. Hussain, and M. Arif. 1994. Fossil evidence for the origin of aquatic locomotion in archaeocete whales. *Science* 263:210–212.

Thewissen, J. G. M., and S. Nummela, eds. 2008. *Sensory Evolution on the Threshold: Adaptations in Secondarily Aquatic Vertebrates*. Berkeley: Univ. California Press.

Thom, M. D., D. D. P. Johnson, and D. W. Macdonald. 2004. The evolution and maintenance of delayed implantation in the Mustelidae (Mammalia: Carnivora). *Evolution* 58:175–183.

Thomas, E. M. 1954. *Wyoming Fur Bearers. Their Life Histories and Importance*. Wyoming Game and Fish Dept. Bull. no. 7. Cheyenne, WY: Wyoming Game and Fish Dept.

Thomas, J., K. Handasyde, M. L. Parrott, and P. Temple-Smith. 2018. The platypus nest: Burrow structure and nesting behaviour in captivity. *Aust. J. Zool.* 65:347–356.

Thompson, S. D. 1988. Thermoregulation in the water opossum (*Chironectes minimus*): An exception that "proves" a rule. *Physiol. Zool.* 61:450–460.

Timmins, W. H. 1971. Observations on breeding the oriental short clawed otter *Amblonyx cinerea* at Chester Zoo. *Int. Zoo. Yearb.* 11:109–111.

Toldt, K. 1933. *Das Haarkleid der Pelztiere*. Deutsche Gesellschaft für Kleintier- und Pelztierzucht, Leipzig, Germany.

Tomas, W. M., and S. M. Salis. 2000. Diet of the marsh deer (*Blastocerus dichotomus*) in the Pantanal wetland, Brazil. *Stud. Neotrop. Fauna Environ.* 35:165–172.

Trayhurn, P., and J. H. Beattie. 2001. Physiological role of adipose tissue: White adipose tissue as an endocrine and secretory organ. *Proc. Nutr. Soc.* 60:329–339.

Traylor-Holzer, K., K. Leus, and K. Bauman. 2019. Integrated Collection Assessment and Planning (ICAP) workshop: Helping zoos move toward the One Plan approach. *Zoo Biol.* 38:95–105.

Tseng, Z. J., A. Pacheco-Castro, O. Carranza-Castañeda, J. J. Aranda-Gómez, X. Wang, and H. Troncoso. 2017. Discovery of the fossil otter *Enhydritherium terraenovae* (Carnivora, Mammalia) in Mexico reconciles a palaeozoogeographic mystery. *Biol. Lett.* 13:20170259.

Tubbs, R. S. 2015. "Anatomy is to physiology as geography is to history; it describes the theatre of events." *Clin. Anat.* 28:151–151.

Turvey, S. T., R. J. Kennerley, J. M. Nuñez-Miño, and R. P Young. 2017. The last survivors: Current status and conservation of the non-volant land mammals of the insular Caribbean. *J. Mammal.* 98:918–936.

Twigg, G. I. 1965. Studies on *Holochilus sciureus berbicensis*, a cricetine rodent from the coastal region of British Guiana. *Proc. Zool. Soc. Lond.* 145:263–283.

Uhen, M. D. 2007. Evolution of marine mammals: Back to the sea after 300 million years. *Anat. Rec.* 290:514–522.

United States Environmental Protection Agency. 2015. *Connectivity of Streams and Wetlands to Downstream Waters: A Review and Synthesis of the Scientific Evidence*. Washington, DC: EPA Office of Research and Development.

VanBik, D., S. H. Lee, M. G. Seo, B. R. Jeon, Y. K. Goo, S. J. Park, M. H. Rhee, O. D. Kwon, T. H. Kim, P. J. L. Geraldino, and D. Kwak. 2017. Borrelia species detected in ticks feeding on wild Korean water deer (*Hydropotes inermis*) using molecular and genotypic analyses. *J. Med. Entomol.* 54:1397–1402.

Vander Wall, S. B. 1990. *Food Hoarding in Animals*. Chicago: Univ. Chicago Press.

van Helvoort, B. E., R. Melisch, I. R. Lubis., and B. O'Callaghan. 1996. Aspects of preying behaviour of smooth-coated otters *Lutrogale perspicillata* from Southeast Asia. *IUCN Otter Spec. Group Bull.* 13:3–6.

Van Rompaey, H., and M. Colyn. 2013. *Genetta piscivora* aquatic genet: Fr. genette aquatique; Ger. Wasserschleichkatze. In *Mammals of Africa*, Vol. 5. edited by J. Kingdon and M. Hoffmann, 239–240. London: Bloomsbury Publishing.

Van Rompaey, H., P. Gaubert, and M. Hoffmann. 2008. *Genetta piscivora*. *IUCN Red List of Threatened Species* 2008:e.T15628A4922427.

van Zyll de Jong, C. G. 1972. A systematic review of the Nearctic and Neotropical river otters (genus *Lutra*, Mustelidae, Carnivora). *Life Sci. Contrib. R. Ont. Mus.* 80:1–104.

Vásquez, C. A. V. 2012. Wild hippos in Colombia. *Aliens Invas. Spec. Bull.* 32:8–12.

Velandia-Perilla, J. H., and C. A. Saavedra-Rodríguez. 2013. Mammalia, Rodentia, Cricetidae, *Neusticomys monticolus*. (Anthony, 1921): Noteworthy records of the montane fish-eating rat in Colombia. *Check List* 9:686–688.

Venzal, J. M., A. Estrada-Pena, O. Castro, C. G. de Souza, M. L. Felix, S. Nava, and A. A. Guglielmone. 2008. *Amblyomma triste* Koch, 1844 (Acari: Ixodidae): Hosts and seasonality of the vector of *Rickettsia parkeri* in Uruguay. *Vet. Parasitol.* 155:104–109.

Veron, G. 2010. Phylogeny of the Viverridae and "viverrid-like" feliforms. In *Carnivoran Evolution. New Views on Phylogeny, Form and Function*, edited by A. Goswami and A. Friscia, 64–91. Cambridge: Cambridge Univ. Press.

Veron, G., P. Gaubert, N. Franklin, A. P. Jennings, and L. I. Grassman. 2006. A reassessment of the distribution and taxonomy of the endangered otter civet *Cynogale bennettii* (Carnivora: Viverridae) of South-east Asia. *Oryx* 40:42–49.

Veron, G., B. D. Patterson, and R. Reeves. 2008. Global diversity of mammals (Mammalia) in freshwater. *Hydrobiologia* 595:607–617.

Verweij, R. J. T., J. Verrelst, P. E. Loth, I. M. A. Heitkönig, and A. M. H. Brunsting. 2006. Grazing lawns contribute to the subsistence of mesoherbivores on dystrophic savannas. *Oikos* 114:108–116.

Vié, J.-C., C. Hilton-Taylor, and S. N. Stuart, eds. 2009. *Wildlife in a Changing World: An Analysis of the 2008 IUCN Red List of Threatened Species*. Gland, Switzerland: IUCN.

Vogel, P. 1983. Contribution à l'écologie et a la zoogéographie de *Micropotamogale lamottei* (Mammalia, Tenrecidae). *Rev. Ecol. Terre Vie*. 38:37–49.

Vogel, P., C. Bodmer, M. Spreng, and J. Aeschimann. 1998. Diving capacity and foraging behaviour of the water shrew. In *Behaviour and Ecology of Riparian Mammals*, edited by N. Dunstone and M. Gorman, 31–48. Cambridge: Cambridge Univ. Press.

Vogt, P. 1995. The European breeding program (EEP) for *Lutra lutra*: Its chances and problems. *Hystrix* 7:247–253.

Voss, R. S. 1988. Systematics and ecology of Ichthyomyine rodents (Muroidea): Patterns of morphological evolution in a small adaptive radiation. *Bull. Amer. Mus. Nat. Hist.* 188:259–493.

———. 2015. Tribe Ichthyomyini Vorontsov, 1959. In *Mammals of South America*, edited by J. L. Patton, U. F. J. Pardiñas, and G. D'Elía, 279–291. Chicago: Univ. Chicago Press.

Wackernagel, M., and W. E. Rees. 1996. Our ecological footprint: Reducing human impact on the earth. *Environ. Urban*. 8:216–216.

Wall, W. P. 1983. The correlation between high limb-bone density and aquatic habits in recent mammals. *J. Paleontol.* 57:197–207.

Walls, G. L. 1942. *The Vertebrate Eye and Its Adaptive Radiation*. Bloomfield Hills, NY: Cranbrook Press.

Wang, Q., and C. Yang. 2013. The phylogeny of the Cetartiodactyla based on complete mitochondrial genomes. *Int. J. Biol.* 5:30–36.

Warren, W. C., L. W. Hillier, J. A. Marshall Graves, E. Birney, C. P. Ponting, F. Grutzner, K. Belov, W. Miller, L. Clarke, A. T. Chinwalla, et al. 2008. Genome analysis of the platypus reveals unique signatures of evolution. *Nature* 453:175–183.

Watland, A. M., E. M. Schauber, and A. Woolf. 2007. Translocation of swamp rabbits in southern Illinois. *Southeast. Nat.* 6:259–270.

Watt, J. 1993. Ontogeny of hunting behaviour of otters (*Lutra lutra* L.) in a marine environment. *Symp. Zool. Soc. Lond.* 65:87–104.

Watts, C. H. S., and H. J. Aslin. 1981. *The Rodents of Australia*. Sydney: Angus and Robertson Publishers.

Watts, C. H. S., and R. W. Braithwaite. 1978. The diet of *Rattus Lutreolus* and five other rodents in Southern Victoria. *Aust. Wildl. Res.* 5:47–57.

Weatherbee, S. D., R. R. Behringer, J. J. Rasweiler, and L. A. Niswander. 2006. Interdigital webbing retention in bat wings illustrates genetic changes underlying amniote limb diversification. *Proc. Natl. Acad. Sci. USA* 103:15103–15107.

Webb, J. B. 1975. Food of the otter (*Lutra lutra*) on the Somerset Levels. *J. Zool.* 177:486–491.

Weber, D. 1990. The end of the otter and of otter reintroduction plans in Switzerland. *IUCN Otter Spec. Group Bull.* 5:45–50.

Weir, V., and K. E. Bannister. 1973. The food of the otter in the Blakeney area. *Trans. Norf. Norw. Nat. Soc.* 22:377–382.

Werdelin, L., N. Yamaguchi, W. E. Johnson, and S. J. O'Brien. 2010. Phylogeny and evolution of cats (Felidae). In *Biology and Conservation of Wild Felids*, edited by D. W. Macdonald and A. J. Loveridge, 59–82. Oxford: Oxford Univ. Press.

Whitaker, J. O., and B. Abrell. 1986. The swamp rabbit, *Sylvilagus aquaticus*, in Indiana, 1984–1985. *Proc. Ind. Acad. Sci.* 95:563–570.

Wilkens, L. A., and M. H. Hofmann. 2008. Electroreception. In *Sensory Evolution on the Threshold: Adaptations in Secondarily Aquatic Vertebrates*, edited by J. G. N. Thewissen and S. Nummela, 325–332. Berkeley: Univ. California Press.

Willemsen, G. F. 1992. A revision of the Pliocene and Quaternary Lutrinae from Europe. *Scr. Geol.* 101:1–115.

———. 2006. *Megalenhydris* and its relationship to *Lutra* reconsidered. *Hell. J. Geosci.* 41:83–87.

Willemsen, G. F., and A. Malatesta. 1987. *Megalenhydris barbaricina* sp. nov., a new otter from Sardinia. *Proc. K. Ned. Akad. Wet. B.* 90:83–92.

Williams, T. D., D. D. Allen, J. M. Groff, and R. L. Glass. 1992. An analysis of California sea otter (*Enhydra lutris*) pelage and integument. *Mar. Mamm. Sci.* 8:1–18.

Williams, T. M. 1983. Locomotion in the North American mink, a semi-aquatic mammal. I. Swimming energetics and body drag. *J. Exp. Biol.* 103:155–168.

———. 1986. Thermoregulation of the North American mink during rest and activity in the aquatic environment. *Physiol. Zool.* 59:293–305.

———. 1989. Swimming by sea otters: Adaptations for low energetic cost locomotion. *J. Comp. Physiol. A* 164:815–824.

———. 1998. Physiological challenges in semi-aquatic mammals: Swimming against the energetic tide. In *Behaviour and Ecology of Riparian Mammals*, edited by N. Dunstone and M. C. Gorman, 17–29. Cambridge: Cambridge Univ. Press.

———. 1999. The evolution of cost efficient swimming in marine mammals: Limits to energetic optimization. *Phil. Trans. R. Soc. Lond. B* 354:193–201.

———. 2001. Intermittent swimming by mammals: A strategy for increasing energetic efficiency during diving. *Am. Zool.* 41:166–176.

Williams, T. M., M. Ben-David, S. Noren, M. Rutishauser, K. McDonald, and W. Heyward. 2002. Running energetics of the North American river otter: Do short legs necessarily reduce efficiency on land? *Comp. Biochem. Physiol. A* 133:203–212.

Williams, T. M., R. W. Davis, L. A. Fuiman, J. Francis, B. J. Le Boeuf, M. Horning, J. Calambokidis, and D. A. Croll. 2000. Sink or swim: Strategies for cost efficient diving by marine mammals. *Science* 288:133–136.

Willner, G. R., J. A. Chapman, and D. Pursley. 1979. Reproduction, physiological responses, food habits, and abundance of nutria on Maryland marshes. *Wildl. Monogr.* 65:3–43.

Willner, G. R., G. A. Feldhamer, E. E. Zucker, and J. A. Chapman. 1980. *Ondatra zibethicus. Mamm. Species* 141:1–8.

Wilson, D., and S. Ruff. 1999. *The Smithsonian Book of North American Mammals*. Washington, DC: Smithsonian Institution Press.

Wilson, D. E., and D. M. Reeder, eds. 2005. *Mammal Species of the World: A Taxonomic and Geographic Reference*. 3rd ed., 2 Vols. Baltimore: Johns Hopkins Univ. Press.

Withers, P. D., C. E. Cooper, S. K. Maloney, F. Bozinovic, and A. P. Cruz-Neto. 2016. *Ecological and Environmental Physiology of Mammals*. Oxford: Oxford Univ. Press.

Wittenberg, R., and M. J. W. Cock, eds. 2001. *Invasive Alien Species: A Toolkit of Best Prevention and Management Practices*. Wallingford, Oxon, UK: CAB International.

Woinarski, J., and A. A. Burbidge. 2016. *Xeromys myoides*. *IUCN Red List of Threatened Species* 2016:e.T23141A22454469.

Wojcik, D. P., C. R. Allen, R. J. Brenner, E. A. Forys, D. P. Jouvenaz, and R. S. Lutz. 2001. Red imported fire ants: Impact on biodiversity. *Am. Entomol.* 47:16–23.

Wolff, J. O., and R. D. Guthrie. 1985. Why are aquatic small mammals so large? *Oikos* 45:365–373.

Wołk, K. 1976. The winter food of the European water-shrew. *Acta Theriol.* 21:117–129.

Won, C. M., and K. G. Smith. 1999. History and current status of mammals of the Korean Peninsula. *Mammal Review* 29:3–33.

Woodburne, M. O., and R. H. Tedford. 1975. The first Tertiary monotreme from Australia. *Am. Mus Novit.* 2588:1–11.

Wourms, J. P., and I. P. Callard. 1992. A retrospect to the symposium on evolution of viviparity in vertebrates. *Am. Zool.* 32:251–255.

Wozencraft, W. C. 2005. Order Carnivora. In *Mammal Species of the World*, 3rd ed., edited by D. E. Wilson and D. M. Reeder, 532–628. Baltimore: Johns Hopkins Univ. Press.

Wright, L., A. Olsson, and B. Kanchanasaka 2008. A working review of the hairy-nosed otter (*Lutra sumatrana*). *IUCN Otter Spec. Group Bull.* 25:38–59.

WWF. 2018. *Living Planet Report—2018: Aiming Higher*, edited by M. Grooten and R. E. A. Almond. Gland, Switzerland: WWF.

Yáber, M. C., and E. A. Herrera. 1994. Vigilance, group size and social status in capybaras. *Anim. Behav.* 48:1301–1307.

Yalden, D. W., M. J. Largen, D. Kock, and J. C. Hillman. 1996. Catalogue of the mammals of Ethiopia and Eritrea. 7. Revised checklist, zoogeography and conservation. *Trop. Zool.* 9:73–164.

Yamaguchi, N., R. J. Sarno, W. E. Johnson, S. J. O'Brien, and D. W. Macdonald. 2004. Multiple paternity and reproductive tactics of free-ranging American minks, *Mustela vison*. *J. Mammal.* 85:432–439.

Youngman, P. M. 1990. *Mustela lutreola*. *Mamm. Species* 362:1–3.

Yuan, B., L. Wang, L., S. Xie, Y. Ren, B. Liu, Y. Jia, H. Shen, D. Sun, and H. Ruan. 2017. Density dependence effects on demographic parameters—a case study of Père David's deer (*Elaphurus davidianus*) in captive and wild habitats. *Biol. Environ.* 117:139–144.

Zabala, J., I. Zuberogoitia, I. Garin, and J. Aihartza. 2003. Landscape features in the habitat selection of European mink (*Mustela lutreola*) in south-western Europe. *J. Zoo. Lond.* 260:415–421.

Zhu, L., C. Deng, X. Zhao, J. Ding, H. Huang, S. Zhu, Z. Wang, S. Qin, Y. Ding, G. Lu, and Z. Yang. 2018. Endangered Père David's deer genome provides insights into population recovering. *Evol. Appl.* 11:2040–2053.

Zuberogoitia, I., J. A. González-Oreja, J. Zabala, and C. Rodríguez-Refojos. 2010. Assessing the control/eradication of an invasive species, the American mink, based on field data; how much would it cost? *Biodivers. Conserv.* 19:1455–1469.

SUBJECT INDEX

Page numbers in *italics* indicate tables and figures.

aquatic tenrec. *See* web-footed tenrec
Archaeopotamus, 72
Archimedes' principle, 214, *214*
Argentine swamp rat, *15*, *101*, 279, *314*, *334*
Artiodactyla, 27–29, 70, *134*, 141, 144, 167–168, 227, 259
Arvicola: amphibius (*see* European water vole); *sapidus* (*see* southern water vole)
Arvicolinae, 58, 142
Asian small-clawed otter, 25–26, 66, *103*, 107, *108*, 115–116, *151*, 166–167, 171, 172, *174*, 177, 243, 247–248, 277, 298, 330, *316*, *331*, *347*, 383
Atilax paludinosus. See marsh mongoose
Australian swamp rat, *18*, 86, *87*, 95, 98, 175, 263, 274, 284, *314*, 335
Australian zoogeographic realm, 95, *98*

Babyrousa: babyrussa (*see* hairy babirusa); *celebensis* (*see* Sulawesi babirusa); *togeanensis* (*see* Togian Islands babirusa)
bacteria, 275, 277–279
Baiyankamys: habbema (*see* mountain water rat); *shawmayeri* (*see* Shaw Mayer's water rat)
basal metabolic rate (BMR), 189–192
basal temperature, 131, 193
beaver beetles, 112, *112*
beaver rats. *See* rakali
beavers. *See* Eurasian beaver; North American beaver
behavior. *See* foraging strategies; predation; reproduction
Bering land bridge, 40, 56, 65–66
biodiversity: conservation of, 381, 383, 397; hotspots, 81, 342, 380; loss of, 386
Blastocerus dichotomus. See marsh deer
blood sweat, 148, 161
body size, 2, 142–143, 186, 188–190, 215
Bornean water shrew, 21, 79, *103*, 305, 330
Bovidae, 27, 29, 73–74
Brachylagus, 53
bradycardia, 200
brain, the, 157, 168, 186, 210–211
Brazilian spotted fever, 279
bristles, 219
brown fat. *See* adipose tissue
Bunolagus monticularis. See riverine rabbit
Bunn's short-tailed bandicoot rat, *18*, 107, *107*, 240, 263, 295, *314*, 328, 330

buoyancy: and locomotion 155, 179–180, 200–201, 214–215, 222–224; and pelage, 74, 148, 154, 156, 187, 216
Burrell, Harry, 291, 294, 395
burrows, 19, 46–47, 118–123, 129–131, 167, 196, 240, 283, 294, 304
Buxolestes 46, *47*

Cabrera's hutia, 13, 56, 80, 104, *106*, 122, *250*, 274, 328–329, *330*
caching, 123, 233, 240
Cape clawless otter. *See* African clawless otter
Cape marsh rat, *18*, 84, *96*, 330, *331*, 391
Capreolinae, 73
Caprolagus, 53
Capromyidae, 13, 56
captive breeding programs, 311, 322, 336, 353, 376, 388–390
capybara: conservation of, 19, 276, *334*, *347*, 391, 394; and disease, 279; distribution of, 91, 98, *100*, 106; foraging, 205, 238–239, 257–258, 260, *261*; locomotion, 224, *225*, *226*; morphology of, 2, 3, 19, 141, *171*, *175*, *178*, 219, 221, 222, *270*; phylogeny of, 56–57; reproduction, 300–301, *301*, 309–310, 312; species interactions, 112–113, 269–270, 283, 285–286; taxonomy of, 6, 18
Carletonomys cailoi, 58
Carnivora, 7, 21–27, 46, 62–69, *134*, 150, 227, 241–242, 252
Castor: canadensis (*see* North American beaver); *fiber* (*see* Eurasian beaver)
castoreum, 163, 346
Castoridae, 13, 54–56, *171*, *174*, *177*
Castorocauda lutrasimilis, 45, *45*
Castoroides, 55, *55*
cat, domestic, 12, 374
caudal vertebrae, 45
Caviidae, 18–19, 57, *171*, *175*, *178*
Caviinae, 56
Cenozoic Era, 8, 40, *41*, *42*, 49, 51, 204
central nervous system, 210–211, 277
Cervidae, 27–30, 73
Cetacea, 27, 71, 74–75, 182
Chaco marsh rat, *15*, *100*, 265, *312*, *334*
Chalicomys, 55
channels, *31*, *32*, 118–122, *120*, 167, 339
chemoreception, 176

geologic time, *41, 42*
Geomyoidea, 56
gestation, 292–295, 303–304, 308–309, *312–317*
giant otter: conservation of, 27, 328–329, *330*, 344, 346, 349–351; distribution of, 27, 91, *96, 101*; foraging, 116, 157, 239, 253, 255–256, 285; locomotion, 225–228; morphology of, 25, 146, 156, *166, 171, 174, 177, 221, 221, 284*; phylogeny of, 66–67; reproduction, 297–300, *316*; species interactions, *274*; taxonomy of, 24–25
giant otter shrew, 10–11, 51, 96, 119, 145, *177, 215, 225, 226, 247, 312, 334, 393*
Glacier Bay water shrew, *21, 87, 99, 306, 337, 338*
golden-bellied water rat. *See* rakali
Goldman's water mouse, *15, 82, 83, 175, 334*
Gomphos elkema, 53
Gomphotheriidae, 52
Gondwana, 39–40
greater cane rat, 13, 56, *96*, 260, *261, 312, 301, 334, 347*
grooming, 55, 150, 160, 165–167, *166, 185–187, 187*, 196, 206, 223, 281

habitat: degradation, 3, 22, 26, 86, 339, 344–345; enhancement, 375, 383–386; fragmentation, 108, 344, 368; loss, 341–343; protection, 381–383
hair. *See* integumentary system
hairy babirusa, 27, 70, *102, 102*, 122, *151, 264, 301, 316, 331*
hairy-nosed otter, 26, *103, 107, 108*, 166, 254, *316*, 321–322, 329, 330, 343, 345–347, 383
Haldanodon exspectatus, 44
hantavirus, 227–228
harderian gland, 161
hearing, 169, 173, 284
heat exchange, countercurrent, 124, 127, 194–196, *195*
heat loss, modes of, 161, 183–185, *184*
Hemigalinae, 64
Herpestidae, 21–23, 63, *174*
Herpestides, 64
Heterosoricidae, 60

heterothermy, regional, 193–194
Hexaprotodon madagascariensis, 72
Himalayan water shrew, *21, 103, 104, 105, 107, 107, 108*, 116, *305, 335, 336, 387*
hindgut fermentation, 209–210, 256
hippo: highways, 118; lawns, 117, 262
Hippopotamidae, 27, 71–72, *171, 176, 178*
hippopotamus. *See* common hippopotamus; pygmy hippopotamus
Hippopotamus: amphibius (*see* common hippopotamus); *lemerlei*, 72, *324*
Holocene, *41, 42*, 50, 58, 72
Holochilus: brasiliensis (*see* web-footed marsh rat); *chacarius* (*see* Chaco marsh rat); *lagigliai* (*see* Lagiglia's marsh rat); *sciureus* (*see* marsh rat)
Holt, Ben, 93–94
Hopkins's groove-toothed swamp rat, *96*
Howell, A. Brazier, 1, 29–30, 144
huillin. *See* southern river otter
hunting. *See* exploitation of species
hut, muskrat, 123, 127, 129, 131
hybridization, and phylogeny, 65, 73
Hydrictis maculicollis. *See* spotted-necked otter
Hydrochoerinae, 56
Hydrochoerus: hydrochaeris (*see* capybara); *isthmius* (*see* lesser capybara)
Hydromyinae, 59
Hydromys: chrysogaster (*see* rakali); *hussoni* (*see* western water rat); *neobritannicus* (*see* New Britain water rat); *ziegleri* (*see* Ziegler's water rat)
Hydropotes inermis. *See* water deer
Hyemoschus aquaticus. *See* water chevrotain
hypothermia, 124, 189, 197

Ichthyomys: hydrobates (*see* crab-eating rat); *pittieri* (*see* Pittier's crab-eating rat); *stolzmanni* (*see* Stolzmann's crab-eating rat); *tweedii* (*see* Tweedy's crab-eating rat)
Insectivora. *See* Eulipotyphla
insulation: and fat storage, 146, 148, 152; and pelage, 127, 155–156, 186–187, 222–223
integumentary system: antlers and horns, 73; claws, nails, and hooves, 22,

Oryzomys: couesi (see Coues's rice rat); palustris (see marsh rice rat)

Otomys lacustris. See Tanzanian vlei rat

otter civet, 6, 21–23, 62, 64, 103, *103*, 150, 157, *157*, 165, 174, 177, 243, 315, 329, 330, 343, 385–386, 394

otters: distribution of, 88–93, *90*, 98, 104, 107, 113–116, *114*, *115*; phylogeny of, 62, 66–67, *68*; taxonomy of, 24–27; trade in, 346–347

oviparous species, 5, 291–292

ovulation, induced, 307, 309

oxygen exchange, 198–204

Oyapock's fish-eating rat, 15, 82, 101, 337, *338*

Pacific water shrew, 21, 99, 240, 306, *314*, 335, 336, *336*

Paenelimnoecus crouzeli, 61

Palaeosinopa, 46

Palasiomys, 57

Palearctic zoogeographic realm, 104, *105*

Paleocene epoch, *41*, *42*, 49, 53–54, 63, 70

Paleolagus, 53

palm oil plantations, 343

Panamanian zoogeographic realm, 104, *106*

Pangea, 38

Pantolestidae, 46

Parageogale, 51

Parahydromys asper. See waterside rat

Paramyidae, 54

Paramys, 54

parasites and parasitoids, 273, 275–281

Pecora, 73

pelage. See fur

Pelomys: fallax. See creek groove-toothed swamp rat; hopkinsi (see Hopkins's groove-toothed swamp rat); isseli (see Issel's groove-toothed swamp rat)

pelycosaurs, 38, 43

Père David's deer, 29, 73, 107, *108*, 152, 164, 168, 209, *261*, *262*, 281, 303, *316*, 324, 326, 378, 390–391

perissodactyls, 7, 70

Permian period, 38, *39*, *41*, 43–44

Peruvian fish-eating rat, 15, 82, 101, 334

Petromyscinae, 60

Phosphatherium escuillei, 52

physiology. See digestive systems; hetero-

thermy; metabolism; nervous system; oxygen exchange, thermoregulation

piloerection, 130, 195

pinnae, 140, 172–173, *174*, 284. See also hearing

Pittier's crab-eating rat, 15, *100*, 331, 392

placental mammals. See eutherian mammals

plateau swamp rat, 15, 101, 337

platypus: conservation of, 332, *333*, 353, 367–368, 394; distribution of, 86, *87*, 95, 98; eggs of, 8, 48, 146, 291–294; engineering by, 119–120, 167; foraging by, 169–170, 205, 249, *250*; locomotion, 203, 213, 225, *226*, 229; morphology of, 145–146, 150, 152, 153, 155, *156*, 158, 162, 167, 171, 172–173, *174*, 176, *177*, 206–207, 211, 218, 220, *221*; phylogeny of, 48–49, *49*; physiology of, 189–190, 192, 195, 199–201; reproduction, 162, 291–294, 304–307, *305*, *312*; in research, 391; species interactions, 274, 278; taxonomy of, 4, 5, 8, 37

Pleistocene epoch, 40, *41*, *42*, 50, 52–53, 55, 58, 62–63, 66–69, *68*, 72–74, 85

Plesictis plesictis, 64

Pliocene epoch, *41*, *42*, 48, 50–51, 55, 58–59, 61–62, 65–66, *68*, 69–74

poaching. See exploitation of species

population, bottleneck, 281, 326

Potamogale velox. See giant otter shrew

Potamogalidae, 10, 51

pouch. See also marsupium

predation: and alarm signals, 146, 284–285; avoidance, 120, 281–288; and synergistic effects, 268–270, 350

Prionailurus: planiceps (see flat-headed cat); viverrinus (see fishing cat)

prions, 276, 278

Proailurus lemanensis, 63

Proboscidea, 7, 52

Procastoroides, 55

Procyon cancrivorus. See crab-eating raccoon

Procyonidae, 21, 69

propulsion. See locomotion

proteins: and brown fat, 148, 197; and development, 149, 153, 294; tannin-binding, 257

Protenrec, 51

Protocetidae, 74
prototherian species, 8–9, 43
Pseudobassaris, 69
Pteronura brasiliensis. *See* giant otter
pygmy hippopotamus, 28, 72, 90–91, *91*,
    97, 122, 145, 181, 200, 261, *261*, 287,
    287, 303, 306, 307, *316*, 328–329, 330,
    343
Pyrenean desman, 20, *21*, 31, *105*, 140, 146,
    150, 157–158, *174*, *177*, 219, 250, 252,
    306, *314*, 330–331, *331*, 344, 365, 387

Quaternary period, *41*, 57–58, *68*

rabies, 277
radiations (species), 38, 40, 51, 58, 63,
    65–66, 70, 72–73, 204
rakali, 8, 17–18, *18*, 79, *80*, 86, *87*, 88,
    95, *98*, 102, *102*, 119, 131, *156*, 189, *191*,
    192–197, *203*, 240, 247, 265, *314*, 335
rats, as introduced species, 80, 155, 246,
    274, 324–325, 345
Rattus lutreolus. *See* Australian swamp rat
Reduncini, 73
reintroductions. *See* conservation
relict species, 81, 84–85
reproduction: and fecundity, *312–317*; and
    gestation, 292–295, 303–304, 308, 309;
    and reproductive systems, 291–295,
    292; and social behavior, 300–307, *301*,
    305. *See also* lactation
research and monitoring, 391–393, 395
reserves, nature, 355, 372–375, 382
reservoirs, as habitats, 33
respiratory system, 198
Rheomys: mexicanus (*see* Mexican water
    mouse); *raptor* (*see* Goldman's water
    mouse); *thomasi* (*see* Thomas's water
    mouse); *underwoodi* (*see* Underwood's
    water mouse)
riverine rabbit, 12
Robert's shaggy rat, *18*, *96*, 330, *331*
Rodentia, 7, 12–19, 54–60, 134, 150, 226
Rodhocetus, 74, *74*
Romerolagus, 53
round-tailed muskrat, 14, *15*, 58, 99, 158,
    175, 260, *261*, 272, 274, 286, 305, 312, 334
Ruminantia, 73
ruminants, 70, 72–74, 206, 208, 209, 235
Russian desman, 20, *20*, *21*, 93, *105*, 140,

146, 150, *156*, 158, *174*, 177, 202, 212,
    219, 227, 250, 252, 274, *314*, 330, 361,
    363, 373–374
Rwenzori otter shrew, 10–11, 51, 85–86,
    96, 118, 150, 249, 250, *312*, 334, 392

Sahara-Arabian zoogeographic realm,
    104, *107*
Saotherium mingoz, 72
Sardolutra ichnusae, 67, *68*
scale-like structures, 153, 160
Scapteromys: aquaticus (*see* Argentine
    swamp rat); *meridionalis* (*see* plateau
    swamp rat); *tumidus* (*see* Waterhouse's
    swamp rat)
Scottish Beaver Trial, 353, 366
sea mink, 24, 65, *142*
semi-aquatic mammals (freshwater): com-
    mon traits in, 1–3, 140–146; definition
    of, 1–3
semi-aquatic mammals (marine}, 4, 224
Semigenetta, 64
sensory systems, 168–178
Shamolagus, 53
Shaw Mayer's water rat, *18*, 102, *151*, 334,
    335
Sigmodontinae, 16, 58
Sigmodontomys alfari. *See* Alfaro's rice
    water rat
Sino-Japanese zoogeographic realm,
    107–108, *108*
sirenians, 3, 7, 52, 164, 176
sitatunga, 29, 73–74, 89, *97*, 122, 161,
    167–168, 209, *261*, 262–263, *283*, *316*,
    335, 336
skin. *See* integumentary system
small-clawed otter (*Aonyx cirereus*). *See*
    Asian small-clawed otter
small-clawed otter (*Aonyx congicus*). *See*
    Congo clawless otter
smell. *See* olfaction
smooth-coated otter, 26, 66–67, 103, 104,
    107, *107*, *108*, 115–116, *115*, 146, 156,
    166, 239, 253–254, 285, 288, 298, *316*,
    331, 346–347, 383
Sorex: alaskanus (*see* Glacier Bay water
    shrew); *bendirii* (*see* Pacific water
    shrew); *palustris* (*see* American water
    shrew)
Soricidae, 20, *21*, 60–61, 158, 160, *174*

Tragulidae, 27–28, 72–73, *178*
trails, 118, 121–122
Transcaucasian water shrew, *21*, *105*, 240,
    251, *306*, *314*, *335*, 392
translocations. *See* conservation
trapping. *See* exploitation of species
trematodes, 280–281, 363
Triassic period, 38, *39*, *41*, *43*, 48
Trinidad water rat, *15*, 16, *100*, *106*, *334*
Trogontherium cuvieri, 55
trypanosomiasis, 269
turbidity, and vision, 170
Tweedy's crab-eating rat, *100*, *338*

Ucayali water rat, *15*, 81, *100*, *151*, 159,
    265, *338*
underwater: hearing, 169, 173; sniffing,
    20, 176, 212, 234; vision, 170–172, 181;
    vocalization, 173
Underwood's water mouse, *15*, *82*, 83, *106*,
    141, *175*, *177*, *334*
ungulates, 27–28, 70–74, 164–165, 167,
    260–262, 279
upper critical temperatures, 192
urine, and communication, 163, 178

Varanosaurus, 43, *43*
vectors of disease, 275, 279
Venezuelan fish-eating rat, *15*, 81–82, *82*,
    101, *106*, 330, *331*
venom, 8, 162, *162*, 240, 251, 286
vibrissae, 157–158, 169, 247, 253
villi, lack of, 207
vision. *See* sensory systems; underwater
    vision
Viverridae, 21–23, 62, 64, 165, *174*, 177
Voss fish-eating rat, *15*, *82*, 337
vulnerable species (IUCN), 329–332, *331*

Waiomys mamasae. *See* Sulawesi water rat
Wallace, Alfred Russel, 76, 93, 95
water chevrotain, 28, 72, 97, *178*, 209, 256,
    261, 265, 270, 286, *306*, *316*, *335*, 347
water deer, 29, 73, *105*, *108*, 209, 256, 261,
    262, 276, *316*, 330, 331, *331*, 346–347,
    383
water mongoose. *See* marsh mongoose
water opossum, 4, 5, 9, 50, 95, 99, *100*,
    106, 149–150, *152*, 153, 167, 179–180,
    190, 205, 218, 225, 226, 229, 249, 250,

292–293, *293*, 295, 305, *307*, *312*, *334*,
    339
water shrew (*Neomys*). *See* Eurasian water
    shrew
water shrew (*Sorex*). *See* American water
    shrew
water vole (*Arvicola*). *See* European water
    vole
water vole (*Microtus*). *See* North Ameri-
    can water vole
water, properties of, 139, 183–184, 213
Waterhouse's swamp rat, *15*, *101*, *314*, *334*
waterproofing, fur, 160–161, *187*, 345
waterside rat, 17, *18*, 79, *102*, *102*, 250, *335*
webbing, interdigital, 16, 145–146,
    149–152, *151*, *152*, 218, 220
web footed marsh rat, *15*, *100*, *179*, 260,
    261, 265, *312*, *334*, 391
web-footed tenrec, 10, 51, 80, 95, 111,
    119, 190, 205, 221, 250, 282, *312*, *331*,
    343–344
Wechiau Community Hippo Sanctuary,
    396
West African shaggy rat, *18*, *83*, 84, 96,
    335, *335*
western Amazonian nectomys, *15*, 16, *100*,
    159, *175*, *334*, 392
western water rat, *18*, 79, 80, *102*, 265, 337,
    338, 339, 349
wetlands, types of, 32–33
Whippomorpha, 71
whiskers. *See* vibrissae
white fat. *See* adipose tissue
wildlife: control, 358, 387–388; parks,
    388; rehabilitation, 370; rescue, 322,
    388
World Wildlife Fund (WWF), *338*, 395

Xeromys myoides. *See* false water rat

Yanoconodon, 46
yapok. *See* water opossum

Zazamys, 56
Ziegler's water rat, *18*, 79, 80, *102*, 250, *338*
zoos: and conservation, 326, 328, 332, 336,
    347, 351, 388, 395–396; and wildlife
    trade, 396
zoogeographic realms, 78, 93–108, *94*
zoonotic diseases, 276, 392

GLYNNIS A. HOOD is a Professor of Environmental Science at the University of Alberta's Augustana Campus in Camrose, Alberta, Canada. Prior to joining the University of Alberta in 2007, she had a nineteen-year career with Parks Canada's warden service in the Rocky Mountains, sub-Arctic, and southern mixed-wood boreal forest. She is an environmental ecologist who examines the interface of freshwater systems and wildlife ecology. Her research program focuses on the ecology and management of beavers and human-wildlife conflicts, as they relate to habitat availability and wildlife management. She is the author of *The Beaver Manifesto*, published by Rocky Mountain Books, and has authored several academic papers on freshwater ecology and wildlife biology. Her research has been highlighted on CBC's *The Nature of Things*, PBS's *Nature*, the Discovery Channel's *Canadian Geographic*, and various international print and radio media. She currently teaches wildlife and freshwater ecology at the University of Alberta.

MEAGHAN BRIERLEY holds a Bachelor of Fine Arts in Design Art and Biology from Concordia University, Montreal, a Master of Science in Biomedical Communications from the University of Toronto, and a PhD in Communication Studies from the University of Calgary. She has served on the BiodiverCity Advisory Committee for the City of Calgary, helping to advance the commitments and procedures identified within Calgary's ten-year biodiversity strategic plan. She is also a member of the Grassroutes Ethnoecological Association, which is dedicated to fostering a sense of place through arts, sciences, and education, and to engendering stewardship and sustainability through strengthening community and environmental connections.